T0295297

PHYSICS RESEARCH AND TECHNOLOGY

NEUTRINOS

PROPERTIES, SOURCES AND DETECTION

PHYSICS RESEARCH AND TECHNOLOGY

Additional books in this series can be found on Nova's website
under the Series tab.

Additional E-books in this series can be found on Nova's website
under the E-books tab.

PHYSICS RESEARCH AND TECHNOLOGY

NEUTRINOS

PROPERTIES, SOURCES AND DETECTION

JOSHUA P. GREENE
EDITOR

Nova Science Publishers, Inc.
New York

For permission to use material from this book please contact us:
Telephone 631-231-7269; Fax 631-231-8175
Web Site: http://www.novapublishers.com

LIBRARY OF CONGRESS CATALOGING-IN-PUBLICATION DATA

Neutrinos : properties, sources, and detection / [edited by] Joshua P. Greene.
p. cm.
Includes bibliographical references and index.
ISBN 978-1-61209-650-6 (hardcover : alk. paper)
1. Neutrinos. I. Greene, Joshua P.
QC793.5.N42N496 2011
539.7'215--dc22
2011002573

Published by Nova Science Publishers, Inc. † New York

CONTENTS

PREFACE

This new book presents current research in the study of the properties, reactions, sources and detection of neutrinos. Topics discussed include the study of differential cross sections and spin asymmetries for the processes of elastic weak, electromagnetic and electroweak scattering of neutrinos by polarized proton targeting; neutrino-nucleus interactions at low and intermediate neutrino energies; double beta decay and the neutrino mass; observable consequences of sterile neutrinos in galactic, cluster, and cosmological scales; evaluation of GT transitions in neutrino-nucleus reactions and nuclear responses to supernova neutrinos for stable molybdenum isotopes. (Imprint: Nova Press)

In Chapter 1 the authors study differential cross sections and spin asymmetries for the processes of elastic weak, electromagnetic and electroweak scattering of the neutrino by polarized (unpolarized) proton target. In the calculations we take into account initial and final neutrino helicities, charge radius and magnetic moment of the Dirac neutrino, electroweak form factors, including neutral weak Pauli magnetic, as well as *C*-, *P*-, *T*/*CP*- symmetries violating anapole and electric dipole proton's electromagnetic form factors.

Calculated angular and energy distributions of the recoil protons for the channels of helicity conserving ($\nu_L \rightarrow \nu_L$) and helicity changing ($\nu_L \rightarrow \nu_R$) neutrino scattering, contain *P*- and *T*-symmetries violating correlations between target proton spin and momenta of the incoming neutrino and recoil proton.

The authors obtain analytical expressions for the longitudinal (target spin parallel vs. antiparallel to incoming neutrino momentum), and for the transverse (target spin is orthogonal up vs. down to reaction plane) asymmetries. Longitudinal asymmetry is caused by binary *P*-odd spin-momentum correlations, which contain contributions from weak, weak-electromagnetic, and anapole electromagnetic interactions. Transverse asymmetry arises due to *T*-odd triple spin-momentum correlation, which is caused by proton's electric dipole moment.

Comparative study of the pure weak, electroweak interference, and electromagnetic spin asymmetries in dependence of angle, energy and form factors parameters provide a possibility to get supplementary new information about the structure of the proton's electroweak interaction with neutrino, and about the electromagnetic properties of the neutrino.

In Chapter 2 the authors review various field theory approaches to the description of neutrino oscillations in vacuum and external fields. First we discuss a relativistic quantum mechanics based approach which involves the temporal evolution of massive neutrinos. To describe the dynamics of the neutrinos system we use exact solutions of wave equations in presence of an external field. It allows one to exactly take into account both the characteristics of neutrinos and the properties of an external field. In particular, we examine flavor oscillations an vacuum and in background matter as well as spin flavor oscillations in matter under the influence of an external electromagnetic field. Moreover we consider the situation of hypothetical nonstandard neutrino interactions with background fermions. In the case of ultrarelativistic particles we reproduce an effective Hamiltonian which is used in the standard quantum mechanical approach for the description of neutrino oscillations. The corrections to the quantum mechanical Hamiltonian are also discussed. Note that within the relativistic quantum mechanics method one can study the evolution of both Dirac and Majorana neutrinos. We also consider several applications of this formalism to the description of oscillations of astrophysical neutrinos emitted by a supernova and compare the behavior of Dirac and Majorana neutrinos. Then we study a spatial evolution of mixed massive neutrinos emitted by classical sources. This method seems to be more realistic since it predicts neutrino oscillations in space. Besides oscillations among different neutrino flavors, we also study transitions between particle and antiparticle states. Finally we use the quantum field theory method, which involves virtual neutrinos propagating between production and detection points, to describe particle-antiparticle transitions of Majorana neutrinos in presence of background matter.

Neutrino studies have been paid considerable attention in nuclear physics, astrophysics and cosmology. It has been pointed out that neutrinos are sensitive probes for investigating astrophysical processes and stellar evolution. The study of solar neutrinos provides important information on nuclear processes inside the sun as well as on matter densities. Moreover, Supernova neutrinos provide sensitive probes for study supernova explosions, neutrino properties and stellar collapse mechanisms. In the present work we have developed a formalism describing (anti)-neutrino-nucleus crosssections in neutral and charged-current processes, taking into account vector and axial vector contributions of the hadronic current. The calculation of the single-particle transition matrix elements, based on the multipole expansion treatment of the relevant hadronic currents, was improved ending to a more compact form. The differential and integrated neutrino-nucleus cross sections were evaluated for low and intermediate neutrino energies ($0\ MeV \leq \varepsilon_\upsilon \leq 100\ MeV$) by examining the dependence of the cross sections on the scattering angle and initial neutrino-energy of Fermi and Gamow- Teller type transition rates. The required many-body nuclear wave-functions were calculated by utilizing the quasi-particle random phase approximation. The results presented refer to the neutral current scattering of electron neutrino on the various nuclear isotopes such as Mo, Fe, O, and Cd which play a significant role in astrophysical neutrino studies. Finally we have explored the nuclear response of some of these isotopes as a supernova neutrino detector assuming Fermi-Dirac distribution for supernova neutrino spectra. In Chapter 3 inelastic neutrino-nucleus reaction cross sections at low and intermediate neutrino energies are studied. The required many-body nuclear wave-functions are calculated in the context of quasi-particle random phase approximation (QRPA) that uses realistic two-body forces. The results presented here refer to the differential, integrated and total cross sections of neutral-current induced reactions of even-even isotopes used recently at low-energy astrophysical neutrino searches.

Even after the impressive confirmation of the non-zero mass of neutrinos by observing flavor oscillations in neutrinos from different sources, neutrino physics is a hot and exciting research area since many unknowns still remain. Double beta decay (DBD) is a rare nuclear process that some nuclei can undergo where a nucleus changes into an isobar with the emission of two electrons and two antineutrinos; although rare, this process has been observed for several nuclei. A non-standard version of this decay without the emission of antineutrinos has been proposed and great efforts are being devoted to its observation due to the outstanding implications of its occurrence, which can shed light on some of the pending questions in the field of neutrino physics: violation of leptonic number, confirmation of the Majorana nature of neutrinos, that is, if neutrinos and antineutrinos are the same particle, and even an estimate of the neutrino mass scale. In Chapter 4, motivation and status of double beta decay searches will be reviewed. In particular, double beta decay processes will be firstly described, emphasizing the connection with the neutrino mass problem. Then, experimental searches of this nuclear process will be also revised, describing their stringent requirements, comparing the different experimental approaches and techniques followed and summarizing the relevant results and prospects.

In Chapter 5, the authors investigate observable consequences of sterile neutrinos, in galactic, cluster, and cosmological scales. They first review some basic physics of the formation of sterile neutrinos from the oscillation of active neutrinos in the early universe. Then they discuss the possibility of having sterile neutrino halos in the Milky Way and galaxy clusters. The radiative decay of the sterile neutrinos gives energy to heat up the interstellar medium which can simultaneously explain the cooling flow problem in galaxy clusters and the hot gas problem in the Milky Way. Also, this model enables them to obtain the possible ranges of rest mass, decay rate and mixing angle of sterile neutrinos. The mixing angle obtained is not as small as we expected before so that the oscillation of active-sterile neutrino may be visible in near future experiments. However, the existence of the sterile neutrinos cannot be treated as the major component of the dark matter in the universe.

In Chapter 6, an improvement of the evaluation of GT transitions in neutrino-nucleus reactions is done. The transition strengths at each finite-momentum transfers are obtained by shell model diagonalizations, and used to evaluate the neutrino-induced reaction cross sections. The method is applied to charge-exchange reaction on ^{56}Fe, and reliable reaction cross sections are obtained up to $E_{\upsilon} \cong 100\,\mathrm{MeV}$

Detection of supernova neutrinos and their properties is very important for astrophysical applications. In Chapter 7 the theoretical framework for neutral-current neutrino-nucleus scattering calculations is reviewed. We then use the formalism to calculate cross sections for the stable (A=92,94,95,96,97,98,100) molybdenum isotopes. Both the coherent and incoherent contributions to the cross sections are considered. The authors use the quasiparticle random-phase approximation (QRPA) to construct the final and initial states of the even-even isotopes and for the odd isotopes the microscopic quasiparticle-phonon model (MQPM) is employed. The computed cross sections are folded with a two-parameter Fermi-Dirac distribution to obtain realistic estimates of the responses to supernova neutrinos.

In Chapter 8 the authors consider a supersymmetric E_6 GUT in a six dimesional $M^4 \otimes T^2$ space. E_6 breaking is achieved by orbifolding the extra two dimensional torus T^2 via a $T^2/(Z_2 \times Z_2' \times Z_2'')$. In effective four dimension we obtain an extended Pati-Salam group with N = 1 SUSY, as a result of orbifold compactification. The authors then discuss the problem of neutrino mass with imposed parity assignment. A light Dirac neutrino is predicted with mass of the order of 10^{-2} eV.

In Chapter 9, measuremets on the anomalous magnetic moment and the electric dipole moment of the tau and its neutrino are calculated through the reactions $e^+e^- \rightarrow \tau^+\tau^-\gamma$ and $e^+e^- \rightarrow \upsilon_\tau\bar{\upsilon}_\tau\gamma$ at the Z_1-pole and in the framework of a left-right symmetric model. The results are based on the recent data reported by the L3 and OPAL Collaborations at CERN LEP. Due to the stringent limit of the model mixing angle ϕ, the effect of this angle on the dipole moments is quite small.

In: Neutrinos: Properties, Sources and Detection
Editor: Joshua P. Greene, pp. 1-22
ISBN: 978-1-61209-650-6

Chapter 1

TESTING OF DISCRETE SYMMETRIES VIOLATION IN THE POLARIZED NEUTRINO-PROTON ELECTROWEAK SCATTERING

B.K. Kerimov and M.Ya. Safin*
Moscow State University, Physics Faculty, Moscow, Russia

Abstract

In this chapter we study differential cross sections and spin asymmetries for the processes of elastic weak, electromagnetic and electroweak scattering of the neutrino by polarized (unpolarized) proton target. In the calculations we take into account initial and final neutrino helicities, charge radius and magnetic moment of the Dirac neutrino, electroweak form factors, including neutral weak Pauli magnetic, as well as *C*-, *P*-, *T*/*CP*- symmetries violating anapole and electric dipole proton's electromagnetic form factors.

Calculated angular and energy distributions of the recoil protons for the channels of helicity conserving ($\nu_L \to \nu_L$) and helicity changing ($\nu_L \to \nu_R$) neutrino scattering, contain *P*- and *T*-symmetries violating correlations between target proton spin and momenta of the incoming neutrino and recoil proton.

We obtain analytical expressions for the longitudinal (target spin parallel vs. antiparallel to incoming neutrino momentum), and for the transverse (target spin is orthogonal up vs. down to reaction plane) asymmetries. Longitudinal asymmetry is caused by binary *P*-odd spin-momentum correlations, which contain contributions from weak, weak-electromagnetic, and anapole electromagnetic interactions. Transverse asymmetry arises due to *T*-odd triple spin-momentum correlation, which is caused by proton's electric dipole moment.

Comparative study of the pure weak, electroweak interference, and electromagnetic spin asymmetries in dependence of angle, energy and form factors parameters provide a possibility to get supplementary new information about the structure of the proton's electroweak interaction with neutrino, and about the electromagnetic properties of the neutrino.

---

* E-mail address: misafin@gmail.com, kerimovbk@gmail.com

Introduction

Studying of possible violation of the fundamental discrete symmetries in the lepton-nucleon scattering, and especially in the processes of polarized electron-proton and neutrino-proton scattering, seems to be exclusively important for revealing new phenomena beyond the Standard Model (SM) in the particles physics. The key role thus belongs to detailed studying of spin-momentum correlations in the scattering on polarized target.

We repeatedly discussed a question on discrete symmetries violation in elastic electromagnetic and electroweak $e^- p(\mathbf{s})$-scattering by studying angular distributions of the scattered electrons [1], as well as energy spectrum ($y = E_k / E$-distribution) of the recoil protons [4, 5, 6]. In elastic weak, electromagnetic and electroweak neutrino-proton scattering *P*-odd effects were considered in energy distribution of the recoil protons [2, 6]. Comparative study of the fundamental symmetries violation in the elastic $e^- p(\mathbf{s})$- and $\nu(\bar{\nu})p(\mathbf{s})$-electroweak scattering were carried out in [7].

Earlier in accelerator experiments the cross section of elastic muon neutrino scattering by unpolarized proton $\nu_\mu(\bar{\nu}_\mu)p \to \nu_\mu(\bar{\nu}_\mu)p$ has been measured in [8] ($E_{\nu_\mu} = 1.25\ GeV$), but analysis of the data obtained was given in [9]. In these papers the question on discrete symmetries violation was not discussed.

To study violation of *P*-, and especially *T*-symmetry in the elastic $\nu(\bar{\nu})p(\mathbf{s})$-scattering one must consider angular spectrum of the recoil protons because of difficulties in the registration of the scattered neutrinos.

In this chapter, which extends and summarizes our works [1,2,7], we give the results of calculations in the tree approximation, obtained for the differential cross sections of elastic weak, electroweak and electromagnetic neutrino scattering on polarized (unpolarized) proton target without and with neutrino helicity changing. We define *P*-violating longitudinal $A_{\nu p}^{(\|)}$ and *T*-violating transverse $A_{\nu p}^{(\perp)}$ asymmetries of the cross sections on the target proton spin $\mathbf{s}$ orientation, and analyze their behavior in dependence of incoming neutrino energy, angle or kinetic energy of the recoil proton. These asymmetries depend certainly on the form factors of electromagnetic and weak neutral proton's currents, as well as on the neutrino charge radius and magnetic moment (r_ν, μ_ν).

Results obtained testify, that investigation of the effects due to symmetries violation, which lead to *P*-odd spin-momentum correlations $(\mathbf{sn})$, $(\mathbf{sm})$, and *T*-odd triple correlation $(\mathbf{s}[\mathbf{nm}])$, entering into the angular dependencies of the cross sections of elastic $\nu p(\mathbf{s})$-scattering, gives one a possibility to get additional new information about structure of the electroweak interaction between proton and neutrino.

We consider here two-particle reaction of the lepton-proton elastic scattering in the lab system: $p = (m_p, 0)$ and $k = (E, \mathbf{k})$ are 4-momenta of the initial proton and neutrino; $s = (0, \mathbf{s})$ is 4-vector of the target proton spin; $p' = (E_p, \mathbf{p})$ and $k' = (E', \mathbf{k}')$ are 4-momenta of the recoil proton and scattered neutrino. After cross section integration on a 3-

momentum of the scattered lepton $d^3k' = k'^2 dk' d\Omega_{k'}$ and on a recoil proton momentum p in neglect of lepton mass, angular distribution of the recoil protons takes the next form:

$$\frac{d\sigma}{d\Omega_p} = \frac{\sqrt{\tau(1+\tau)}(1+\tau)}{16\pi^2 m_p^2 \omega^2 \omega_0} |M|^2 . \tag{1}$$

Here $d\Omega_p = \sin\theta_p d\theta_p d\varphi_p$ is solid angle of the recoil proton; $\cos\theta_p = \frac{\mathbf{kp}}{|\mathbf{k}||\mathbf{p}|}$, $\omega = \frac{E}{m_p}$, E is energy of the incident neutrino in the lab system $(E >> m_\nu)$, $\omega_0 = \frac{\omega+1}{\omega}$, 4-momentum transfer $q = k - k' = p' - p$. It follows from the conservation laws $\mathbf{k} = \mathbf{k}' + \mathbf{p}$, $E + m_p = E' + E_p$, that

$$p = |\mathbf{p}| = \frac{2m_p\omega_0 \cos\theta_p}{\omega_0^2 - \cos^2\theta_p} = 2m_p\sqrt{\tau(1+\tau)}, \; E_p = m_p \frac{\omega_0^2 + \cos^2\theta_p}{\omega_0^2 - \cos^2\theta_p} = m_p(1+2\tau),$$

$$\cos\theta_p = \omega_0\sqrt{\frac{\tau}{1+\tau}}, \; \tau = \frac{Q^2}{4m_p^2} = \frac{\cos^2\theta_p}{\omega_0^2 - \cos^2\theta_p}, \; Q^2 = -q^2 > 0,$$

$$\tau_{\max} = \tau\big|_{\theta_p=0} = \frac{\omega^2}{2\omega+1}. \tag{2}$$

In case of assumed geometry of experiment shown in Figure 1, and full proton polarization, spin-momentum correlations arising in the cross sections are given by expressions:

$$\begin{aligned} (\mathbf{sn}) &= \cos\theta_s , \\ (\mathbf{sm}) &= \cos\theta_p \cos\theta_s + \sin\theta_p \sin\theta_s \cos\varphi_s , \\ (\mathbf{s}[\mathbf{nm}]) &= \sin\theta_s \sin\varphi_s , \end{aligned} \tag{3}$$

where θ_s and φ_s are polar and azimuthal angles of the unit vector of proton spin $\mathbf{s}$ with respect to unit vector $\mathbf{n}$ along the momentum of incoming lepton; $\mathbf{m}$ is unit vector along the momentum of recoil proton.

If proton is polarized orthogonally to reaction plane $(\mathbf{n},\mathbf{m})$, then $\theta_s = \frac{\pi}{2}$, and $(\mathbf{sn}) = 0$, $(\mathbf{sm}) = 0$. Thus $(\mathbf{s}[\mathbf{nm}]) = +1$, when proton is polarized along axis Y $\left(\varphi_s = +\frac{\pi}{2}\right)$, and $(\mathbf{s}[\mathbf{nm}]) = -1$, when proton is polarized against axis Y $\left(\varphi_s = -\frac{\pi}{2}\right)$.

Let's define asymmetry on orientation of the target proton spin with respect to reaction plane – transverse asymmetry:

$$A^{(\perp)} = \frac{d\sigma(\mathbf{s} \uparrow\uparrow [\mathbf{nm}]) - d\sigma(\mathbf{s} \downarrow\uparrow [\mathbf{nm}])}{d\sigma(\mathbf{s} \uparrow\uparrow [\mathbf{nm}]) + d\sigma(\mathbf{s} \downarrow\uparrow [\mathbf{nm}])} =$$
$$= \frac{d\sigma(\omega, \theta_p, \theta_s = \pi/2, \varphi_s = \pi/2) - d\sigma(\omega, \theta_p, \theta_s = \pi/2, \varphi_s = -\pi/2)}{d\sigma(\omega, \theta_p, \theta_s = \pi/2, \varphi_s = \pi/2) + d\sigma(\omega, \theta_p, \theta_s = \pi/2, \varphi_s = -\pi/2)}. \quad (4)$$

Here $\varphi_s = \pi/2$ corresponds to orientation of the target proton spin $\mathbf{s}$ along axis *Y*, while $\varphi_s = -\pi/2$ corresponds to its orientation against axis *Y* (see Figure 1).

Asymmetry on orientation of the target proton spin with respect to momentum of the incident neutrino (longitudinal asymmetry, $(\mathbf{s}[\mathbf{nm}]) = 0$) is defined as

$$A^{(\|)} = \frac{d\sigma(\mathbf{s} \uparrow\uparrow \mathbf{n}) - d\sigma(\mathbf{s} \downarrow\uparrow \mathbf{n})}{d\sigma(\mathbf{s} \uparrow\uparrow \mathbf{n}) + d\sigma(\mathbf{s} \downarrow\uparrow \mathbf{n})} = \frac{d\sigma(\omega, \theta_p, \theta_s = 0, \varphi_s) - d\sigma(\omega, \theta_p, \theta_s = \pi, \varphi_s)}{d\sigma(\omega, \theta_p, \theta_s = 0, \varphi_s) + d\sigma(\omega, \theta_p, \theta_s = \pi, \varphi_s)}. \quad (5)$$

When $\theta_s = 0$ and $\theta_s = \pi$ we get $(\mathbf{sn}) = +1$, $(\mathbf{sm}) = \cos\theta_p$ and $(\mathbf{sn}) = -1$, $(\mathbf{sm}) = -\cos\theta_p$ respectively.

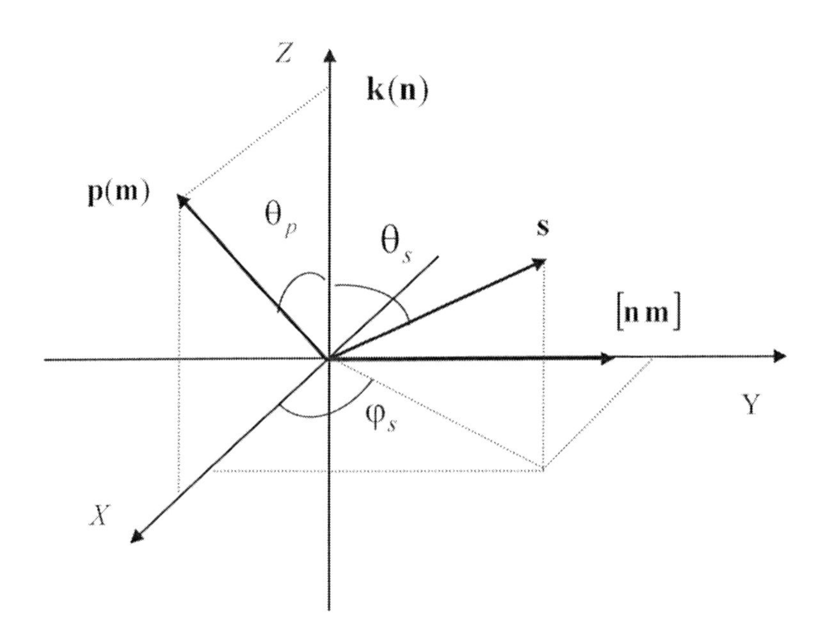

Figure 1. Coordinate system, in which we consider elastic electroweak neutrino scattering off polarized proton. Neutrino beam is directed up along Z axis. Reaction plane formed by momenta of incident neutrino and recoil proton, lies in the plane *XZ*.

Elastic Electroweak Scattering of the Neutrino on Polarized/Unpolarized Proton Target

Amplitude for the process of electroweak neutrino-proton scattering

$$\nu + p(\mathbf{s}) \xrightarrow{Z^0,\gamma} \nu + p\,,\ \nu = \nu_e, \nu_\mu, \nu_\tau\,, \tag{6}$$

including simultaneously amplitudes, corresponding to interaction of the weak neutral currents $M^{\nu p}_{weak}$ and electromagnetic interaction $M^{\nu p}_{em}$, is given by [2]

$$M^{\nu p} = M^{\nu p}_{weak} + M^{\nu p}_{em} = \frac{G_F}{\sqrt{2}} m^{\nu p}_{weak} + \frac{4\pi\alpha}{q^2} m^{\nu p}_{em} = \frac{G_F}{\sqrt{2}}\left(m^{\nu p}_{weak} - \rho_p m^{\nu p}_{em}\right), \tag{7}$$

where $\rho_p = -\frac{4\pi\alpha\sqrt{2}}{G_F q^2} = \frac{\pi\alpha\sqrt{2}}{G_F m_p^2}\frac{1}{\tau} = \frac{\rho_{0p}}{\tau}$, and parameter $\rho_{0p} = \frac{\pi\alpha\sqrt{2}}{G_F m_p^2} = 3.18\cdot 10^3$.

Amplitude of weak interaction, describing process (6) in the framework of Standard Model, can be written as [2]:

$$m^{\nu p}_{weak} = j^{\nu;\,weak}_{\alpha} J^{weak}_{\alpha} =$$

$$= \bar{u}_\nu(k')\,\gamma_\alpha(1+\gamma_5)u_\nu(k)\,\bar{u}_p(p')\left[\gamma_\alpha\left(g_{Mp} + g_{Ap}\gamma_5\right) - \frac{P_\alpha}{2m_p} f_{Vp}\right]u_p(p,s). \tag{8}$$

Here k and p are 4-momenta of the initial neutrino and proton; k' and p' are 4-momenta of the final neutrino and proton ($m_\nu << m_e, E_\nu >> m_\nu$); $P_\alpha = (p+p')_\alpha$; s is 4-vector of the initial proton spin; form factors $g_{Mp} = g_{Vp} + f_{Vp}$ and $g_{Ep} = g_{Vp} + \frac{q^2}{4m_p^2} f_{Vp}$ are known as neutral weak magnetic and electric form factors, while f_{Vp} is known as neutral weak Pauli form factor of the proton (term $\bar{u}_p(p')\left(\frac{1}{2m_p} f_{Vp}\sigma_{\alpha\beta}q_\beta\right)u_p(p)$ in the proton's weak neutral current, describing distribution of the neutral weak magnetism).

Amplitude for direct electromagnetic interaction (γ- exchange) of Dirac neutrino with proton is given by expression [2, 3, 10]:

$$m^{\nu p}_{em} = j^{\nu;\,em}_{\alpha} J^{em}_{\alpha} = \bar{u}_\nu(k')\left[f_{1\nu}\gamma_\alpha + \frac{1}{2m_e} f_{2\nu}\sigma_{\alpha\beta}q_\beta\right]u_\nu(k)J^{em}_\alpha$$

$$= \bar{u}_\nu(k')\left[f_{m\nu}\gamma_\alpha - \frac{K_\alpha}{2m_e} f_{2\nu}\right]u_\nu(k)\times\bar{u}_p(p')\left[\gamma_\alpha\left(G_{Mp} + \frac{q^2}{m_p^2}G_{1p}\gamma_5\right) - \frac{P_\alpha}{2m_p}\left(F_{2p} - iG_{2p}\gamma_5\right)\right]u_p(p,s). \tag{9}$$

Here $K_\alpha = (k+k')_\alpha$; $F_{1p}(q^2)$, $F_{2p}(q^2)$, $G_{1p}(q^2)$ и $G_{2p}(q^2)$ are form factors, describing respectively distributions of the charge, anomalous magnetic, anapole and electric dipole magnetic moments of the proton; $G_{Ep}(q^2) = F_{1p}(q^2) - \tau F_{2p}(q^2)$, $G_{Mp}(q^2) = F_{1p}(q^2) + F_{2p}(q^2)$ are Sachs electromagnetic proton's form factors; $f_{m\nu}(q^2) = f_{1\nu}(q^2) + \frac{m_\nu}{m_e} f_{2\nu}(q^2)$, $f_{1\nu}(q^2)$ and $f_{2\nu}(q^2)$ are Dirac charge and Pauli magnetic dipole form factors of the neutrino. Expression J_α^{em} in (9) takes into account electron current conservation $q_\alpha j_\alpha^{em} = 0$. Charge radius $r_\nu = \sqrt{\langle r_\nu^2 \rangle}$ and magnetic moment μ_ν of the neutrino are defined in the following way: $f_{1\nu}(q^2) = -\frac{1}{6} r_\nu^2 q^2$ and $\mu_\nu = f_{2\nu}(0)\mu_B$ with $\mu_B = \frac{e\hbar}{2m_e c}$ being Bohr magneton. Laboratory upper limit on the value of μ_ν extracted from $\bar{\nu}e$- and νe- scattering data is $\mu_\nu < 10^{-10}\mu_B$ [15].

Presence of form factors G_{1p} and G_{2p} in (9) causes violation of *P*- and *T/CP*-symmetries in νp-interactions.

Keeping in mind (7)-(9), we find from (1) following expression for dependence of differential cross section of the process (6) from recoil proton angle θ_p:

$$\frac{d\sigma^{\nu p}}{d\Omega_p} = \sum_{i=weak,\,\mathrm{int},\,em} \frac{d\sigma_i^{\nu p}}{d\Omega_p} = \frac{G_F^2 m_p^2}{2\pi^2 \omega^2 \omega_0} \sqrt{\tau(1+\tau)}(1+\tau) \frac{\left|m_{weak}^{\nu p}\right|^2 - 2\rho_p m_{int}^{\nu p} + \rho_p^2 \left|m_{em}^{\nu p}\right|^2}{16 m_p^4}, \quad (10)$$

where interference amplitude reads as $m_{int}^{\nu p} = \mathrm{Re}\left(m_{em}^{\nu p^*} m_{weak}^{\nu p}\right)$.

Taking into account helicities of the incoming and scattered neutrinos ($\zeta = \mp 1$ and $\zeta' = \mp 1$), we can split differential cross section of the electroweak scattering of Dirac neutrino $\nu(\zeta) + \mathrm{p}(\mathbf{s}) \xrightarrow{Z^0,\gamma} \nu(\zeta') + \mathrm{p}$ $(\nu = \nu_{e L,eR}, \nu_{\mu L,\mu R}, \nu_{\tau L,\tau R})$ into two parts:

$$\frac{d\sigma^{\nu p}}{d\Omega_p} = \frac{1+\zeta\zeta'}{2} \frac{d\sigma_+^{\nu p}}{d\Omega_p} + \frac{1-\zeta\zeta'}{2} \frac{d\sigma_-^{\nu p}}{d\Omega_p}. \quad (11)$$

For cross section of helicity conserving neutrino scattering ($\zeta' = \zeta = -1$, $\nu_L \to \nu_L$) we find

$$\frac{d\sigma_+^{\nu p}}{d\Omega_p}(\mathbf{s}; \omega, \theta_p, \varphi_p) = \frac{2G_F^2 m_p^2}{\pi^2} \frac{\sqrt{\tau(1+\tau)}}{\omega_0} \left\{A_+^{\nu p} + B_+^{\nu p}(\mathbf{sn}) + C_+^{\nu p}(\mathbf{sm}) + D_+^{\nu p}(\mathbf{s}[\mathbf{nm}])\right\}, \quad (12.1)$$

and for cross section of neutrino scattering with helicity flip ($\zeta' = -\zeta = +1,\ \nu_L \to \nu_R$) we get

$$\frac{d\sigma_-^{\nu p}}{d\Omega_p}\left(\mathbf{s};\omega,\theta_p,\varphi_p\right) = \frac{\alpha^2}{m_e^2} f_{2\nu}^2 \frac{1}{\omega_0}\sqrt{\frac{1+\tau}{\tau}}\left\{A_-^{\nu p} + B_-^{\nu p}(\mathbf{sn}) + C_-^{\nu p}(\mathbf{sm}) + D_-^{\nu p}(\mathbf{s}[\mathbf{nm}])\right\}. \quad (12.2)$$

Correlation factors in (12.1) are described by expressions

$$A_+^{\nu p} = A_{weak}^{\nu p} - \beta_\nu A_{int}^{\nu p} + \frac{1}{4}\beta_\nu^2 A_{em}^{\nu p},\ B_+^{\nu p} = B_{weak}^{\nu p} - \beta_\nu B_{int}^{\nu p} + \frac{1}{4}\beta_\nu^2 B_{em}^{\nu p},$$
$$C_+^{\nu p} = C_{weak}^{\nu p} - \beta_\nu C_{int}^{\nu p} + \frac{1}{4}\beta_\nu^2 C_{em}^{\nu p},\ D_+^{\nu p} = -\beta_\nu\left(D_{int}^{\nu p} - \frac{1}{4}\beta_\nu D_{em}^{\nu p}\right). \quad (13)$$

In these formulas and in what follows parameter $\beta_\nu = \rho_p f_{1\nu}$, defines relative contribution of the interference weak and direct electromagnetic interaction of neutrino with proton in the cross section for reaction (6). Using for neutrino charge form factor expression $f_{1\nu}(q^2) = \frac{2}{3} m_p^2 r_\nu^2\ \tau$ and relation $\rho_p = \frac{\rho_{0p}}{\tau}$, we find

$$\beta_\nu = \rho_p f_{1\nu} = \frac{2}{3} m_p^2 r_\nu^2\ \rho_{0p} = \frac{2\pi\alpha\sqrt{2}}{3}\frac{r_\nu^2}{G_F}. \quad (14)$$

With $r_\nu = 10^{-17}\ cm$ this parameter takes value $\beta_\nu = 4.79\cdot 10^{-4}$.

We find the next expressions for pure weak contributions in (13):

$$A_{weak}^{\nu p} = g_{Ep}^2 f_-(\omega,\tau) + \left(\tau g_{Mp}^2 + (1+\tau) g_{Ap}^2\right) f_+(\omega,\tau) + 4\frac{\tau}{\omega} f_0(\omega,\tau) g_{Mp} g_{Ap},$$
$$B_{weak}^{\nu p} = 2 g_{Ep}\left[\frac{\tau}{\omega}(1+\tau) g_{Mp} + f_0(\omega,\tau) g_{Ap}\right], \quad (15)$$
$$C_{weak}^{\nu p} = -2\sqrt{\tau(1+\tau)}\left[\frac{\tau}{\omega} g_{Mp} + \left(1 - \frac{\tau}{\omega}\right) g_{Ap}\right]\left[\omega_0 g_{Ep} - \left(1 - \frac{\tau}{\omega}\right) g_{Mp} - \frac{1+\tau}{\omega} g_{Ap}\right],$$
$$D_{weak}^{\nu p} = 0.$$

We should underline, that weak interaction by itself does not lead to violating *T*-invariance term $D_{weak}^{\nu p}$.

For interference contributions due to Z^0- and γ- exchange $(\sim \beta_\nu)$ in the formulas (13) we get:

$$A_{int}^{\nu p} = g_{Ep}G_{Ep}f_{-}(\omega,\tau)+\tau g_{Mp}\left[f_{+}(\omega,\tau)\,G_{Mp} - 8\frac{\tau}{\omega}f_0(\omega,\tau)\,G_{1p}\right]$$
$$+2\tau(1+\tau)g_{Ap}\left[\frac{1}{\omega}\left(1-\frac{\tau}{\omega}\right)G_{Mp} - 2f_{+}(\omega,\tau)\,G_{1p}\right],$$
$$B_{int}^{\nu p} = (1+\tau)\left\{G_{Ep}\left[\frac{\tau}{\omega}g_{Mp} + \left(1-\frac{\tau}{\omega}\right)g_{Ap}\right] + \tau g_{Ep}\left[\frac{1}{\omega}G_{Mp} - 4\left(1-\frac{\tau}{\omega}\right)G_{1p}\right]\right\},$$
$$C_{int}^{\nu p} = -\sqrt{\tau(1+\tau)}\left\{\omega_0\tau g_{Ep}\left[\frac{1}{\omega}G_{Mp} - 4\left(1-\frac{\tau}{\omega}\right)G_{1p}\right]\right. \tag{16}$$
$$+\tau g_{Mp}\left[\frac{\omega_0}{\omega}G_{Ep} - \frac{2}{\omega}\left(1-\frac{\tau}{\omega}\right)G_{Mp} + 4f_{+}(\omega,\tau)\,G_{1p}\right]$$
$$\left.+g_{Ap}\left[\omega_0\left(1-\frac{\tau}{\omega}\right)G_{Ep} - f_{+}(\omega,\tau)G_{Mp} + 8\frac{\tau}{\omega}f_0(\omega,\tau)\,G_{1p}\right]\right\},$$
$$D_{int}^{\nu p} = \sqrt{\tau(1+\tau)}f_0(\omega,\tau)g_{Ap}G_{2p}.$$

It is easy to see, that correlation functions $B_{int}^{\nu p}$ and $C_{int}^{\nu p}$, responsible for *P*-symmetry violation, contain contributions of proton's anapole moment in the form of correlations between following form factors $f_{1\nu}g_{Ep}G_{1p}$, $f_{1\nu}g_{Mp}G_{1p}$, $f_{1\nu}g_{Ap}G_{1p}$, while function $D_{int}^{\nu p}$, responsible for *T*-symmetry violation, includes contribution of proton's EDM in the form of correlation of G_{2p} with neutrino charge form factor (radius) and weak axial form factor $f_{1\nu}g_{Ap}G_{2p}$.

For the pure electromagnetic contributions $\left(\sim\beta_\nu^2\right)$ in (13) we find following expressions:

$$A_{em}^{\nu p} = f_{-}(\omega,\tau)\left[G_{Ep}^2 + \tau(1+\tau)G_{2p}^2\right] - 16\frac{\tau^2}{\omega}f_0(\omega,\tau)G_{Mp}G_{1p} + \tau f_{+}(\omega,\tau)\left[G_{Mp}^2 + 16\tau(1+\tau)G_{1p}^2\right],$$
$$B_{em}^{\nu p} = 2\tau(1+\tau)G_{Ep}\left[\frac{1}{\omega}G_{Mp} - 4(1-\frac{\tau}{\omega})G_{1p}\right],$$
$$C_{em}^{\nu p} = 2\tau\sqrt{\tau(1+\tau)}\left\{\frac{1}{\omega}\left(1-\frac{\tau}{\omega}\right)\left[G_{Mp}^2 + 16\tau(1+\tau)G_{1p}^2\right]\right. \tag{17.1}$$
$$\left.-4f_{+}(\omega,\tau)G_{Mp}G_{1p} - \omega_0 G_{Ep}\left[\frac{1}{\omega}G_{Mp} - 4\left(1-\frac{\tau}{\omega}\right)G_{1p}\right]\right\},$$
$$D_{em}^{\nu p} = 2\tau(1+\tau)\sqrt{\tau(1+\tau)}\left[\frac{1}{\omega}G_{Mp} - 4\left(1-\frac{\tau}{\omega}\right)G_{1p}\right]G_{2p}.$$

Correlation factors in cross section (12.2), describing angular distribution of the recoil protons in helicity changing $\nu_L \to \nu_R$ neutrino scattering, are defined by directly neutrino magnetic moment $\mu_\nu = f_{2\nu}\mu_B$ interaction with electromagnetic field of the target proton, possessing not only form factors $G_{Ep,Mp}$, but anapole and electric dipole proton form factors $G_{1p,2p}$:

$$A_-^{\nu p} = \left(1-\frac{\tau}{\omega}\right)^2 \left[G_{Ep}^2 + \tau(1+\tau)G_{2p}^2\right] + \tau f_-(\omega,\tau)\left[G_{Mp}^2 + 16\tau(1+\tau)G_{1p}^2\right],$$

$$B_-^{\nu p} = -8\tau(1+\tau)\left(1-\frac{\tau}{\omega}\right)G_{Ep}G_{1p}, \qquad (17.2)$$

$$C_-^{\nu p} = 8\tau\sqrt{\tau(1+\tau)}\left[\omega_0\left(1-\frac{\tau}{\omega}\right)G_{Ep} - f_-(\omega,\tau)G_{Mp}\right]G_{1p},$$

$$D_-^{\nu p} = -8\tau(1+\tau)\sqrt{\tau(1+\tau)}\left(1-\frac{\tau}{\omega}\right)G_{1p}G_{2p}.$$

Here are the designations used in the formulas (15)-(17):

$$f_\pm(\omega,\tau) = \left(1-\frac{\tau}{\omega}\right)^2 \pm \frac{\tau}{\omega}\frac{1+\tau}{\omega}, \quad f_0(\omega,\tau) = (1+\tau)\left(1-\frac{\tau}{\omega}\right). \qquad (18)$$

Expressions (12.1), (13)-(16) are the analogs of the formulas (8), (9), (15)-(17) in [2] for energy distributions of the recoil protons in elastic electroweak $\nu\mathbf{p}(\mathbf{s})$-scattering.

Using (12.1) we get the next expression for summary differential cross section with account of neutrino charge radius, describing neutrino scattering off unpolarized proton ($(\mathbf{sn}) = 0, (\mathbf{sm}) = 0, (\mathbf{s}[\mathbf{nm}]) = 0$) without changing neutrino helicity :

$$\frac{d\sigma_+^{\nu p}}{d\Omega_p}(\omega,\theta_p;\nu_L \to \nu_L) = \frac{d\sigma_{weak}^{\nu p}}{d\Omega_p} + \frac{d\sigma_{int}^{\nu p}}{d\Omega_p} + \frac{d\sigma_{em}^{\nu p}}{d\Omega_p} = \frac{2G_F^2 m_\mathrm{p}^2}{\pi^2}\frac{\sqrt{\tau(1+\tau)}}{\omega_0}\left\{A_{weak}^{\nu p} - \beta_\nu A_{int}^{\nu p} + \frac{1}{4}\beta_\nu^2 A_{em}^{\nu p}\right\}$$

$$= \frac{2G_F^2 m_\mathrm{p}^2}{\pi^2}\frac{\sqrt{\tau(1+\tau)}}{\omega_0}\left\{f_-(\omega,\tau)\left[\left(g_{Ep} - \frac{1}{2}\beta_\nu G_{Ep}\right)^2 + \frac{1}{4}\beta_\nu^2\tau(1+\tau)G_{2p}\right]\right.$$

$$+ f_+(\omega,\tau)\left[\tau\left(g_{Mp} - \frac{1}{2}\beta_\nu G_{Mp}\right)^2 + (1+\tau)(g_{Ap} + 2\beta_\nu\tau G_{1p})^2\right]$$

$$\left. + 4\frac{\tau}{\omega} f_0(\omega,\tau)\left(g_{Mp} - \frac{1}{2}\beta_\nu G_{Mp}\right)(g_{Ap} + 2\beta_\nu\tau G_{1p})\right\} \qquad (19)$$

In case of pure weak $(\beta_\nu = 0)$ neutrino scattering off unpolarized proton $\nu_L + p \xrightarrow{Z^0} \nu_L + p$ we get from (12.1), (13), (15)-(17.1) angular distribution of the recoil protons, following from Standard Model

$$\frac{d\sigma_+^{\nu p}}{d\Omega_p}(\omega, \theta_p; \nu_L \to \nu_L) = \frac{d\sigma_{weak}^{\nu p}}{d\Omega_p} = \frac{2G_F^2 m_p^2}{\pi^2} \frac{\sqrt{\tau(1+\tau)}}{\omega_0}$$
$$\times \left\{ f_-(\omega,\tau) g_{Ep}^2 + f_+(\omega,\tau)\left(\tau g_{Mp}^2 + (1+\tau) g_{Ap}^2\right) + 4\frac{\tau}{\omega} f_0(\omega,\tau) g_{Mp} g_{Ap} \right\}. \tag{20}$$

This cross section can be written in more traditional for weak interactions sector form:

$$\frac{d\sigma_{weak}^{\nu p}}{d\Omega_p} = \frac{G_F^2 m_p^2}{\pi^2} \frac{1}{\omega_0} \sqrt{\tau(1+\tau)}(1+\tau) \left\{ g_{Lp}^2 + \left(1 - \frac{2\tau}{\omega}\right)^2 g_{Rp}^2 - \frac{2\tau}{\omega^2} g_{Lp} g_{Rp} \right.$$
$$\left. + 2\tau f_{Vp}\left[\left(1 - \frac{2\tau}{\omega} + \frac{\tau}{\omega^2}\right) f_{Vp} + \frac{2}{\omega}\left(g_{Lp} - \left(1 - \frac{2\tau}{\omega}\right) g_{Rp}\right)\right]\right\}, \tag{21}$$

which includes chiral parameters $g_{Lp} = g_{Vp} + g_{Ap}$, $g_{Rp} = g_{Vp} - g_{Ap}$ of the proton's weak neutral current and form factor $f_{Vp}(q^2)$, describing neutral weak magnetism of the proton. The last allows to get information about strange quark contribution $(\bar{s}s)$ to magnetic form factor of the proton when studying spin asymmetries in the elastic electroweak $e^-_{L,R}p$-scattering [1, 5, 6, 11, 12], $e^-p(\mathbf{s})$-scattering [1, 4, 5, 6] and $\nu(\bar{\nu})p(\mathbf{s})$- scattering [2, 6].

Making in (21) replacement $g_{Lp} \leftrightarrow g_{Rp}$ $(g_{Ap} \to -g_{Ap})$ we get appropriate expression for the cross section $d\sigma_{weak}^{\bar{\nu}p}/d\Omega_p$ of antineutrino-proton scattering $\bar{\nu}_{lR}\, p \xrightarrow{Z^0} \bar{\nu}_{lR}\, p$, $l = e, \mu, \tau$.

With the help of formulas (12.2) and (17.2) we find, that cross section for the helicity changing scattering of the neutrino, possessing magnetic moment $\mu_\nu = f_{2\nu}\mu_B$, off unpolarized target proton is given by formula, containing all releavant electromagnetic form factors:

$$\frac{d\sigma_-^{\nu p}}{d\Omega_p}(\omega, \theta_p; \nu_L \to \nu_R)$$
$$= \frac{\alpha^2}{m_e^2}\left(\frac{\mu_\nu}{\mu_B}\right)^2 \frac{1}{\omega_0}\sqrt{\frac{1+\tau}{\tau}} \left\{ \left(1 - \frac{\tau}{\omega}\right)^2 \left[G_{Ep}^2 + \tau(1+\tau) G_{2p}^2\right] + \tau f_-(\omega,\tau)\left[G_{Mp}^2 + 16\tau(1+\tau) G_{1p}^2\right] \right\}, \tag{22}$$

Electroweak Asymmetries of the Recoil Proton Angular Distributions

Inserting (12.1) into (4) with account of (13), we find formula for transverse electroweak asymmetry of the cross section of the neutrino helicity conserving scattering $\nu_L + \mathrm{p}(\mathbf{s}) \xrightarrow{Z^0,\gamma} \nu_L + \mathrm{p}$:

$$A_{\nu p}^{(\perp)}(\omega, \theta_p; \nu_L \to \nu_L) = -\beta_\nu \frac{D_{int}^{\nu p} - \frac{1}{4}\beta_\nu D_{em}^{\nu p}}{A_{weak}^{\nu p} - \beta_\nu A_{int}^{\nu p} + \frac{1}{4}\beta_\nu^2 A_{em}^{\nu p}}. \qquad (23.1)$$

Corresponding asymmetry for the neutrino scattering with helicity flip $\nu_L + \mathrm{p}(\mathbf{s}) \xrightarrow{\gamma} \nu_R + \mathrm{p}$ is caused by electromagnetic interaction only, and can be obtained from (4) and (12.2):

$$A_{\nu p}^{(\perp)}(\omega, \theta_p; \nu_L \to \nu_R) = \frac{D_-^{\nu p}}{A_-^{\nu p}}. \qquad (23.2)$$

Analogically, from (5) we get the next expression for longitudinal weak and electroweak asymmetry of the cross section of neutrino helicity conserving scattering, containing contributions of $G_{1\mathrm{p}}$ together with other form factors:

$$A_{\nu p}^{(\parallel)}(\omega, \theta_p; \nu_L \to \nu_L) = \frac{B_{weak}^{\nu p} - \beta_\nu B_{int}^{\nu p} + \frac{1}{4}\beta_\nu^2 B_{em}^{\nu p} + \left(C_{weak}^{\nu p} - \beta_\nu C_{int}^{\nu p} + \frac{1}{4}\beta_\nu^2 C_{em}^{\nu p}\right)\cos\theta_p}{A_{weak}^{\nu p} - \beta_\nu A_{int}^{\nu p} + \frac{1}{4}\beta_\nu^2 A_{em}^{\nu p}}. \qquad (24.1)$$

Pure electromagnetic longitudinal asymmetry of the cross section of neutrino helicity changing scattering according to (12.2) takes form:

$$A_{\nu p}^{(\parallel)}(\omega, \theta_p; \nu_L \to \nu_R) = \frac{B_-^{\nu p} + C_-^{\nu p}\cos\theta_p}{A_-^{\nu p}}. \qquad (24.2)$$

As it follows from (23.1) and (16), (17.1), T-odd transverse electroweak asymmetry $A_{\nu p}^{(\perp)}$ for scattering without neutrino helicity change, is determined mainly by interference correlation of the neutrino charge form factor $f_{1\nu}$ (radius r_ν) and proton electric dipole form factor G_{2p} with weak axial-vector form factor g_{Ap} and electromagnetic form factors G_{Mp}, G_{1p}. While transverse asymmetry for electromagnetic scattering with neutrino helicity flip, according to (23.2) and (17.2) is determined exclusively by correlation of proton anapole and electric dipole form factors $G_{1p}G_{2p}$.

In limiting cases of backward $(\theta_p = 0)$ and forward $\left(\theta_p = \frac{\pi}{2}\right)$ neutrino scattering we find from (23) for helicity conserving neutrino scattering $\left(\tau_{max} = \frac{\omega^2}{2\omega+1}\right)$:

$$A_{\nu p}^{(\perp)}(\omega, \theta_p = 0; \nu_L \to \nu_L) =$$

$$= -\beta_\nu\, \omega\,\omega_0 G_{2p}(\tau) \frac{2\omega_0 g_{Ap}(\tau) + \beta_\nu\left[G_{Mp}(\tau) - 4\omega_0\tau G_{1p}(\tau)\right]}{\left\{2\left[g_{Mp}(\tau) + \omega_0 g_{Ap}(\tau)\right] - \beta_\nu\left[G_{Mp}(\tau) - 4\omega_0\tau\, G_{1p}(\tau)\right]\right\}^2}\Bigg|_{\tau=\tau_{max}},$$

$$A_{\nu p}^{(\perp)}\left(\omega, \theta_p = \frac{\pi}{2}; \nu_L \to \nu_L\right) = 0, \quad (25.1)$$

and for helicity changing neutrino scattering

$$A_{\nu p}^{(\perp)}(\omega, \theta_p = 0; \nu_L \to \nu_R) = -8\frac{\omega\,\omega_0^2\tau^2}{G_{Ep}^2(\tau) + \omega_0^2\tau^2 G_{2p}^2(\tau)} G_{1p}(\tau)G_{2p}(\tau)\Bigg|_{\tau=\tau_{max}},$$

$$A_{\nu p}^{(\perp)}\left(\omega, \theta_p = \frac{\pi}{2}; \nu_L \to \nu_R\right) = 0. \quad (25.2)$$

The same way we get from (24) for helicity conserving neutrino scattering

$$A_{\nu p}^{(\parallel)}(\omega, \theta_p = 0; \nu_L \to \nu_L) = 1,$$

$$A_{\nu p}^{(\parallel)}\left(\omega, \theta_p = \frac{\pi}{2}; \nu_L \to \nu_L\right) = \frac{2g_{Ap}(0)\left[g_{Ep}(0) - \frac{1}{2}\beta_\nu G_{Ep}(0)\right]}{g_{Ap}^2(0) + \left[g_{Ep}(0) - \beta_\nu G_{Ep}(0)\right]^2}, \quad (26.1)$$

and for helicity changing neutrino scattering

$$A_{\nu p}^{(\parallel)}(\omega, \theta_p = 0; \nu_L \to \nu_R) = 0,$$

$$A_{\nu p}^{(\parallel)}\left(\omega, \theta_p = \frac{\pi}{2}; \nu_L \to \nu_R\right) = 0. \quad (26.2)$$

One can see from formulas (25)-(26), that values of these both asymmetries at the left $(\theta_p = 0)$ and right $\left(\theta_p = \frac{\pi}{2}\right)$ bounds of the angular spectrum do not depend upon incident neutrino energy, except the values of transverse asymmetry $A_{\nu p}^{(\perp)}$ at the left bound.

Energy Distribution of the Recoil Protons

If one is interested in energy spectrum of recoil protons in the process under discussion, he ought to limit consideration by *P*-odd spin-momentum correlations only. To obtain energy distribution we must integrate cross section $d\sigma/d\Omega_p = d\sigma/d\varphi_p d\cos\theta_p$ on azimuthal angle φ_p of the recoil proton. After integration in accordance with (3), contribution from triple spin-momentum correlation $(\mathbf{s}[\mathbf{nm}])$ vanishes, and there remain only contributions from pseudoscalar correlations

$$(\mathbf{sn}) = \cos\theta_s\,,\ (\mathbf{sm}) = \cos\theta_p \cos\theta_s = \cos\theta_p (\mathbf{sn}). \tag{27}$$

Having used expression (2) for $\cos\theta_p$ through τ, and relation $\tau = \omega y/2$, we find formulas of transition from angular distribution to τ- and energy y-distribution of the recoil protons ($y = E_k/E$, E_k-kinetic energy of the recoil proton):

$$\frac{d\sigma}{d\tau} = \frac{\pi\omega_0}{(1+\tau)\sqrt{\tau(1+\tau)}}\frac{d\sigma}{d\Omega_p},\ \frac{d\sigma}{dy} = \frac{\omega}{2}\frac{d\sigma}{d\tau}. \tag{28}$$

As the result, we get from (12) next expressions for energy distributions of the recoil protons in polarized neutrino-proton electroweak scattering without and with neutrino helicity change:

$$\frac{d\sigma^{\nu p}}{dy}(\mathbf{s};\,\omega, y; \nu_L \to \nu_L) = \frac{G_F^2 m_p^2 \omega}{\pi}\frac{2}{2+\omega y}\left[A_+^{\nu p} + F_+^{\nu p}(\mathbf{sn})\right], \tag{29.1}$$

$$\frac{d\sigma^{\nu p}}{dy}(\mathbf{s};\,\omega, y; \nu_L \to \nu_R) = \frac{\pi\alpha^2}{m_e^2}\frac{2 f_{2\nu}^2}{y(2+\omega y)}\left[A_-^{\nu p} + F_-^{\nu p}(\mathbf{sn})\right]. \tag{29.2}$$

Here we introduce correlation functions

$$F_\pm^{\nu p} = B_\pm^{\nu p} + \omega_0\sqrt{\frac{\tau}{1+\tau}}C_\pm^{\nu p}, \tag{30}$$

in right hand of which one must substitute $\tau = \frac{\omega y}{2}$, with this substitution functions (18) become:

$$f_-(\omega, y) = 1 - y - \frac{y}{2\omega},\; f_+(\omega, y) = 1 - y + \frac{y^2}{2} + \frac{y}{2\omega},\; f_0(\omega, y) = \left(1 + \frac{\omega y}{2}\right)\left(1 - \frac{y}{2}\right). \quad (31)$$

Energy distributions of the recoil protons in neutrino scattering off unpolarized proton ($\nu_L p \xrightarrow{Z^0,\gamma} \nu_L p, \nu_L p \xrightarrow{\gamma} \nu_R p$) according to (29) are given by

$$\frac{d\sigma^{\nu p}}{dy}(\omega, y; \nu_L \to \nu_L) = \frac{G_F^2 m_p^2 \omega}{\pi} \frac{2}{2 + \omega y} A_+^{\nu p}, \quad (32.1)$$

$$\frac{d\sigma^{\nu p}}{dy}(\omega, y; \nu_L \to \nu_R) = \frac{\pi\alpha^2}{m_e^2} \frac{2 f_{2\nu}^2}{y(2 + \omega y)} A_-^{\nu p}. \quad (32.2)$$

From (13) and (15)-(17) with the help of (31) we get

$$A_+^{\nu p} = \left(1 - y - \frac{y}{2\omega}\right)\left[\left(g_{Ep} - \frac{\beta_\nu}{2} G_{Ep}\right)^2 + \frac{\omega y}{8}\left(1 + \frac{\omega y}{2}\right)\beta_\nu^2 G_{2p}^2\right]$$
$$+ 2y\left(1 - \frac{y}{2}\right)\left(1 + \frac{\omega y}{2}\right)\left(g_{Ap} + \omega y \beta_\nu G_{1p}\right)\left(g_{Mp} - \frac{\beta_\nu}{2} G_{Mp}\right)$$
$$+ \left(1 - y + \frac{y^2}{2} + \frac{y}{2\omega}\right)\left[\left(g_{Ap} + \omega y \beta_\nu G_{1p}\right)^2 + \frac{\omega y}{2}\left(g_{Mp} - \frac{\beta_\nu}{2} G_{Mp}\right)^2\right], \quad (33.1)$$

$$A_-^{\nu p} = \left(1 - \frac{y}{2}\right)^2$$
$$\times \left[G_{Ep}^2 + \frac{\omega y}{2}\left(1 + \frac{\omega y}{2}\right) G_{2p}^2\right] + \frac{\omega y}{2}\left(1 - y - \frac{y}{2\omega}\right)\left[G_{Mp}^2 + 8\omega y\left(1 + \frac{\omega y}{2}\right) G_{1p}^2\right]. \quad (33.2)$$

Expressions (32), (33) generalize the result, obtained in [13], to the case when proton's anapole and electric dipole moments are taken into account.

From energy spectra (29) one can get only longitudinal asymmetries:

$$A_{\nu p}^{(\parallel)}(\omega, y; \nu_L \to \nu_L) = \frac{F_+^{\nu p}}{A_+^{\nu p}}, \quad (34.1)$$

$$A_{\nu p}^{(\parallel)}(\omega, y; \nu_L \to \nu_R) = \frac{F_-^{\nu p}}{A_-^{\nu p}}. \tag{34.2}$$

Using (13), (15)-(17) and (31) we find final expressions for correlation functions:

$$\begin{aligned} F_+^{\nu p} = 2\left\{\left(1-\frac{y}{2}\right)g_{Ap} + \frac{\omega y}{2}\left[\frac{1}{\omega}g_{Mp} - \frac{\beta_\nu}{2}\left(\frac{1}{\omega}G_{Mp} - \left(1-\frac{y}{2}\right)G_{1p}\right)\right]\right\} \\ \times\left\{\left(1+\frac{\omega y}{2}\right)\left(g_{Ep} - \frac{\beta_\nu}{2}G_{Ep}\right) + \omega_0\frac{\omega y}{2}\left[\frac{1}{\omega}\left(1+\frac{\omega y}{2}\right)\left(g_{Ap} + \beta_\nu \omega y G_{1p}\right)\right.\right. \\ \left.\left. -\omega_0\left(g_{Ep} + \frac{\beta_\nu}{2}G_{Ep}\right) + \left(1-\frac{y}{2}\right)\left(g_{Mp} - \frac{\beta_\nu}{2}G_{Mp}\right)\right]\right\}, \end{aligned} \tag{35.1}$$

$$F_-^{\nu p} = -4\omega y\left(1-y-\frac{y}{2\omega}\right)\left[\left(1-\frac{y}{2}\right)G_{Ep} + \omega_0\frac{\omega y}{2}G_{Mp}\right]G_{1p}. \tag{35.2}$$

Let's pay attention to that asymmetry (34.2) according to (35.2) is distinct from zero only if proton's anapole moment is not zero $G_{1p} \neq 0$, in opposite case, when $G_{1p} = 0$, cross section for the neutrino scattering with helicity flip due to pure electromagnetic interaction is not depended from target proton spin orientation.

Analysis of the Asymmetries in Neutrino-proton Scattering

Let's point out, that in the elastic neutrino-proton scattering without neutrino helicity flip, electromagnetic interaction is a correction to weak interaction, which is directly responsible for *P*-symmetry violation. Weak-electromagnetic interference thus contributes to appropriate *P*-odd correlations and leads to arising new *CP*/*T*-odd correlations, which significance is determined by parameter $\beta_\nu = 4.79 \cdot 10^{-4}$ (we take here for neutrino charge radius value $r_\nu = 10^{-17}\ cm$ [16]).

Below, in the numeric calculations we use well known value for electroweak mixing parameter $x = \sin^2\theta_W = 0.23$ and dipole parametrization $G_{D;V,A}(Q^2) = \left(1 + Q^2/M_{V,A}^2\right)^{-2}$ for electromagnetic (G_{Ep}, G_{Mp}, G_{Mn}) and weak (g_{Ep}, g_{Mp}, g_{Ap}) nucleon form factors, which was used in our previous works [1, 2, 4, 5], but for neutron electric form factor G_{En} we take parametrization, given in [14] (see also the formula (29) in [1]).

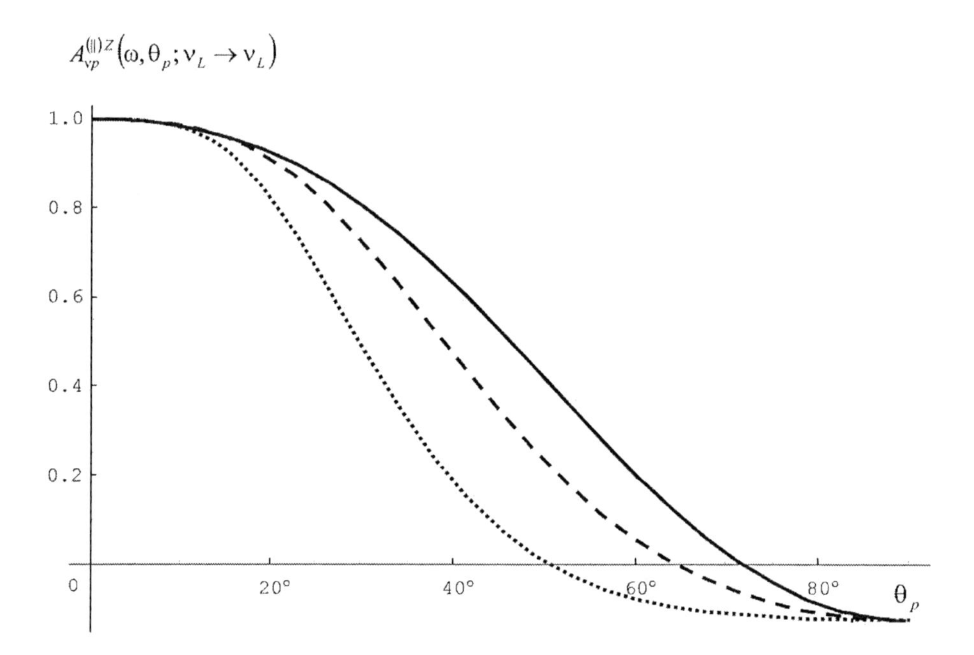

Figure 2. Angular dependences of weak longitudinal asymmetry $A_{\nu p}^{(\|)Z}(\omega,\theta_p;\nu_L \to \nu_L)$ for incident neutrino energies $E = 200\,MeV$ (solid line), $E = 500\,MeV$ (broken line), $E = 1\,GeV$ (dotted line), and $G_{1p}(0) = \delta_{0p}$.

We accept the next dipole form for Q^2 – dependences of anapole and electric dipole proton form factors

$$G_{1p}(Q^2) = G_{1p}(0)G_{D;A}(Q^2),\; G_{2p}(Q^2) = G_{2p}(0)G_{D;V}(Q^2), \tag{36}$$

and we will consider that normalization constants are of order not higher than $G_{1p}(0) \sim \delta_{0p}$, $G_{2p}(0) \sim \delta_{0p}$ $\left(\delta_{0p} = \dfrac{G_F m_p^2}{\pi\alpha\sqrt{2}} = 3.14 \cdot 10^{-4}\right)$. Experimental upper limit on the value of proton electric dipole moment $d_p < 0.54 \cdot 10^{-23}$ e$\cdot cm$ is given in [15].

Figure 2 shows dependences of the longitudinal pure weak asymmetry

$$A_{\nu p}^{(\|)Z}(\omega,\theta_p;\nu_L \to \nu_L) = A_{\nu p}^{(\|)}(\omega,\theta_p;\nu_L \to \nu_L)|_{\beta_\nu = 0} \tag{37}$$

of the cross section of neutrino-proton scattering $\nu_L + \mathrm{p(s)} \xrightarrow{Z^0} \nu_L + \mathrm{p}$ on the recoil proton angle θ_p, which are calculated from (24.1) without account of interference $(\sim \beta_\nu)$ and pure electromagnetic $(\sim \beta_\nu^2)$ contributions for energies of incident neutrino $E = 200, 500\,MeV$ and $1\,GeV$. As one can see, pure weak asymmetry in accordance with

(26.1) starts with positive maximum value $A_{\nu p}^{(\|)Z} = 1$ at $\theta_p = 0°$ and monotonously decreases to fixed negative value $A_{\nu p}^{(\|)Z} = \dfrac{2g_{Ap}(0)g_{Ep}(0)}{g_{Ap}^2(0)+g_{Ep}^2(0)} = -0.126$ at angle $\theta_p = 90°$.

Figure 3 shows calculated from (24.1) influence of interference and pure electromagnetic contributions on the longitudinal asymmetry

$$\Delta_{\nu p}^{(\|)Z\gamma,\gamma}(\omega,\theta_p;\nu_L \to \nu_L) = A_{\nu p}^{(\|)}(\omega,\theta_p;\nu_L \to \nu_L) - A_{\nu p}^{(\|)Z}(\omega,\theta_p;\nu_L \to \nu_L) \quad (38)$$

in dependence of recoil proton angle, for the same energies, as on Figure 2. It is seen, that this influence for neutrino energies up to $1\,GeV$, used values of proton form factors parameters and neutrino charge radius, is positive, i.e. increases pure weak asymmetry, but not exceed 10^{-4}.

Figure 4 shows calculated from (24.2) angular dependences of electromagnetic longitudinal asymmetry $A_{\nu p}^{(\|)}(\omega,\theta_p;\nu_L \to \nu_R)$ of the cross section of elastic neutrino-proton scattering $\nu_L + p(s) \xrightarrow{\gamma} \nu_R + p$ with neutrino helicity flip for the same energies, as on Figure 2. One can see, that this asymmetry is negative, its value reaches a maximum in the region of recoil proton angles $30° \le \theta_p \le 40°$ and vanishes at the bounds of angular spectra.

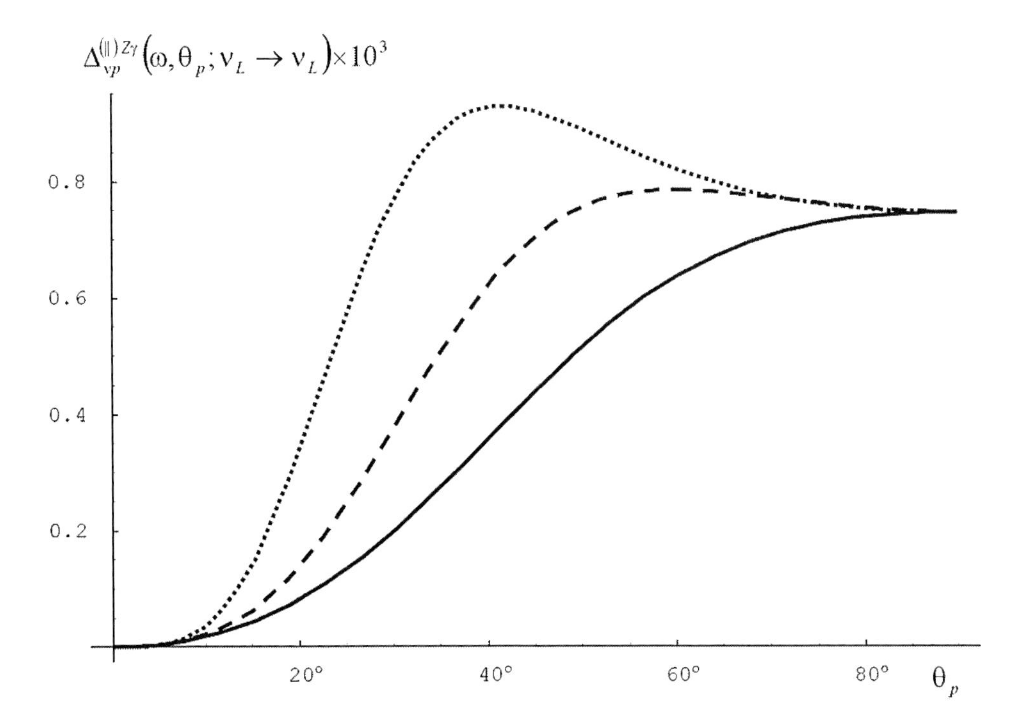

Figure 3. Influence of interference and pure electromagnetic contributions on angular dependence of longitudinal asymmetry $\Delta_{\nu p}^{(\|)Z\gamma,\gamma} = A_{\nu p}^{(\|)} - A_{\nu p}^{(\|)Z}$ for the same energies as on Figure 2.

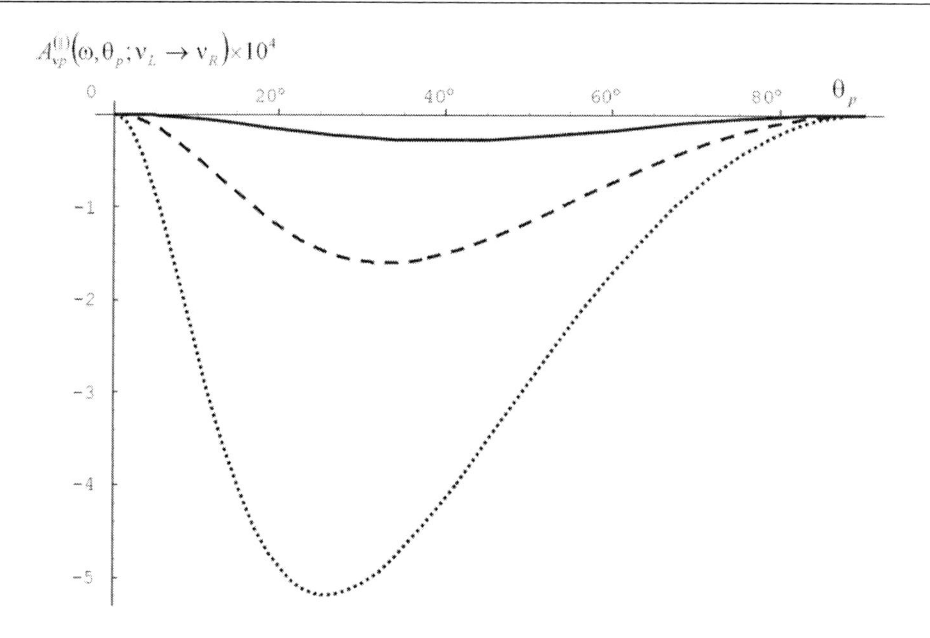

Figure 4. Angular dependences of electromagnetic longitudinal asymmetry $A_{\nu p}^{(\parallel)}(\omega,\theta_p;\nu_L \to \nu_R)$ for the same energies as on Figure 2.

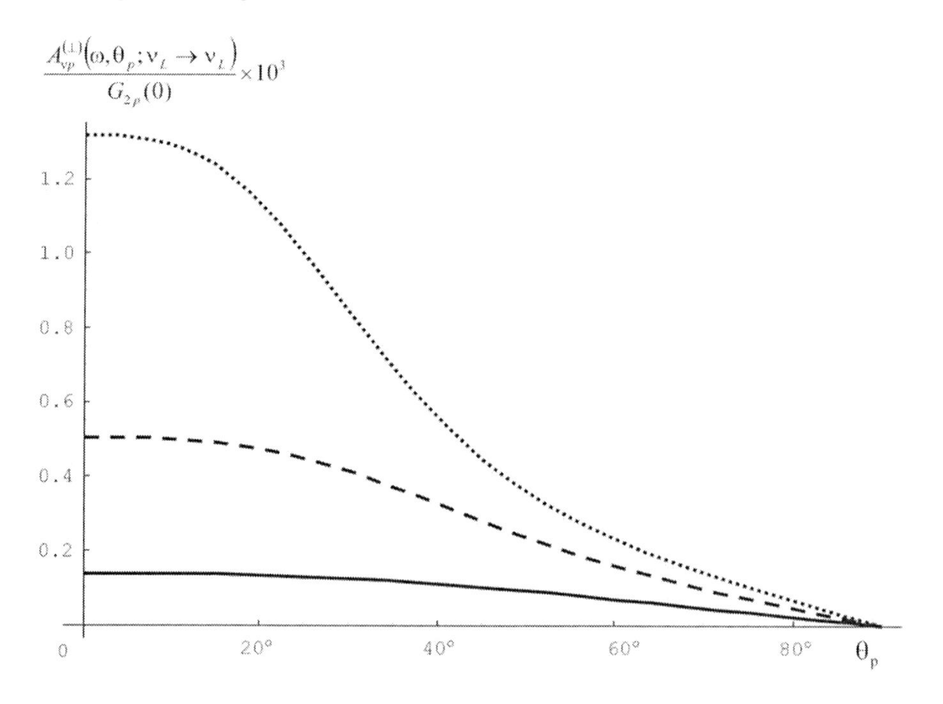

Figure 5. Angular dependences of electroweak reduced transverse T-odd asymmetry $A_{\nu p}^{(\perp)}(\omega,\theta_p;\nu_L \to \nu_L)/G_{2p}(0)$ for the same energies as on Figure 2, and $G_{1p}(0)=\delta_{0p}$, $r_\nu = 10^{-17}\ cm$.

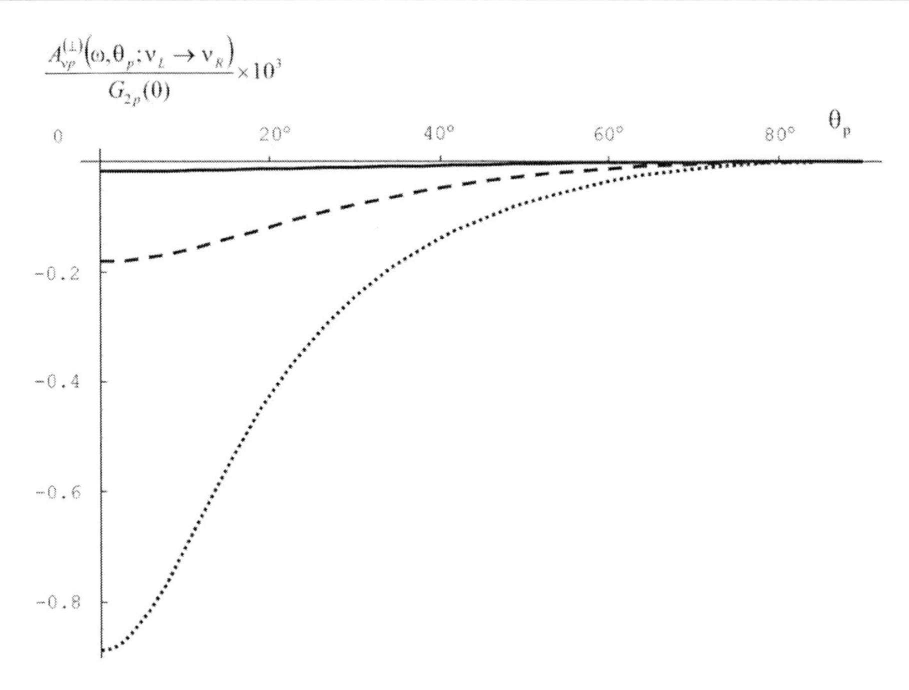

Figure 6. Angular dependences of electromagnetic reduced transverse *T*-odd asymmetry $A_{\nu p}^{(\perp)}(\omega,\theta_p;\nu_L \to \nu_R)/G_{2p}(0)$ for the same energies as on Figure 5.

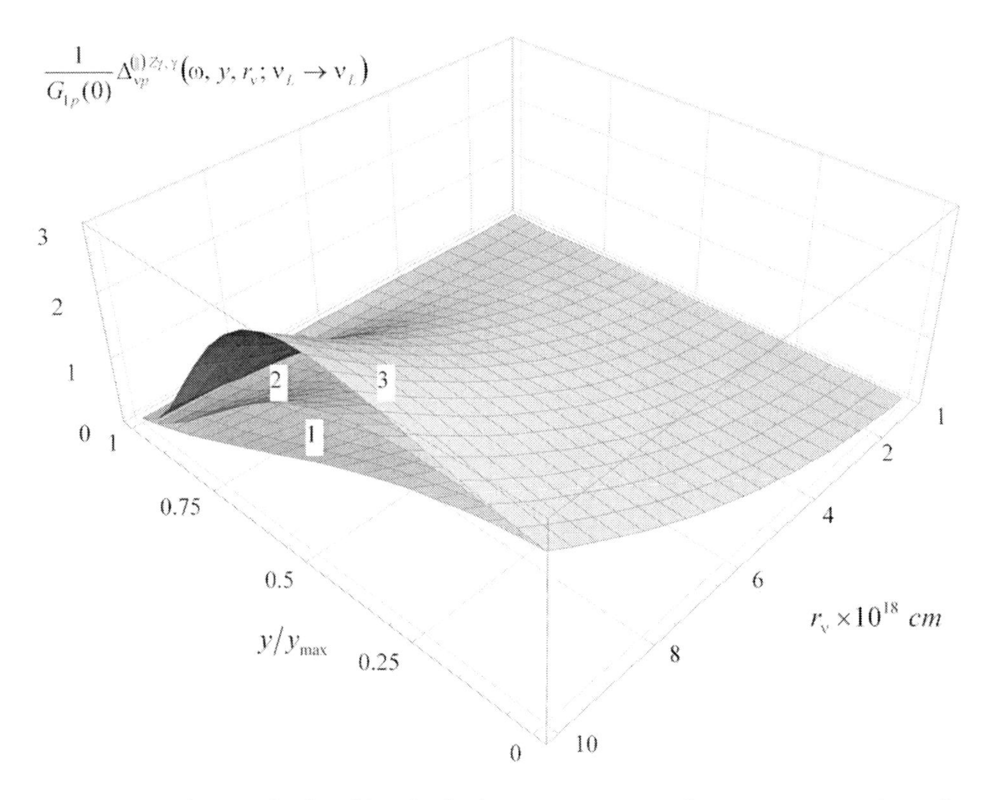

Figure 7. Dependences of reduced longitudinal asymmetry on recoil proton energy and neutrino charge radius: 1 – $E_\nu = 200\ MeV$, 2 – $E_\nu = 500\ MeV$, 3 – $E_\nu = 1\ GeV$.

On Figures 5 and 6 are shown calculated from (23) angular dependences of electroweak and electromagnetic reduced transverse T-odd asymmetries for neutrino scattering without and with its helicity flip for incident neutrino energies $E = 200, 500\ MeV$ and $E = 1\ GeV$ with $G_{1p}(0) = \delta_{0p}$, $r_\nu = 10^{-17}\ cm$. This two asymmetries have opposite signs, so that if the cross sections for neutrino scattering without and with helicity change were comparable by values, then effective transverse asymmetry would be significantly smaller.

On Figure 7 we show calculated from (33.1)-(35.1) and (38) reduced contribution $\frac{1}{G_{1p}(0)}\Delta_{\nu p}^{(\|) Z\gamma,\gamma}(\omega, y, r_\nu; \nu_L \to \nu_L)$ of interference and pure electromagnetic interaction into longitudinal asymmetry for neutrino scattering without helicity change in dependence of kinetic energy of recoil proton ($y = \frac{E_k}{E_\nu}$, $y_{\max} = \frac{2E_\nu}{2E_\nu + m_p}$) and neutrino charge radius r_ν. We see, that when $r_\nu \le 10^{-18}\ cm$ influence of this contribution to asymmetry is quite small $\Delta_{\nu p}^{(\|) Z\gamma,\gamma} << G_{1p}(0)$, whereas when $r_\nu \sim 10^{-17}\ cm$ this contribution at $y = 0$ (forward scattering of the neutrino) approaches to independent of energy of incident neutrino value $\approx 2.4\ G_{1p}(0)$.

Conclusion

In this chapter we have obtained full analytical expressions for differential cross sections of elastic neutrino scattering by polarized and unpolarized proton target. These cross sections include contributions from pure weak (Standard Model) interaction, from direct electromagnetic interaction of the neutrino, possessing charge radius and magnetic moment, with proton, and from interference between these two interactions. We show that cross sections can be divided into two parts, describing possible channels of neutrino scattering: without $(\nu_L \to \nu_L)$ and with $(\nu_L \to \nu_R)$ its helicity change. Remarkable feature of such cross section splitting is that scattering channel, conserving neutrino helicity, is caused by neutrino charge radius, while the channel with neutrino helicity flip is caused by neutrino magnetic moment.

We have defined and analyze P-odd asymmetries of angular and energy distributions of recoil protons due to pseudoscalar spin-momentum correlations, as well as T-odd asymmetry due to triple spin-momentum correlation.

As show calculations in the case of neutrino-proton scattering without helicity flip ($\nu_L \to \nu_L$), the contributions of electroweak interference and pure electromagnetic interaction $\Delta_{\nu p}^{(\|) Z\gamma,\gamma}$ to longitudinal asymmetry, are determined by neutrino charge radius (parameter β_ν) and are most essential in the region of recoil proton angles $\theta_p > 20°$. It appears also that anapole moment contribution to this asymmetry in comparison with ordinary electromagnetic proton form factors, increases with energy of the incident neutrino, and is destructive – reduces size of asymmetry with growth of $G_{1p}(0)$ value. In forward neutrino scattering ($\theta_p = \pi/2$ or $y = 0$) this contribution is not dependent of incoming neutrino energy and is equal to $\approx 2.4\ G_{1p}(0)$ – see figures 3 and 7.

Neutrino-proton scattering with helicity flip ($\nu_L \to \nu_R$) is caused by neutrino magnetic moment. Longitudinal asymmetry in this case appears due to proton anapole moment G_{1p}, it is negative and with $G_{1p}(0) \approx \delta_{0p}$ has order of magnitude 10^{-4}.

What concerns transverse *T*-odd asymmetry, one can see from corresponding formulas and figures 5 and 6, that its order of magnitude is strictly determined by the value of proton electric dipole moment G_{2p}. We pay attention, that asymmetries for neutrino scattering without and with neutrino helicity flip are close by absolute value, but have opposite signs $\left(A_{\nu p}^{(\perp)}(\nu_L \to \nu_L) > 0,\ A_{\nu p}^{(\perp)}(\nu_L \to \nu_R) < 0\right)$. So that, if intensities for this two channels of neutrino scattering were close on size, then effective transverse asymmetry would be considerably suppressed.

The results, obtained in this chapter, could be used in analysis of experimental data on neutrino-proton scattering, as well as in further searches of possible violations of *P*- and *T/CP*-symmetries, caused by electroweak structure of the proton and nonstandard electromagnetic properties of the neutrino.

References

[1] Kerimov B. K.; Safin M. Ya. Izv. *Akad. Nauk. Ser. Fiz.* 2010, 74, № 6, 854 [Bull. Rus. Acad. Sci.: Phys (Engl. Transl.). 2010, 74, No 6, 817].

[2] Kerimov B. K.; Safin M. Ya. *Yadernaya Fizika.* 2009, 72, № 11, [Phys. At. Nucl. (Engl. Transl.), 2009, 72, No 11, 1960]; SpringerLink DOI 10.1134/ S1063778809110179.

[3] Kerimov B. K.; Kasumov Yu. M.; Sadykhov F. S. *Izv. Vyssh. Uchebn. Zaved., Fiz.* 1970, № 8, 59, and references therein.

[4] Kerimov B. K.; Safin M. Ya. *Izv. Akad. Nauk. Ser. Fiz.* 2008, 72, № 11, 1601 [Bull. Rus. Acad. Sci.: Phys (Engl. Transl.). 2008, 72, No 11, 1516].

[5] Kerimov B. K.; Safin M. Ya. *Izv. Akad. Nauk. Ser. Fiz.* 2006, 70, № 5, 698.

[6] Kerimov B. K.; Safin M. Ya. *Izv. Akad. Nauk. Ser. Fiz.* 2004, 68, № 2, 184; 2002, 66, № 10, 1465; 2001, 65, № 11, 1594.

[7] Kerimov B. K. Safin M. Ya. *Book of Abstracts of LX International Conference "Nucleus* 2010", S-Petersburg, Jule 6-9, 2010, 242.

[8] Ahrens L. A. et al. *Phys. Rev.* **D35**, 785, 1987.

[9] Garvey G. T., Louis W.C., White D. H. *Phys. Rev.* **C48**, 761, 1993.

[10] Kerimov B. K.; Safin M. Ya. *Izv. Akad. Nauk. Ser. Fiz.* 2005, 69, № 11, 1681.

[11] Spayde D.T. et al. *Phys. Rev. Lett.* 2004, B58,. 79.

[12] Beise E. J., Pitt M. L., Spayde D. T. *Progr. Part. Nucl. Phys*. 2005, 54, 289 and references therein.

[13] Kerimov B. K.; Safin M. Ya. Vestn. Mosk. Univ. Ser. 3 *Fizika. Astronomia. 1995*, 36, № 3, 80.
[14] Kelly J. *J. Phys. Rev. C.* 2004, 70, 068202.
[15] Review of Particle Physics. Yao W.-M. et al. *J. Phys. G.* 2006, 33, 476, 955.
[16] Degrassi G., Sirlin A., Marciano W. *J. Phys. Rev. D.* 1989, 39, 287.

In: Neutrinos: Properties, Sources and Detection
Editor: Joshua P. Greene, pp. 23-90
ISBN 978-1-61209-650-6

Chapter 2

FIELD THEORY DESCRIPTION OF NEUTRINO OSCILLATIONS

Maxim Dvornikov*
Departamento de Física, Universidad Técnica Federico Santa María,
Casilla 110-V, Valparaíso, Chile and
IZMIRAN, 142190, Troitsk, Moscow Region, Russia

Abstract

We review various field theory approaches to the description of neutrino oscillations in vacuum and external fields. First we discuss a relativistic quantum mechanics based approach which involves the temporal evolution of massive neutrinos. To describe the dynamics of the neutrinos system we use exact solutions of wave equations in presence of an external field. It allows one to exactly take into account both the characteristics of neutrinos and the properties of an external field. In particular, we examine flavor oscillations an vacuum and in background matter as well as spin flavor oscillations in matter under the influence of an external electromagnetic field. Moreover we consider the situation of hypothetical nonstandard neutrino interactions with background fermions. In the case of ultrarelativistic particles we reproduce an effective Hamiltonian which is used in the standard quantum mechanical approach for the description of neutrino oscillations. The corrections to the quantum mechanical Hamiltonian are also discussed. Note that within the relativistic quantum mechanics method one can study the evolution of both Dirac and Majorana neutrinos. We also consider several applications of this formalism to the description of oscillations of astrophysical neutrinos emitted by a supernova and compare the behavior of Dirac and Majorana neutrinos. Then we study a spatial evolution of mixed massive neutrinos emitted by classical sources. This method seems to be more realistic since it predicts neutrino oscillations in space. Besides oscillations among different neutrino flavors, we also study transitions between particle and antiparticle states. Finally we use the quantum field theory method, which involves virtual neutrinos propagating between production and detection points, to describe particle-antiparticle transitions of Majorana neutrinos in presence of background matter.

PACS numbers: 14.60.Pq, 03.65.Pm, 13.40.Em, 14.60.St

Keywords: neutrino oscillations, exact solutions of wave equations, astrophysical neutrinos

*E-mail address: maxim.dvornikov@usm.cl

1. Introduction

Neutrino physics is one of the most rapidly developing area of high energy physics especially after the great success in experimental studies of neutrino properties. In the first place we should mention the investigation of astrophysical neutrinos since they play an important role in the evolution of various astronomical objects like stars, supernovas, quasars etc. In the course of recent experiments for the detection of solar neutrinos [1] it was revealed that transitions between different neutrino flavors, neutrino oscillations, are the most plausible theoretical explanation of the solar neutrinos deficit [2]. Flavor oscillations were also observed in the experimental studies of atmospheric neutrinos [3]. It is worth noticing that there are numerous attempts to detect neutrinos from outside the solar system (see, e.g., Ref. [4]), however, presently only SN1987A supernova neutrinos were observed [5].

Besides natural sources, one can use neutrinos produced by an accelerator or a nuclear reactor to study oscillations of these particles [6]. In this case we control the flavor content of both initial and final fluxes, i.e. it is the best strategy to examine neutrino oscillations. Besides the studies of neutrino oscillations, some accelerator based experiments are dedicated to the investigation of neutrino interactions (see, e.g., Ref. [7]).

The existence of neutrino oscillations is a direct indication to the facts that neutrinos are massive particles and there is a mixing between different neutrino generations. There are multiple theoretical mechanisms to generate masses and mixing of neutrinos [8] to fit the data of aforementioned neutrino oscillations experiments.

It was found that besides the possibility of flavor conversions in vacuum, various external fields, like interaction with background matter [9] or with an external magnetic field [10], can significantly influence neutrino oscillations. For example, the resonant enhancement of neutrino oscillations in matter, Mikheev-Smirnov-Wolfenstein (MSW) effect, plays an important role in the solution of the solar neutrino problem [11]. It should be noted that the standard model of electroweak interactions does not imply the mixing between different neutrino flavors when particles propagate in background matter. However the possibility of nonstandard neutrino interactions, which can cause transitions between neutrino flavors even at the absence of vacuum mixing, was recently discussed [12].

Since neutrinos are unlikely to have a nonzero electric charge [13], the interaction with an external electromagnetic field may be implemented due to the presence of anomalous magnetic moments. Note that, unlike an electron which has a vacuum magnetic moment, electromagnetic moments of a neutrino always arise from the loop corrections. The contributions to neutrino electromagnetic moments in various extensions of the standard model are reviewed in Ref. [14].

It should be noted that neutrino electromagnetic properties have rather complicated structure. First, in the system of mixed neutrinos there can be both diagonal and transition magnetic moments. In presence of an external electromagnetic field the former are responsible for the helicity flip within one neutrino generation and the latter cause the change of neutrino flavor (spin flavor oscillations). Second, electromagnetic properties of Dirac and Majorana neutrinos are completely different. In general case Dirac neutrinos can have all kinds of magnetic moments, whereas Majorana neutrinos possess only transition magnetic moments which are antisymmetric in neutrino flavors [15]. Nowadays there is no universally recognized experimental confirmation of the nature of neutrinos [16]. Thus it is important to study the evolution of both Dirac and Majorana neutrinos in an external electromagnetic field.

The indication to the Majorana nature of neutrinos should be experimental confirmation of the existence of neutrinoless double β-decay ($0\nu2\beta$), which is the manifestation of oscillations between neutrinos and antineutrinos [17]. This kind of transitions is possible only if neutrinos are Majorana particles. Despite numerous attempts to detect ($0\nu2\beta$)-decay in a laboratory were made [18], no confirmed events are known presently.

The great variety of evidences of neutrino oscillations requires a rigorous theoretical explanation of this phenomenon. The conventional approach to the description of neutrino oscillations is based on the quantum mechanical evolution of neutrino flavor eigenstates governed by an effective Hamiltonian [19]. This intuitive approach may be easily extended to the description of neutrino oscillations in background matter [9] and spin flavor oscillations in an external magnetic field [20] giving one a reasonable description of neutrino oscillations in these external fields.

Neutrinos are supposed to propagate as plane waves in the quantum mechanical approach. However, due to uncertainties in production and detection processes, neutrinos seem to have some distribution of their momenta, i.e. they propagate like wave packets [21]. The Gaussian distribution of the neutrino momentum is typically discussed [22]. However the actual form of the distribution strongly depends on the production and detection process (see, e.g., Ref. [23]). The necessity of the wave packets treatment of neutrino oscillations is discussed in Ref. [24]. The details of this approach including its extension to the Dirac theory were considered in Ref. [25].

The attempts to reproduce the quantum mechanical transition probability formula in vacuum were made in Refs. [26, 27] treating massive neutrinos as virtual particles propagating between production and detection points. Using this quantum field theory method one deals with the observables of charged leptons rather than with the characteristics of mixed neutrinos. The special Gaussian form of the source and detector within this quantum field theory approach was discussed in Ref. [28]. The analysis of relativistic effects in neutrino oscillations using this approach was made in Ref. [29].

In Ref. [30] neutrino flavor oscillations in vacuum are described on the basis of time evolution of the Fock states of flavor neutrinos. The time dependent transition probability formula, obtained within the S-matrix approach, was studied in Ref. [31]. The quantum mechanics analysis of the neutrino detection process and the possibility to obtain the non-standard transition probability formulas is considered in Ref. [32].

We contributed to the field theoretical description of neutrino oscillations in Refs. [33, 34, 35, 36, 37, 38, 39, 40]. We developed an approach based on relativistic quantum mechanics and applied it for the description of neutrino flavor oscillations in vacuum [33], in background matter [34, 36], and to spin flavor oscillations in an external electromagnetic field [35, 37, 39, 40]. This method is based on the first quantized neutrino wave packets which was also studied in Ref. [41] to study neutrino oscillations in vacuum and in an external electromagnetic field.

In frames of the relativistic quantum mechanics method we formulate the initial condition problem for the system of flavor neutrinos and study the subsequent time evolution of neutrino wave packets. When we discuss neutrino propagation in an external field, we use an exact solution of the corresponding wave equation in presence of this external field. With help of this method we could reproduce the Schrödinger type evolution equation, which is used in the standard quantum mechanical description of Dirac and Majorana neutrino oscillations, and discuss the correction to the standard approach.

In Ref. [38] we studied neutrino flavor oscillation in vacuum within the external classical sources method. Note that the analogous approach to the description of neutrino oscillations was also discussed in Ref. [42]. Within this approach we studied the spatial evolution of flavor neutrino waves emitted by external classical sources. We could describe the evolution of both Dirac and Majorana neutrinos in vacuum.

In the present work we review our recent achievements in theoretical description of neutrino flavor oscillations. This work is organized as follows. In Secs. 2.-5. we discuss oscillations of Dirac neutrinos in frames of relativistic quantum mechanics method. We start with the description of neutrino flavor oscillations in vacuum (Sec. 2.). Then we consider oscillations in background matter (Sec. 3.), where we also study oscillations in case of nonstandard neutrino interactions with matter. In Sec. 4. we apply the relativistic quantum mechanics method to the description of spin flavor oscillations of Dirac neutrinos in an external magnetic field. Finally, in Sec. 5., we discuss the most general situation of spin flavor oscillations in matter under the influence of an external magnetic field. In Secs. 6. and 7. we examine applications of the relativistic quantum mechanics method to the description of propagation and oscillations of astrophysical neutrinos in a magnetized envelope after the supernova explosion.

In Secs. 8. and 9. we analyze the evolution of massive mixed Majorana neutrinos in vacuum as well as in matter and magnetic field in frames of the relativistic quantum mechanics approach. Besides oscillations among different neutrino flavors we also consider transitions between neutrino and antineutrino states which are allowed if neutrinos are Majorana particles. In Sec. 10. we apply the general results to the studies of oscillations of astrophysical neutrinos in the supernova envelope supposing that neutrinos are Majorana particles and compare them with the Dirac neutrino case studied in Sec. 6..

Then, in Sec. 11., we formulate an alternative formalism for the description of neutrino flavor oscillations which is based on the mixed massive neutrinos emission by the classical sources. We examine the spatial distribution of flavor neutrino wave functions in case of localized sources. Note that vacuum oscillations of both Dirac (Sec. 11.1.) and Majorana (Sec. 11.2.) neutrinos are studied. When we discuss Majorana neutrinos case, we also consider the possibility of transitions between particles and antiparticles.

Finally, in Sec. 12., we consider the transitions among neutrinos and antineutrinos in frames of the quantum field theory, treating neutrinos as virtual particles propagating between macroscopically separated production and detection points (see also Refs. [26, 27, 28, 29]). In particular we are interested in the influence of background matter on the oscillations process since this kind of transition typically happens inside a nucleus (see, e.g., Ref. [17]) and one cannot neglect the presence of dense nuclear matter in neutrino oscillations.

In Sec. 13. we summarize our results. Several technical issues are considered in Appendices A.-D. in order not to encumber the description of the main results in Secs. 2.-12..

2. Dirac Neutrinos in Vacuum

In this section we discuss the evolution of mixed flavor neutrinos in vacuum, i.e. at the absence of external fields, using the relativistic quantum mechanics approach [33]. First we formulate the initial condition problem for the system of flavor neutrinos. We exactly solve this problem and find the time dependent wave functions and the transition probability. We also discuss the validity of the developed formalism.

Without loss of generality we study the situation of only two particles $\nu = (\nu_\alpha, \nu_\beta)$, where α and β can stay for electron, muon or τ-neutrinos. In various theoretical models (see, e.g., Ref. [43]) bigger number of flavor neutrino fields $N_\nu > 3$ is proposed. The case of arbitrary number of neutrino fields can be considered in our formalism by straightforward generalization. The Lorentz invariant Lagrangian for this system has the following form:

$$\mathcal{L} = \sum_{\lambda=\alpha,\beta} \bar{\nu}_\lambda \mathrm{i}\gamma^\mu \partial_\mu \nu_\lambda - \sum_{\lambda\lambda'=\alpha,\beta} m_{\lambda\lambda'} \bar{\nu}_\lambda \nu_{\lambda'}, \tag{1}$$

where $\gamma^\mu = (\gamma^0, \boldsymbol{\gamma})$ are Dirac matrices and $(m_{\lambda\lambda'})$ is the mass matrix which is not diagonal in the flavor neutrinos basis. The nondiagonal elements of this matrix $m_{\alpha\beta} = m_{\beta\alpha}$ correspond to the mixing between different neutrino flavors.

To describe the time evolution of the system (1) we formulate the initial condition problem for flavor neutrinos ν,

$$\nu_\lambda(\mathbf{r}, t = 0) = \nu_\lambda^{(0)}(\mathbf{r}), \quad \lambda = \alpha, \beta, \tag{2}$$

where $\nu_\lambda^{(0)}(\mathbf{r})$ are known functions. Eq. (2) means that initial field distributions of flavor neutrinos are known and we will search for their wave functions at subsequent moments of time: $\nu_\lambda(\mathbf{r}, t > 0)$. The situation when only one of the flavor neutrinos, i.e. belonging to the type "β", is present initially corresponds to a typical neutrino oscillations experiment: one looks for the initially absent neutrino flavor "α" in a beam consisting of neutrinos of the flavor "β".

To solve the initial condition problem (2) for the system (1) we introduce the mass eigenstates neutrinos ψ_a, $a = 1, 2$,

$$\nu_\lambda = \sum_{a=1,2} U_{\lambda a} \psi_a, \tag{3}$$

where the matrix $(U_{\lambda a})$ is chosen in such a way to diagonalize the mass matrix $(m_{\lambda\lambda'})$. The eigenvalues of the matrix $(m_{\lambda\lambda'})$, which are real and positive, have the meaning of the masses of the fields ψ_a. We define them as m_a.

The Lagrangian formulated in terms of the flavor neutrino fields does not provide any information about the nature of neutrinos, i.e. whether neutrinos are Dirac or Majorana particles, since in general case it is written using the two component left and right handed spinors [44, 45]. Only when we introduce the mass eigenstates and analyze the structure of resulting mass matrices, we can reveal the nature of neutrinos. We suppose that in our case the fields ψ_a are Dirac particles.

For the case of only two neutrinos system the mixing matrix $(U_{\lambda a})$ in Eq. (3) has the form,

$$(U_{\lambda a}) = \begin{pmatrix} \cos\theta & -\sin\theta \\ \sin\theta & \cos\theta \end{pmatrix}, \tag{4}$$

where θ is the vacuum mixing angle.

The Lagrangian (1) expressed in terms of the fields ψ_a reads

$$\mathcal{L} = \sum_{a=1,2} \bar{\psi}_a (\mathrm{i}\gamma^\mu \partial_\mu - m_a)\psi_a. \tag{5}$$

The Lagrangian (5) should be supplied with the initial condition,

$$\psi_a(\mathbf{r}, t=0) = \psi_a^{(0)}(\mathbf{r}), \quad \psi_a^{(0)} = (U_{a\lambda}^{-1})\nu_\lambda^{(0)}, \tag{6}$$

which follows from Eqs. (2) and (3).

The Dirac equations,

$$\mathrm{i}\dot{\psi}_a = (\boldsymbol{\alpha}\mathbf{p} + \beta m_a)\psi_a, \tag{7}$$

which result from the Lagrangian (5), reveal that the fields ψ_a decouple. In Eq. (7) we use the standard definitions for Dirac matrices $\boldsymbol{\alpha} = \gamma^0\boldsymbol{\gamma}$ and $\beta = \gamma^0$. The solution of Eq. (7) can be found in the following way:

$$\psi_a(\mathbf{r}, t) = \int \frac{\mathrm{d}^3\mathbf{p}}{(2\pi)^{3/2}} e^{\mathrm{i}\mathbf{p}\mathbf{r}} \sum_{\zeta=\pm 1} \left[a_a^{(\zeta)} u_a^{(\zeta)} e^{-\mathrm{i}E_a t} + b_a^{(\zeta)} v_a^{(\zeta)} e^{\mathrm{i}E_a t} \right], \tag{8}$$

where $E_a = \sqrt{|\mathbf{p}|^2 + m_a^2}$ is the energy of a massive neutrino in vacuum, $\zeta = \pm 1$ is the helicity of massive neutrinos, and $u_a^{(\zeta)}$ and $v_a^{(\zeta)}$ are the basis spinors corresponding to a definite helicity.

In the relativistic quantum mechanics approach to the description of neutrinos evolution [33] the coefficients $a_a^{(\zeta)}$ and $b_a^{(\zeta)}$ are c-number quantities rather than operators acting in the Fock space. Our task is to find these coefficients. Since the fields ψ_a are independent the values of the coefficients $a_a^{(\zeta)}$ and $b_a^{(\zeta)}$ depends only on the initial condition (6).

Using the orthonormality of the basis spinors,

$$u_a^{(\zeta')\dagger} u_a^{(\zeta)} = v_a^{(\zeta')\dagger} v_a^{(\zeta)} = \delta_{\zeta'\zeta}, \quad u_a^{(\zeta')\dagger} v_a^{(\zeta)} = 0, \tag{9}$$

we find the coefficients $a_a^{(\zeta)}$ and $b_a^{(\zeta)}$ in the form,

$$a_a^{(\zeta)} = \frac{1}{(2\pi)^{3/2}} u_a^{(\zeta)\dagger} \psi_a^{(0)}(\mathbf{p}), \quad b_a^{(\zeta)} = \frac{1}{(2\pi)^{3/2}} v_a^{(\zeta)\dagger} \psi_a^{(0)}(\mathbf{p}), \tag{10}$$

where

$$\psi_a^{(0)}(\mathbf{p}) = \int \mathrm{d}^3\mathbf{r} e^{-\mathrm{i}\mathbf{p}\mathbf{r}} \psi_a^{(0)}(\mathbf{r}), \tag{11}$$

is the Fourier transform of the initial condition (6).

With help of Eqs. (8)-(11) we arrive to the expression for the wave function of the neutrino mass eigenstates,

$$\psi_a(\mathbf{r}, t) = \int \mathrm{d}^3\mathbf{r}' S_a(\mathbf{r}' - \mathbf{r}, t)(-\mathrm{i}\gamma^0)\psi_a^{(0)}(\mathbf{r}), \tag{12}$$

where

$$S_a(\mathbf{r}, t) = (\mathrm{i}\gamma^\mu \partial_\mu + m_a) D_a(\mathbf{r}, t), \tag{13}$$

is the Pauli-Jourdan function for a spinor particle and

$$D_a(\mathbf{r},t) = \int \frac{d^3\mathbf{p}}{(2\pi)^3} e^{i\mathbf{pr}} \frac{\sin E_a t}{E_a}, \tag{14}$$

is the Pauli-Jourdan function for a scalar particle.

It is convenient to rewrite Eq. (12) in the form,

$$\psi_a(\mathbf{r},t) = \int \frac{d^3\mathbf{p}}{(2\pi)^3} e^{i\mathbf{pr}} S_a(-\mathbf{p},t)(-i\gamma^0)\psi_a^{(0)}(\mathbf{p}), \tag{15}$$

where

$$\begin{aligned} S_a(-\mathbf{p},t) &= \sum_{\zeta=\pm 1} \left(u_a^{(\zeta)} \otimes u_a^{(\zeta)\dagger} e^{-iE_a t} + v_a^{(\zeta)} \otimes v_a^{(\zeta)\dagger} e^{iE_a t} \right) (i\gamma^0) \\ &= \left[\cos E_a t - i \frac{\sin E_a t}{E_a} (\boldsymbol{\alpha}\mathbf{p} + \beta m_a) \right] (i\gamma^0), \end{aligned} \tag{16}$$

is the Fourier transform of the Pauli-Jourdan function (13). To derive Eq. (16) we use the summation over the helicity index formulas [46],

$$\sum_{\zeta=\pm 1} u_a^{(\zeta)} \otimes u_a^{(\zeta)\dagger} = \frac{1}{2} + \frac{1}{2E_a}(\boldsymbol{\alpha}\mathbf{p} + \beta m_a), \quad \sum_{\zeta=\pm 1} v_a^{(\zeta)} \otimes v_a^{(\zeta)\dagger} = \frac{1}{2} - \frac{1}{2E_a}(\boldsymbol{\alpha}\mathbf{p} + \beta m_a), \tag{17}$$

which are consistent with the normalization of the basis spinors (9).

Now let us specify the initial condition (2). We suggest that initially very broad wave packet is present, i.e. the coordinate dependence of the wave functions $\nu_\lambda^{(0)}(\mathbf{r})$ is close to a plane wave corresponding to the initial momentum $\mathbf{k}$. Moreover we choose the situation when only one flavor neutrino is present,

$$\nu_\alpha^{(0)}(\mathbf{r}) = 0, \quad \nu_\beta^{(0)}(\mathbf{r}) = e^{i\mathbf{kr}} \nu_\beta^{(0)}(\mathbf{k}), \tag{18}$$

where $\nu_\alpha^{(0)}(\mathbf{k})$ is the coordinate independent normalization spinor, $|\nu_\beta^{(0)}(\mathbf{k})|^2 = 1$.

Using Eqs. (3), (4), and (14)-(18) we obtain the wave function of the initially absent flavor neutrino ν_α as

$$\begin{aligned} \nu_\alpha(\mathbf{r},t) =& e^{i\mathbf{kr}} \sin\theta\cos\theta \Big\{ \cos E_1 t - \cos E_2 t \\ & - i \frac{\sin E_1 t}{E_1} (\boldsymbol{\alpha}\mathbf{k} + \beta m_1) + i \frac{\sin E_2 t}{E_2} (\boldsymbol{\alpha}\mathbf{k} + \beta m_2) \Big\} \nu_\beta^{(0)}(\mathbf{k}), \end{aligned} \tag{19}$$

where the energies are the functions of the initial momentum, $E_a = \sqrt{|\mathbf{k}|^2 + m_a^2}$.

With help of Eq. (19) we get the transition probability $P_{\nu_\beta \to \nu_\alpha}(t) = |\nu_\alpha(\mathbf{r},t)|^2$ as

$$\begin{aligned} P_{\nu_\beta \to \nu_\alpha}(t) =& \sin^2(2\theta) \Big[\sin^2(\Phi t) \\ & - \sin(\Phi t)\cos(\sigma t) \frac{1}{2} \left(\frac{m_1^2}{k^2} \sin E_1 t - \frac{m_2^2}{k^2} \sin E_2 t \right) \Big] + \mathcal{O}\left(\frac{m_a^4}{k^4} \right), \end{aligned} \tag{20}$$

where

$$\Phi = \frac{E_1 - E_2}{2} \approx \frac{\delta m^2}{4k} + \cdots, \quad \sigma = \frac{E_1 + E_2}{2} \approx k + \frac{m_1^2 + m_2^2}{4k} + \cdots, \tag{21}$$

and $\delta m^2 = m_1^2 - m_2^2$. The quantity Φ has the meaning of the phase of neutrino oscillations in vacuum. Note that the coordinate dependence is washed out from Eq. (20) since we study a very broad initial wave packet.

In the untrarelativistic limit (21) the main term in Eq. (20) resembles the usual transition probability for neutrino oscillations in vacuum. The correction to the main result is suppressed by the factor $m_a^2/k^2 \ll 1$. Using Eq. (21) we can represent Eq. (20) in the following form:

$$P_{\nu_\beta \to \nu_\alpha}(t) = \sin^2(2\theta)\left[\sin^2\left(\frac{\delta m^2}{4k}t\right) - \frac{\delta m^2}{4k^2}\sin\left(\frac{\delta m^2}{4k}t\right)\sin(2kt)\right] + \cdots, \tag{22}$$

where we drop small terms $\sim m_a^4/k^4$.

Now the leading term in Eq. (22) reproduces the well known transition probability for neutrino oscillations in vacuum derived in frames of the quantum mechanical approach [47]. The correction to the leading term, which is a rapidly oscillating function on the frequency $\sim k$, was first studied in Ref. [30] in frames of the quantum field theory approach to neutrino oscillations. We obtained analogous result using the relativistic quantum mechanics method (see also Ref. [33]). This correction to the leading term in the transition probability results from the accurate account of the Lorentz invariance.

Note that Eq. (22) is invariant under the $m_1 \leftrightarrow m_2$ transformation. It means that one cannot obtain the information about neutrino mass hierarchy studying neutrino oscillations in vacuum even taking into account the correction to the leading term in the transition probability.

Now let us discuss the possibility of initial conditions which differ from the plane wave (18). If the initial wave function is localized in a spatial region with a typical size L_0 and we measure a signal in the wave zone $|\mathbf{r}| \gg L_0$, the dependence on the particle masses m_a is washed out from the Pauli-Jourdan functions (14) and (13) (see Refs. [33, 48]). Thus neutrinos with spatially localized initial wave packets evolve in the wave zone like massless particles, which are known not to reveal flavor oscillations. Therefore the initial wave packet should be sufficiently broad.

3. Dirac Neutrinos in Background Matter

In this section we use the formalism developed in Sec. 2. to study the evolution of the system of mixed flavor neutrinos $\nu = (\nu_\alpha, \nu_\beta)$ in background matter [34, 36]. We formulate the initial condition problem for this system and solve it for ultrarelativistic neutrinos. The case of the standard model neutrino interactions is studied in details. Then we also analyze the dynamics of neutrino oscillations in presence of the nonstandard interactions which mix neutrino flavors.

The neutrino interaction with matter can be represented in the form of an external axial-vector field $f^{\mu}_{\lambda\lambda'}$ [49, 50]. As in Sec. 2., we start from the Lorentz invariant Lagrangian,

$$\mathcal{L} = \sum_{\lambda=\alpha,\beta} \bar{\nu}_\lambda \mathrm{i}\gamma^\mu \partial_\mu \nu_\lambda - \sum_{\lambda,\lambda'=\alpha,\beta} \left[m_{\lambda\lambda'} \bar{\nu}_\lambda \nu_{\lambda'} + f^{\mu}_{\lambda\lambda'} \bar{\nu}_\lambda \gamma^{\mathrm{L}}_\mu \nu_{\lambda'} \right], \tag{23}$$

where $\gamma^{\mathrm{L}}_\mu = \gamma_\mu (1-\gamma^5)/2$, $\gamma^5 = \mathrm{i}\gamma^0\gamma^1\gamma^2\gamma^3$, and keep the same notation for the mass matrix as in Sec. 2..

Note that in general case the axial vector field $f^{\mu}_{\lambda\lambda'}$ can be nondiagonal in the neutrino flavor basis. The appearance of nondiagonal elements of the matrix $(f^{\mu}_{\alpha\beta})$ is the indication to the presence of nonstandard neutrino interactions which mix neutrino flavors since in the standard model of electroweak interactions only diagonal elements of this matrix can appear. The time component of diagonal elements of this matrix, $f^{0}_{\lambda\lambda}$, is proportional to the density of background matter and spatial component, $\mathbf{f}_{\lambda\lambda}$, to the mean velocity and polarization of background fermions. the details of the averaging over the background fermions are presented in Ref. [50]

To describe the evolution of the system (23) we formulate the initial condition problem for flavor neutrinos, with the initial wave functions having an analogous form as in Eq. (2). To solve the initial condition problem we introduce the set of neutrino mass eigenstates ψ_a [see Eqs. (3) and (4)] to diagonalize the mass matrix $(m_{\lambda\lambda'})$. As in Sec. 2. we suggest that mass eigenstates ψ_a are Dirac particles.

The Lagrangian (23) expressed in terms of the fields ψ_a has the form,

$$\mathcal{L} = \sum_{a=1,2} \bar{\psi}_a (\mathrm{i}\gamma^\mu \partial_\mu - m_a)\psi_a - \sum_{a,b=1,2} g^{\mu}_{ab} \bar{\psi}_a \gamma^{\mathrm{L}}_\mu \psi_b, \tag{24}$$

where

$$(g^{\mu}_{ab}) = U^\dagger (f^{\mu}_{\lambda\lambda'}) U = \begin{pmatrix} g^\mu_1 & g^\mu \\ g^\mu & g^\mu_2 \end{pmatrix}, \tag{25}$$

is the external axial-vector field expressed in the mass eigenstates basis.

One can derive Dirac equations for the neutrino mass eigenstates,

$$\begin{aligned} \mathrm{i}\dot{\psi}_a &= \mathcal{H}_a \psi_a + \mathcal{V}\psi_b, \quad a,b = 1,2, \quad a \neq b, \\ \mathcal{H}_a &= \boldsymbol{\alpha}\mathbf{p} + \beta m_a + \beta\gamma^{\mathrm{L}}_\mu g^\mu_a, \quad \mathcal{V} = \beta\gamma^{\mathrm{L}}_\mu g^\mu, \end{aligned} \tag{26}$$

directly from the Lagrangian (24). It should be noted that Dirac equations for different neutrino mass eigenstates are coupled because of the presence of the interaction $\mathcal{V}$. Studying an exact solution of the Dirac equation for a massive neutrino in background matter we can exactly take into account the contribution of the term $\beta\gamma^{\mathrm{L}}_\mu g^\mu_a$ into the dynamics of a particle. On the contrary, the term proportional to $\mathcal{V}$, which mixes different mass eigenstates, should be studied perturbatively. Nevertheless one can account for all terms in the perturbative expansion for ultrarelativistic neutrinos.

We will study the case of nonmoving and unpolarized matter which corresponds to $\mathbf{g}_{ab} = 0$. The matrix (g^{μ}_{ab}) has only time component now,

$$(g_{ab}) \equiv (g^{0}_{ab}) = \begin{pmatrix} g_1 & g \\ g & g_2 \end{pmatrix}, \tag{27}$$

where we introduce the new notations, $g_a \equiv g^0_{aa}$ and $g \equiv g^0_{12} = g^0_{21}$. If the background matter is nonmoving and unpolarized, the Hamiltonian $\mathcal{H}_a$ commutes with the helicity operator $(\boldsymbol{\Sigma}\mathbf{p})/|\mathbf{p}|$, where $\boldsymbol{\Sigma} = \gamma^0\gamma^5\boldsymbol{\gamma}$, and we can classify the states of massive neutrinos with help of the eigenvalues of the helicity operator $\zeta = \pm 1$.

The general solution of Eq. (26) can be presented in the following way:

$$
\begin{aligned}
\psi_a(\mathbf{r},t) = & e^{-\mathrm{i}g_a t/2} \int \frac{\mathrm{d}^3\mathbf{p}}{(2\pi)^{3/2}} e^{\mathrm{i}\mathbf{p}\mathbf{r}} \\
& \times \sum_{\zeta=\pm 1} \left[a_a^{(\zeta)}(t) u_a^{(\zeta)} \exp\left(-\mathrm{i}E_a^{(\zeta)} t\right) + b_a^{(\zeta)}(t) v_a^{(\zeta)} \exp\left(+\mathrm{i}E_a^{(\zeta)} t\right) \right],
\end{aligned}
\tag{28}
$$

where $a_a^{(\zeta)}$ and $b_a^{(\zeta)}$ are the undetermined nonoperator coefficients [see Eq. (8)], which are, however, time dependent now because of the presence of the term $\mathcal{V}$ in Eq. (26).

The energy spectrum $E_a^{(\zeta)}$ in Eq. (28) was found in Ref. [51],

$$
E_a^{(\zeta)} = \sqrt{(|\mathbf{p}| - \zeta g_a/2)^2 + m_a^2},
\tag{29}
$$

for the case of nonmoving and unpolarized neutrinos. The basis spinors $u_a^{(\zeta)}$ and $v_a^{(\zeta)}$ in Eq. (28) are the eigenvectors of the helicity operator $(\boldsymbol{\Sigma}\mathbf{p})/|\mathbf{p}|$, with the eigenvalues ζ. As an example, we present here the basis spinors which correspond to an ultrarelativistic particle propagating along the z-axis,

$$
u^- = \frac{1}{\sqrt{2}} \begin{pmatrix} 0 \\ -1 \\ 0 \\ 1 \end{pmatrix}, \quad
u^+ = \frac{1}{\sqrt{2}} \begin{pmatrix} 1 \\ 0 \\ 1 \\ 0 \end{pmatrix}, \quad
v^- = \frac{1}{\sqrt{2}} \begin{pmatrix} 0 \\ 1 \\ 0 \\ 1 \end{pmatrix}, \quad
v^+ = \frac{1}{\sqrt{2}} \begin{pmatrix} 1 \\ 0 \\ -1 \\ 0 \end{pmatrix}.
\tag{30}
$$

where we omit the subscript "a" since we neglect the neutrino mass in Eq. (30). Basis spinors corresponding to arbitrary energies were also found in the explicit form in Ref. [51].

Now we should specify the initial condition. We can choose it as in Eq. (18), with $\mathbf{k} = (0,0,k)$ and $k \gg m_a$. It is also convenient to take $\nu_\beta^{(0)} = u^-$ [see Eq. (30)]. Such an initial wave function corresponds to a neutrino propagating along the z-axis, with the spin directed opposite to the particle momentum, i.e. it describes a left polarized neutrino.

If we put the *ansatz* (28) in the wave equations (26), we get the following ordinary differential equations for the functions $a_a^{(\zeta)}(t)$ and $b_a^{(\zeta)}(t)$:

$$
\begin{aligned}
\mathrm{i}\dot{a}_a^{(\zeta)} = & e^{\mathrm{i}(g_a - g_b)t/2} \exp\left(\mathrm{i}E_a^{(\zeta)} t\right) u^{(\zeta)\dagger} \mathcal{V} \\
& \times \sum_{\zeta'=\pm 1} \left[a_b^{(\zeta')} u^{(\zeta')} \exp\left(-\mathrm{i}E_b^{(\zeta')} t\right) + b_b^{(\zeta')} v^{(\zeta')} \exp\left(\mathrm{i}E_b^{(\zeta')} t\right) \right], \\
\mathrm{i}\dot{b}_a^{(\zeta)} = & e^{\mathrm{i}(g_a - g_b)t/2} \exp\left(\mathrm{i}E_a^{(\zeta)} t\right) v^{(\zeta)\dagger} \mathcal{V} \\
& \times \sum_{\zeta'=\pm 1} \left[a_b^{(\zeta')} u^{(\zeta')} \exp\left(-\mathrm{i}E_b^{(\zeta')} t\right) + b_b^{(\zeta')} v^{(\zeta')} \exp\left(\mathrm{i}E_b^{(\zeta')} t\right) \right].
\end{aligned}
\tag{31}
$$

To obtain Eq. (31) we use the orthonormality of the basis spinors (30) [see Eq. (9)]. We should supply the Eq. (31) with the initial condition,

$$a_1^{(\zeta)}(0) = \frac{\sin\theta}{(2\pi)^{3/2}} u^{(\zeta)\dagger}\nu_\beta^{(0)}, \quad a_2^{(\zeta)}(0) = \frac{\cos\theta}{(2\pi)^{3/2}} u^{(\zeta)\dagger}\nu_\beta^{(0)}, \tag{32}$$

that result from Eqs. (6) and (28). If we study an arbitrary wave packet initial condition rather than the plane wave distribution for $\nu_\beta^{(0)}(\mathbf{r})$ (18), we have to replace $\nu_\beta^{(0)}$ in Eq. (32) with the Fourier transform of the initial wave function $\nu_\beta^{(0)}(\mathbf{r})$.

Taking into account the fact that $\langle u^{(\zeta)}|\mathcal{V}|v^{(\zeta')}\rangle = 0$, we get that equations for $a_a^{(\zeta)}(t)$ and $b_a^{(\zeta)}(t)$ decouple, i.e. the interaction $\mathcal{V}$ does not mix positive and negative energy eigenstates. In the following we will consider the evolution of only $a_a^{(\zeta)}(t)$ since the dynamics of $b_a^{(\zeta)}(t)$ is studied analogously.

The only nonzero matrix elements of the potential $\mathcal{V}$ in Eq. (31) are $\langle u^-|\mathcal{V}|u^-\rangle = \langle v^+|\mathcal{V}|v^+\rangle = g$, which result from Eq. (30). Finally Eq. (31) are reduced to the ordinary differential equations only for the functions $a_a^-(t)$,

$$\mathrm{i}\dot{a}_a^- = a_b^- g \exp\left[\mathrm{i}\{E_a^- - E_b^- + (g_a - g_b)/2\}t\right], \quad a, b = 1, 2, \quad a \neq b. \tag{33}$$

It follows from Eq. (33) that equations for the functions a_a^- and a_a^+ (not shown here), decouple since the interaction with background matter conserves the particle helicity and we have chosen the initial condition corresponding to a left polarized particle. Indeed one can obtain from Eq. (32) that the functions $a_a^+(t)$ are equal to zero at $t = 0$.

The solution of Eq. (33) can be expressed in the form (see Appendix A.),

$$\begin{aligned} a_1^-(t) &= F a_1^-(0) + G a_2^-(0), & F &= \left[\cos\Omega_- t - \mathrm{i}\frac{\omega_-}{2\Omega_-}\sin\Omega t\right]\exp\left(\mathrm{i}\omega_- t/2\right), \\ a_2^-(t) &= F^* a_2^-(0) - G^* a_1^-(0), & G &= -\mathrm{i}\frac{g}{\Omega_-}\sin\Omega_- t \exp\left(\mathrm{i}\omega_- t/2\right), \end{aligned} \tag{34}$$

where $\Omega_- = \sqrt{g^2 + (\omega_-/2)^2}$ and $\omega_- = E_1^- - E_2^- + (g_1 - g_2)/2$.

Using the identity $\left(v^+ \otimes v^{+\dagger}\right)\nu_\beta^{(0)} = 0$ [see Eq. (30)] as well as Eqs. (3), (4), (28), and (34) we arrive to the wave function of the flavor neutrino ν_α,

$$\begin{aligned} \nu_\alpha(z, t) = &- \mathrm{i}\exp\left(-\mathrm{i}\sigma_- t + \mathrm{i}kz\right)\frac{\sin\Omega_- t}{\Omega_-} \\ &\times \left[g\cos 2\theta + (\omega_-/2)\sin 2\theta\right]\nu_\beta^{(0)} + \mathcal{O}\left(\frac{m_a}{k}\right), \end{aligned} \tag{35}$$

where $\sigma_- = (E_1^- + E_2^-)/2 + (g_1 + g_2)/2$. Note that Eq. (35) is the most general one which takes into account the nonstandard interactions of relativistic neutrinos with nonmoving and unpolarized matter of arbitrary density.

Now let us discuss the standard model neutrino interactions with background matter. In this case the matrix $(f_{\lambda\lambda'}^\mu)$ is diagonal: $f_{\lambda\lambda'}^\mu = f_\lambda^\mu \delta_{\lambda\lambda'}$. Since we study the nonmoving and unpolarized matter the spatial components of the four vector f_λ^μ are equal zero. If we study the matter composed of electrons, neutrons and protons, the zero-th component $f^0_\lambda \equiv f_\lambda$ is (see, e.g., Ref. [49])

$$f_\lambda = \sqrt{2}G_\mathrm{F} \sum_{f=e,p,n} n_f q_f^{(\lambda)}, \quad q_f^{(\lambda)} = \left(I_{3\mathrm{L}}^{(f)} - 2Q^{(f)} \sin^2\theta_W + \delta_{fe}\delta_{\lambda\nu_e} \right), \tag{36}$$

where n_f is the number density of background particles, $I_{3\mathrm{L}}^{(f)}$ is the third isospin component of the matter fermion f, $Q^{(f)}$ is its electric charge, θ_W is the Weinberg angle and G_F is the Fermi constant.

Using Eqs. (4), (25), and (27) and we get that matrix (g_{ab}) has the form,

$$(g_{ab}) = \begin{pmatrix} f_\alpha \cos^2\theta + f_\beta \sin^2\theta & \sin 2\theta \Delta V_\mathrm{eff}/2 \\ \sin 2\theta \Delta V_\mathrm{eff}/2 & f_\alpha \sin^2\theta + f_\beta \cos^2\theta \end{pmatrix}, \tag{37}$$

where $\Delta V_\mathrm{eff} = f_\beta - f_\alpha$ is the difference between the effective potentials of the flavor neutrinos interaction with background matter. With help of Eq. (36) we present ΔV_eff in the form,

$$\Delta V_\mathrm{eff} = \sqrt{2}G_\mathrm{F} \times \begin{cases} n_e, & \text{for } \nu_e \to \nu_{\mu,\tau}, \\ 0, & \text{for } \nu_{\mu,\tau} \to \nu_{\tau,\mu}, \end{cases} \tag{38}$$

for various oscillations channels.

In the following we will discuss the low density matter limit, $g_a \ll k$, which is fulfilled for all realistic neutrino momenta and densities of background matter. Indeed, even for the background matter in the center of a neutron star where $n_n = 10^{38}\,\mathrm{cm}^{-3}$, using Eq. (36) we get $g_a \sim 10\,\mathrm{eV}$, which is much less than any reasonable neutrino energy. With help of Eq. (29) we get that in this approximation $\omega_-/2 \approx \Phi - \cos 2\theta \Delta V_\mathrm{eff}/2$, where Φ and δm^2 are defined in Eq. (21).

The transition probability for the process $\nu_\beta \to \nu_\alpha$ can be calculated on the basis of Eq. (35) as

$$P_{\nu_\beta \to \nu_\alpha}(t) = |\nu_\alpha(z,t)|^2 \approx P_\mathrm{max} \sin^2\left(\frac{\pi}{L_\mathrm{osc}} t \right), \tag{39}$$

where

$$\begin{aligned} P_\mathrm{max} &= \frac{\Phi^2 \sin^2(2\theta)}{(\Phi \cos 2\theta - \Delta V_\mathrm{eff}/2)^2 + \Phi^2 \sin^2(2\theta)}, \\ \frac{\pi}{L_\mathrm{osc}} &= \sqrt{(\Phi \cos 2\theta - \Delta V_\mathrm{eff}/2)^2 + \Phi^2 \sin^2(2\theta)}, \end{aligned} \tag{40}$$

are the maximal transition probability and the oscillations length.

One can see that Eqs. (39) and (40) reproduce the well known formula for the neutrino oscillations probability in the background matter (see Ref. [9]). If the background density has the resonance value determined by, $\Phi \cos 2\theta = \Delta V_\mathrm{eff}^{(\mathrm{res})}/2$, the maximal transition probability reaches big values ~ 1. This resonance enhancement of neutrino oscillations in matter is known as the MSW effect [9].

Now let us discuss the modification of Eqs. (39) and (40) which include a hypothetical nonstandard neutrino interaction. One of the possibilities to include this kind of interactions is to study the nondiagonal element of the matrix $(f^{\mu}_{\lambda\lambda'})$. This interaction can produce the neutrino flavor conversion in presence of background matter. If still we discuss the nonmoving and unpolarized matter, we define the additional nonzero element as $f \equiv f^{0}_{\alpha\beta} \neq 0$. We can express the nonstandard interaction as $f = \epsilon_{\alpha\beta}(f_{\alpha} + f_{\beta})/2$ although the exact form of the nonstandard interaction dependence on the densities of background fermions is still open. The experimental constraint on the $\epsilon_{\alpha\beta}$ parameters reads $|\epsilon_{\alpha\beta}| \lesssim 0.4$ [52].

Using the same technique as to get Eqs. (39) and (40) we arrive to the modified maximal transition probability and the oscillations length,

$$
\begin{aligned}
P_{\max} &= \frac{(\Phi \sin 2\theta + f)^2}{(\Phi \cos 2\theta - \Delta V_{\text{eff}}/2)^2 + (\Phi \sin 2\theta + f)^2}, \\
\frac{\pi}{L_{\text{osc}}} &= \sqrt{(\Phi \cos 2\theta - \Delta V_{\text{eff}}/2)^2 + (\Phi \sin 2\theta + f)^2},
\end{aligned} \tag{41}
$$

Eq. (41) is an exact one valid for arbitrary magnitude of the nonstandard interaction f in contrast to the perturbative formulas derived in Ref. [53].

As one can see in Eq. (41) that in the majority of cases the nonstandard neutrino interaction of the considered type does not generate any additional resonances in neutrino oscillations. The small new interaction can just slightly change the shape of the transition probability.

We can however notice that the new interaction produces flavor oscillations even for massless neutrinos. Indeed, if we suggest that $m_a = 0$ (or $\Phi = 0$), we get that the parameters of transition probability formula, given in Eq. (41), formally coincide with that in Eq. (40), derived for the massive neutrinos, if replace $\Phi \sin 2\theta \to f$ there. However we cannot expect the appearance of the usual MSW resonance in this model since $\Phi = 0$. The amplification of neutrino oscillation can happen only if $\Delta V_{\text{eff}} = 0$. For example, it is the case for $\nu_{\mu} \leftrightarrow \nu_{\tau}$ oscillations [see Eq. (38)].

Note that we have chosen the plane wave initial condition corresponding to ultrarelativistic particles to study neutrino oscillations in background matter. It allowed us to exactly take into account the contribution of the field g in Eq. (31). It is, however, possible to study the neutrino evolution with arbitrary initial condition in low density matter [34]. It was shown in Ref. [34] that the dynamics of neutrino oscillations is consistent with the results of Ref. [9].

4. Dirac Neutrinos in an External Magnetic Field

In this section we apply the formalism developed in Secs. 2. and 3. for the description of neutrino evolution in an external electromagnetic field [35]. In contrast to the previous sections we examine the situation when the helicity of a neutrino changes together with its flavor, i.e. we study so called neutrino spin flavor oscillations, $\nu_{\beta}^{\text{L,R}} \leftrightarrow \nu_{\alpha}^{\text{R,L}}$. We derive the new transition probability formulas which account for arbitrary magnetic moments matrix.

Neutrinos are known to be uncharged particles. The constraint on the neutrino electric charge is at the level of $10^{-13}\,e$ [13]. Nevertheless there is a possibility for them to interact with an external electromagnetic field $F_{\mu\nu} = (\mathbf{E}, \mathbf{B})$ via the anomalous magnetic moments. The experimental constraint on the neutrino magnetic moments is $\sim 10^{-10}\mu_{\text{B}}$ [52, 54], where μ_{B} is the Bohr magneton. Despite the smallness of magnetic moments, its interaction with strong electromagnetic fields can produces sizeable effects (see, e.g., Secs. 6. and 7. below).

The Lagrangian for the considered system of two flavor neutrinos $\nu = (\nu_\alpha, \nu_\beta)$ is expressed in the following way:

$$\mathcal{L} = \sum_{\lambda=\alpha,\beta} \bar{\nu}_\lambda \mathrm{i}\gamma^\mu \partial_\mu \nu_\lambda - \sum_{\lambda\lambda'=\alpha,\beta} \bar{\nu}_\lambda \left(m_{\lambda\lambda'} + \frac{1}{2} M_{\lambda\lambda'} \sigma_{\mu\nu} F^{\mu\nu} \right) \nu_{\lambda'}, \tag{42}$$

where $\sigma_{\mu\nu} = (\mathrm{i}/2)(\gamma_\mu\gamma_\nu - \gamma_\nu\gamma_\mu)$. The magnetic moments matrix $(M_{\lambda\lambda'})$ in Eq. (42) is defined in the flavor eigenstates basis. In general case this matrix is independent from the mass matrix $(m_{\lambda\lambda'})$, i.e. the diagonalization of the mass matrix does not necessarily imply the diagonal form of the magnetic moments matrix.

To analyze the dynamics of the system (42) we formulate the initial condition problem [see Eqs. (2), (6), and (18)] and introduce the mass eigenstates ψ_a [see Eqs. (3) and (4)]. However, in contrast to the previous sections we should choose the normalization spinor $\nu_\beta^{(0)}$ in Eq. (18) in a specific form.

When a neutrino with anomalous magnetic moment propagate in an external electromagnetic field its helicity changes. Therefore we can impose the additional constraint on the initial spinor,

$$P_\pm \nu_\beta^{(0)} = \nu_\beta^{(0)}, \quad P_\pm = \left(1 \pm \frac{(\mathbf{\Sigma} \cdot \mathbf{k})}{|\mathbf{k}|} \right), \tag{43}$$

which means that one has neutrinos of the specific helicity initially. Here $\mathbf{\Sigma} = \gamma^5 \boldsymbol{\alpha}$ is the Dirac matrix. If we act with the operator $P_\mp$ on the final state $\nu_\alpha(\mathbf{r}, t)$, we can study the appearance of the opposite helicity eigenstates among neutrinos of the flavor "α", i.e this situation corresponds to the neutrino spin flavor oscillations $\nu_\beta^{\mathrm{L,R}} \leftrightarrow \nu_\alpha^{\mathrm{R,L}}$. For the sake of definiteness we choose the initial wave function corresponding to a left polarized neutrino and the final one to a right polarized particle.

Now we express the Lagrangian (42) using the mass eigenstates ψ_a, which diagonalize the mass matrix,

$$\mathcal{L} = \sum_{a=1,2} \bar{\psi}_a (\mathrm{i}\gamma^\mu \partial_\mu - m_a) \psi_a - \frac{1}{2} \sum_{ab=1,2} \mu_{ab} \bar{\psi}_a \sigma_{\mu\nu} \psi_b F^{\mu\nu}, \tag{44}$$

where

$$(\mu_{ab}) = U^\dagger (M_{\lambda\lambda'}) U = \begin{pmatrix} \mu_{11} & \mu_{12} \\ \mu_{21} & \mu_{22} \end{pmatrix}, \tag{45}$$

is the magnetic moment matrix presented in the mass eigenstates basis which, as we mentioned above, not necessarily to be diagonal.

As in Secs. 2. and 3. we will study the situation of mass eigenstates neutrinos ψ_a which are Dirac particles. It means that the magnetic moments matrix (μ_{ab}) can have both diagonal and nondiagonal elements. The diagonal elements of this matrix correspond to usual magnetic moments and the nondiagonal to the transition ones. The transition magnetic moments are responsible for the transitions between left and right polarized particles of different species.

Let us assume that the magnetic field is constant, uniform and directed along the z-axis, $\mathbf{B} = (0, 0, B)$, and that the electric field vanishes, $\mathbf{E} = 0$. In this case we write down the Pauli-Dirac equations for ψ_a, resulting from Eq. (44), as follows:

$$\begin{aligned} \mathrm{i}\dot{\psi}_a =& \mathcal{H}_a\psi_a + \mathcal{V}\psi_b, \quad a, b = 1, 2, \quad a \neq b, \\ \mathcal{H}_a =& (\boldsymbol{\alpha}\mathbf{p}) + \beta m_a - \mu_a\beta\Sigma_3 B, \quad \mathcal{V} = -\mu\beta\Sigma_3 B \end{aligned} \tag{46}$$

where $\mu_a = \mu_{aa}$, and $\mu = \mu_{ab} = \mu_{ba}$ are the elements of the matrix (μ_{ab}) (45).

We should notice that, as in Sec. 3., the wave equations (46) are coupled due to the presence of the interaction $\mathcal{V}$. Therefore we have to use a sort of the perturbative approach to account for this term, whereas analogous diagonal magnetic interaction $-\mu_a\beta\Sigma_3 B$ will be taken into account exactly from the very beginning.

We will study the propagation of neutrinos in the transverse magnetic field. Therefore it convenient to choose the initial momentum along the x-axis $\mathbf{k} = (k, 0, 0)$ and the initial spinor as $\nu_\beta^{(0)\mathrm{T}} = (1/2)(1, -1, -1, 1)$. It is easy to see that the wave function $\nu_\beta^{(0)}(\mathbf{r})$ describes an ultrarelativistic particle propagating along the x-axis with its spin directed opposite to the x-axis, i.e. a left polarized neutrino. The contribution of the longitudinal magnetic field to the dynamics of neutrino oscillations is suppressed by the factor $m_a/k \ll 1$ [49, 55, 56].

The general solution of Eq. (46) can be presented as follows:

$$\psi_a(\mathbf{r}, t) = \int \frac{\mathrm{d}^3\mathbf{p}}{(2\pi)^{3/2}} e^{\mathrm{i}\mathbf{p}\mathbf{r}} \sum_{\zeta=\pm 1} \left[a_a^{(\zeta)}(t) u_a^{(\zeta)} \exp\left(-\mathrm{i}E_a^{(\zeta)} t\right) + b_a^{(\zeta)}(t) v_a^{(\zeta)} \exp\left(+\mathrm{i}E_a^{(\zeta)} t\right) \right]. \tag{47}$$

Our main goal is to determine the coefficients $a_a^{(\zeta)}$ and $b_a^{(\zeta)}$ consistent with both the initial condition (18) and (43) and the evolution equation (46). As in Sec. 3. these coefficients are in general functions of time.

We have already mentioned that the helicity of a neutral particle with an anomalous magnetic moment is not conserved in an external magnetic field. Therefore to classify the states of massive neutrinos in Eq. (47) one has to use the operator [35, 57, 40],

$$\Pi_a = m_a\Sigma_3 + \mathrm{i}\gamma^0\gamma^5(\boldsymbol{\Sigma} \times \mathbf{p})_3 - \mu_a B, \tag{48}$$

which commutes with the Hamiltonian $\mathcal{H}_a$ in Eq. (46) and thus characterizes the spin direction with respect to the magnetic field. The quantum number $\zeta \pm 1$ is the sign of the eigenvalue of the operator (48).

The energy levels in Eq. (47) have the form [35, 57, 40],

$$E_a^{(\zeta)} = \sqrt{p_3^2 + \mathcal{E}_a^{(\zeta)2}}, \quad \mathcal{E}_a^{(\zeta)} = \mathcal{K}_a - \zeta\mu_a B, \quad \mathcal{K}_a = \sqrt{m_a^2 + p_1^2 + p_2^2}. \tag{49}$$

For our choice of the external magnetic field $\mathbf{B} = (0, 0, B)$ and the initial momentum $\mathbf{k} = (k, 0, 0)$, Eq. (49) reads

$$E_a^{(\zeta)} = \mathcal{K}_a - \zeta\mu_a B \approx k + \frac{m_a^2}{2k} - \zeta\mu_a B, \tag{50}$$

for ultrarelativistic neutrinos with $k \gg m_a$. Here $\mathcal{K}_a = \sqrt{k^2 + m_a^2}$ is the kinetic energy of massive neutrinos.

The exact form for the basis spinors $u_a^{(\zeta)}$ and $v_a^{(\zeta)}$ in Eq. (47) for arbitrary neutrino momentum is presented in Refs. [35, 57]. We reproduce the basis spinors for ultrarelativistic neutrinos,

$$u^- = \frac{1}{\sqrt{2}} \begin{pmatrix} 0 \\ 1 \\ 1 \\ 0 \end{pmatrix}, \quad u^+ = \frac{1}{\sqrt{2}} \begin{pmatrix} 1 \\ 0 \\ 0 \\ 1 \end{pmatrix}, \quad v^- = \frac{1}{\sqrt{2}} \begin{pmatrix} 1 \\ 0 \\ 0 \\ -1 \end{pmatrix}, \quad v^+ = \frac{1}{\sqrt{2}} \begin{pmatrix} 0 \\ 1 \\ -1 \\ 0 \end{pmatrix}, \tag{51}$$

since we will be interested in the evolution of such particles. In Eq. (51) we omit the index "a" since we examine the case of $k \gg m_a$. The basis spinors u^- and v^- correspond to the negative eigenvalue of the operator (48) (neutrino spin is directed oppositely to the magnetic field), and u^+ and v^+ to the positive one (neutrino spin is parallel to the magnetic field).

Using the general solution (47) of the Pauli-Dirac equation (46) containing the undetermined functions $a_a^{(\zeta)}$ and $b_a^{(\zeta)}$ and taking into account the orthonormality of the basis spinors (51) we get the system of ordinary differential equations for these functions $a_a^{(\zeta)}$ and $b_a^{(\zeta)}$:

$$\begin{aligned} \mathrm{i}\dot{a}_a^{(\zeta)} &= \exp(+\mathrm{i}E_a^{(\zeta)}t) u_a^{(\zeta)\dagger} \mathcal{V} \sum_{\zeta'=\pm 1} \left[a_b^{(\zeta')} u_b^{(\zeta')} \exp(-\mathrm{i}E_b^{(\zeta')}t) + b_b^{(\zeta')} v_b^{(\zeta')} \exp(+\mathrm{i}E_b^{(\zeta')}t) \right], \\ \mathrm{i}\dot{b}_a^{(\zeta)} &= \exp(-\mathrm{i}E_a^{(\zeta)}t) v_a^{(\zeta)\dagger} \mathcal{V} \sum_{\zeta'=\pm 1} \left[a_b^{(\zeta')} u_b^{(\zeta')} \exp(-\mathrm{i}E_b^{(\zeta')}t) + b_b^{(\zeta')} v_b^{(\zeta')} \exp(+\mathrm{i}E_b^{(\zeta')}t) \right], \end{aligned} \tag{52}$$

which should be supplied with the initial condition (32) but with different initial wave function $\nu_\beta^{(0)\mathrm{T}} = (1/2)(1, -1, -1, 1)$ (see above).

With help of the obvious identities $\langle u_a^{\pm} | \mathcal{V} | u_b^{\pm} \rangle = \mp \mu B$ and $\langle u_a^{\pm} | \mathcal{V} | v_b^{\mp} \rangle = 0$, which result from Eq. (51), one can cast Eq. (52) into the form

$$\mathrm{i}\dot{a}_a^{\pm} = \mp a_b^{\pm} \mu B \exp[\mathrm{i}(E_a^{\pm} - E_b^{\pm})t], \tag{53}$$

which is analogous to Eq. (33) studied in Sec. 3.. Note that the ordinary differential equations for the functions $a_a^{(\zeta)}$ and $b_a^{(\zeta)}$ again decouple.

On the basis of the results of Appendix A. we are able to write down the solution of Eq. (53) as

$$a_1^{\pm}(t) = F^{\pm} a_1^{\pm}(0) + G^{\pm} a_2^{\pm}(0), \quad a_2^{\pm}(t) = F^{\pm *} a_2^{\pm}(0) - G^{\pm *} a_1^{\pm}(0), \tag{54}$$

where

$$\begin{aligned} F^{\pm} &= \left[\cos \Omega_{\pm} t - \mathrm{i} \frac{\omega_{\pm}}{2\Omega_{\pm}} \sin \Omega_{\pm} t \right] \exp(\mathrm{i}\omega_{\pm} t/2), \\ G^{\pm} &= \pm\, \mathrm{i} \frac{\mu B}{\Omega_{\pm}} \sin \Omega_{\pm} t \exp(\mathrm{i}\omega_{\pm} t/2), \end{aligned} \tag{55}$$

and

$$\Omega_\pm = \sqrt{(\mu B)^2 + (\omega_\pm/2)^2}, \quad \omega_\pm = E_1^\pm - E_2^\pm. \tag{56}$$

The details of the derivation of Eqs. (54)-(56) from Eqs. (53) are also presented in Ref. [35].

Using Eq. (47) and Eqs. (52)-(56) and the identity $\left(v^{(\zeta)} \otimes v^{(\zeta)\dagger}\right) \nu_\beta^{(0)} = 0$ [see Eq. (51)] we obtain the wave functions ψ_a, $a = 1, 2$, as,

$$\begin{aligned}
\psi_1(x,t) = & \exp(-\mathrm{i}E_1^+ t) \left(u^+ \otimes u^{+\dagger}\right) [F^+ \psi_1(x,0) + G^+ \psi_2(x,0)] \\
& + \exp(-\mathrm{i}E_1^- t) \left(u^- \otimes u^{-\dagger}\right) [F^- \psi_1(x,0) + G^- \psi_2(x,0)], \\
\psi_2(x,t) = & \exp(-\mathrm{i}E_2^+ t) \left(u^+ \otimes u^{+\dagger}\right) [F^{+*} \psi_2(x,0) - G^{+*} \psi_1(x,0)] \\
& + \exp(-\mathrm{i}E_2^- t) \left(u^- \otimes u^{-\dagger}\right) [F^{-*} \psi_2(x,0) - G^{-*} \psi_1(x,0)],
\end{aligned} \tag{57}$$

which satisfy the chosen initial condition since $G^\pm(0) = 0$, $F^\pm(0) = 1$ [see Eqs. (55)] and $[(u^+ \otimes u^{+\dagger}) + (u^- \otimes u^{-\dagger})]\psi_a(x,0) = \psi_a(x,0)$ [see Eq. (51)].

To study the appearance of right polarized neutrinos of the type "α" we should act with the operator $P_+ = (1 + \Sigma_1)/2$ defined in Eq. (43) on the final wave function $\nu_\alpha(x,t)$,

$$\nu_\alpha^{\mathrm{R}}(x,t) = \frac{1}{2}(1 + \Sigma_1)\left[\cos\theta\psi_a(x,t) - \sin\theta\psi_a(x,t)\right], \tag{58}$$

where $\psi_a(x,t)$ are shown in Eq. (57).

With help of Eqs. (3), (4), (57), and (58) we receive for the right polarized component of ν_α the expression

$$\begin{aligned}
\nu_\alpha^{\mathrm{R}}(x,t) = & \frac{1}{2}\big\{ \sin\theta\cos\theta\big[e^{-\mathrm{i}\mathcal{K}_1 t}(e^{\mathrm{i}\mu_1 Bt}F^+ - e^{-\mathrm{i}\mu_1 Bt}F^-) \\
& - e^{-\mathrm{i}\mathcal{K}_2 t}(e^{\mathrm{i}\mu_2 Bt}F^{+*} - e^{-\mathrm{i}\mu_2 Bt}F^{-*})\big] \\
& + \cos^2\theta e^{-\mathrm{i}\mathcal{K}_1 t}(e^{\mathrm{i}\mu_1 Bt}G^+ - e^{-\mathrm{i}\mu_1 Bt}G^-) \\
& + \sin^2\theta e^{-\mathrm{i}\mathcal{K}_2 t}(e^{\mathrm{i}\mu_2 Bt}G^{+*} - e^{-\mathrm{i}\mu_2 Bt}G^{-*})\big\} e^{\mathrm{i}kx}\nu_\alpha^{(0)\mathrm{R}},
\end{aligned} \tag{59}$$

where $\left(\nu_\alpha^{(0)\mathrm{R}}\right)^{\mathrm{T}} = (1/2)(1,1,1,1)$ is the normalized spinor representing the right polarized final neutrino state.

Finally, taking into account Eqs. (55) and (56) it is possible to express the wave function in Eq. (59) in the form

$$\begin{aligned}
\nu_\alpha^{\mathrm{R}}(x,t) = & \Bigg\{ \sin\theta\cos\theta\frac{1}{2\mathrm{i}}\left[\frac{\omega_+}{\Omega_+}\sin(\Omega_+ t)\exp(\mathrm{i}\bar{\mu}Bt) - \frac{\omega_-}{\Omega_-}\sin(\Omega_- t)\exp(-\mathrm{i}\bar{\mu}Bt)\right] \\
& + \mathrm{i}\mu B\left[\frac{\sin(\Omega_+ t)}{\Omega_+}\cos^2\theta - \frac{\sin(\Omega_- t)}{\Omega_-}\sin^2\theta\right]\cos(\bar{\mu}Bt)\Bigg\} \\
& \times \exp(-\mathrm{i}\sigma t + \mathrm{i}kx)\nu_\alpha^{(0)\mathrm{R}},
\end{aligned} \tag{60}$$

where $\sigma = (\mathcal{K}_1 + \mathcal{K}_1)/2$ and $\bar{\mu} = (\mu_1 + \mu_2)/2$. The magnetic moments μ and μ_a are defined in Eq. (45).

The transition probability for the process $\nu_\beta^{\mathrm{L}} \to \nu_\alpha^{\mathrm{R}}$ can be directly obtained as the squared modulus of $\nu_\alpha^{\mathrm{R}}(x,t)$ from Eq. (59) or Eq. (60), that is $P_{\nu_\beta^{\mathrm{L}} \to \nu_\alpha^{\mathrm{R}}}(t) = |\nu_\alpha^{\mathrm{R}}(x,t)|^2$. Notice that the probability is a function of time alone with no dependence on spatial coordinates. This is of course obvious as we have taken the initial wave function as a plane wave and the the magnetic field spatially constant.

Let us now apply the general results Eq. (59) or Eq. (60) to two special cases. We first consider the situation where $\mu_{1,2} \gg \mu$, i.e. the case when the transition magnetic moment is small compared with the diagonal ones. Using Eqs. (55) and (56) we find that, in this case $F^{\pm} \approx 1$ and $\Omega_{\pm} \approx \omega_{\pm}/2$, and Eq. (59) takes the form

$$\begin{aligned}\nu_\alpha^{\mathrm{R}}(x,t) \approx \mathrm{i} \Bigg\{ & \sin\theta\cos\theta \left[e^{-\mathrm{i}\mathcal{K}_1 t} \sin\mu_1 B t - e^{-\mathrm{i}\mathcal{K}_2 t} \sin\mu_2 B t \right] \\ & + \cos 2\theta \frac{\mu B}{2} \left(e^{-\mathrm{i}\Sigma_+ t} \frac{\sin\Delta_+ t}{\Delta_+} + e^{-\mathrm{i}\Sigma_- t} \frac{\sin\Delta_- t}{\Delta_-} \right) \Bigg\} e^{ikx} \nu_\alpha^{(0)\mathrm{R}},\end{aligned} \quad (61)$$

where

$$\Sigma_{\pm} = \sigma \pm \bar{\mu} B, \quad \Delta_{\pm} = \Phi \pm \delta\mu B, \quad \delta\mu = \frac{\mu_1 - \mu_2}{2},$$

and the phase of vacuum oscillations Φ was defined in Eq. (21). Eq. (61) was obtained in Ref. [35] using the perturbative methods. Assuming that $\mu \ll \mu_a$ the perturbation theory was developed in that work. Now we rederive the same result as a particular case of the more general result.

As another application of our general result we will study the situation, where the transition magnetic moments are much larger than the diagonal ones, that is $\mu \gg \mu_{1,2}$. In this case Eqs. (55) gives $F^+ \approx F^-$ and $G^+ \approx -G^-$, and we receive from Eq. (60) for the wave function ν_α^{R} the expression

$$\nu_\alpha^{\mathrm{R}}(x,t) \approx \mathrm{i} \exp(-\mathrm{i}\sigma t + \mathrm{i}kx) \cos(2\theta) \frac{\mu B}{\Omega_{\mathrm{B}}} \sin(\Omega_{\mathrm{B}} t) \nu_\alpha^{(0)\mathrm{R}}, \quad (62)$$

where

$$\Omega_{\mathrm{B}} = \sqrt{(\mu B)^2 + \Phi^2}. \quad (63)$$

The transition probability for the process $\nu_\beta^{\mathrm{L}} \to \nu_\alpha^{\mathrm{R}}$ is then given by

$$P_{\nu_\beta^{\mathrm{L}} \to \nu_\alpha^{\mathrm{R}}}(t) = \cos^2(2\theta) \left(\frac{\mu B}{\Omega_{\mathrm{B}}} \right)^2 \sin^2(\Omega_{\mathrm{B}} t). \quad (64)$$

The behavior of the system in this case is schematically illustrated in Fig 1. It should be noticed that the analog of Eq. (64) was obtained in Ref. [20] where the authors studied the resonant spin flavor precession of Dirac and Majorana neutrinos in matter under the influence of an external magnetic field in frames of the quantum mechanical approach.

Using Eq. (60) one can describe spin flavor oscillations of Dirac neutrinos with arbitrary magnetic moments matrix. It is the new result which was obtained using the relativistic quantum mechanics approach. Nevertheless this result is consistent with the conventional quantum mechanical description of spin flavor oscillations. We will demonstrate the consistency in Sec. 5. where more general case of neutrinos propagating in matter and external magnetic field is studied.

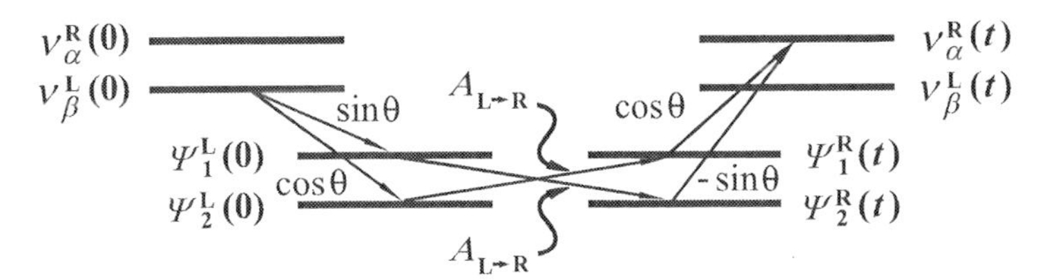

Figure 1. The schematic illustration of the system evolution in the case $\mu \gg \mu_{1,2}$. The horizontal lines of the figure correspond to various neutrino eigenstates at different moments of time ($t = 0$ and t). The expressions next to arrows correspond to the appropriate factors in the formula (62) of the wave function. The arrows from $\nu_\beta^L(0)$ to $\psi_1^L(0)$ and $\psi_2^L(0)$, for example, indicate the vacuum mixing matrix transformation at $t = 0$ and the arrow from $\psi_1^L(0)$ to $\psi_2^R(t)$ the evolution of the mass eigenstates with the helicity change. The transitions $\psi_a^L(0) \to \psi_b^R(t)$ can be described by the formula, $\psi_{1,2}^R(t) = A_{L\to R}\psi_{2,1}^L(0)$, where $A_{L\to R} = i(\mu B/\Omega)\sin\Omega t$. This figure is taken from Ref. [35].

Spin flavor oscillations of Dirac neutrinos with arbitrary initial condition, not necessarily corresponding to ultrarelativistic particles, were studied in Ref. [35] using the perturbative approach. To effectively apply the perturbation theory one has to study the situation of small transition magnetic moment, $\mu \ll \mu_a$. One could derive Eq. (61) using the results of Ref. [35] in the limit $k \gg m_a$.

5. Dirac Neutrinos in Matter under the Influence of a Magnetic Field

In this section, using the relativistic quantum mechanics method, we study the general case of the mixed flavor neutrinos propagating in background matter and interacting with an external electromagnetic field [40]. We formulate the initial condition problem for neutrino spin flavor oscillations. Then we derive the effective Hamiltonian which governs spin flavor oscillations and show the consistence of our approach to the usual quantum mechanics method. The corrections to the standard effective Hamiltonian are also obtained.

The Lagrangian for the system of two mixed flavor neutrinos $\nu = (\nu_\alpha, \nu_\beta)$ interacting with background matter and external electromagnetic field has the form,

$$\mathcal{L} = \sum_{\lambda=\alpha,\beta} \bar{\nu}_\lambda i\gamma^\mu\partial_\mu\nu_\lambda - \sum_{\lambda\lambda'=\alpha,\beta} \bar{\nu}_\lambda \left(m_{\lambda\lambda'} + \gamma_\mu^L f_{\lambda\lambda'}^\mu + \frac{1}{2} M_{\lambda\lambda'}\sigma_{\mu\nu}F^{\mu\nu} \right) \nu_{\lambda'}, \tag{65}$$

where the mass matrix $(m_{\lambda\lambda'})$, matter interaction matrix $(f_{\lambda\lambda'}^\mu)$, and the magnetic moments matrix $(M_{\lambda\lambda'})$ are defined in Secs. 2., 3., and 4. respectively.

In the following we will be interested in the standard model neutrino interaction with matter which corresponds to the diagonal matrix $f_{\lambda\lambda'}^\mu = \delta_{\lambda\lambda'} f_\lambda^\mu$. Moreover we will study the situation of nonmoving and unpolarized matter. In this case only zero-th component of the four vector f_λ^μ is not equal to zero. The explicit form of this component $f_\lambda \equiv f_\lambda^0$ for the background matter composed of electrons, protons, and neutrons is given in Eq. (36).

We choose the configuration of the electromagnetic field $F_{\mu\nu} = (\mathbf{E}, \mathbf{B})$ in Eq. (65) in the same form as in Sec. 4.. Namely, we suppose that the electric field is absent $\mathbf{E} = 0$ and magnetic field is constant and directed along the z-axis, $\mathbf{B} = (0, 0, B)$.

To study the time evolution of flavor neutrinos we should supply the Lagrangian (65) with some initial condition. We choose the initial wave functions in the same form as in Sec. 4., i.e. we suppose that $\nu_\alpha(\mathbf{r}, 0) = 0$ and $\nu_\beta(\mathbf{r}, 0) = e^{ikx}\nu_\beta^{(0)}$, where the spinor $\nu_\beta^{(0)}$ corresponds to either left or right polarized neutrinos. The explicit form of $\nu_\beta^{(0)}$ can be defined with help of the operators $P_\pm$ in Eq. (43). It means that initially we have a beam of neutrinos of the flavor "β" with a specific polarization propagating along the x-axis. If we study the appearance of neutrinos of the flavor "α" of the opposite polarization, it will correspond to a typical situation of neutrino spin flavor oscillations in matter and transversal magnetic field, $\nu_\beta^{\mathrm{L,R}} \leftrightarrow \nu_\alpha^{\mathrm{R,L}}$.

Then we introduce the mass eigenstates ψ_a using Eqs. (3) and (4) to diagonalize the mass matrix $(m_{\lambda\lambda'})$ in Eq. (65). These mass eigenstates are again supposed to be Dirac particles. Now the Lagrangian (65) expressed via the new mass eigenstates has the form,

$$\mathcal{L} = \sum_{a=1,2} \bar{\psi}_a (\mathrm{i}\gamma^\mu \partial_\mu - m_a)\psi_a - \sum_{ab=1,2} \bar{\psi}_a \left(g_{ab}^\mu \gamma_\mu^{\mathrm{L}} + \frac{1}{2}\mu_{ab}\sigma_{\mu\nu}F^{\mu\nu} \right) \psi_b, \tag{66}$$

where (g_{ab}^μ) and (μ_{ab}) are the matrix of neutrino interaction with matter and the neutrino magnetic moments matrix expressed in the mass eigenstates basis, which are defined in Eqs. (25) and (45). We remind that in case of nonmoving and unpolarized matter the matrix (g_{ab}^μ) has only zero-th component (27).

On the basis of the mass eigenstates Lagrangian (66), we can derive the corresponding wave equations which have the following form:

$$\begin{aligned} \mathrm{i}\dot{\psi}_a =& \mathcal{H}_a \psi_a + \mathcal{V}\psi_b, \quad a = 1, 2, \quad a \neq b, \\ \mathcal{H}_a =& (\boldsymbol{\alpha}\mathbf{p}) + \beta m_a - \mu_a \beta \Sigma_3 B + g_a (1 - \gamma^5)/2, \\ \mathcal{V} =& - \mu\beta\Sigma_3 B + g(1 - \gamma^5)/2. \end{aligned} \tag{67}$$

Note that we cannot directly solve the wave equations (67) because of the nondiagonal interaction $\mathcal{V}$ which mixes different mass eigenstates (see also Secs. 3. and 4.). Nevertheless we can point out an exact solution of the wave equation $\mathrm{i}\dot{\psi}_a = \mathcal{H}_a\psi_a$, for a single mass eigenstate ψ_a, that exactly accounts for the influence of the external fields g_a and $\mu_a B$. The contribution of the mixing potential $\mathcal{V}$ can be then taken into account using the perturbation theory, with all the terms in the expansion series being accounted for exactly.

We look for the solution of Eq. (67) in the following form [40]:

$$\begin{aligned} \psi_a(\mathbf{r}, t) =& e^{-\mathrm{i}g_a t/2} \int \frac{\mathrm{d}^3\mathbf{p}}{(2\pi)^{3/2}} e^{\mathrm{i}\mathbf{p}\mathbf{r}} \\ & \times \sum_{\zeta=\pm 1} \left[a_a^{(\zeta)}(t) u_a^{(\zeta)} \exp\left(-\mathrm{i}E_a^{(\zeta)} t\right) + b_a^{(\zeta)}(t) v_a^{(\zeta)} \exp\left(\mathrm{i}E_a^{(\zeta)} t\right) \right], \end{aligned} \tag{68}$$

where the energy levels, which were found in Ref. [37], have the form,

$$E_a^{(\zeta)} = \sqrt{\mathcal{M}_a^2 + m_a^2 + p^2 - 2\zeta R_a^2}, \quad \mathcal{M}_a = \sqrt{(\mu_a B)^2 + g_a^2/4}, \tag{69}$$

where $R_a^2 = \sqrt{p^2\mathcal{M}_a^2 + (\mu_a B)^2 m_a^2}$.

The basis spinors in Eq. (68) can be found in the limit of the small neutrino mass [37],

$$
u_a^{(\zeta)} = \frac{1}{2\sqrt{2\mathcal{M}_a(\mathcal{M}_a - \zeta g_a/2)}}
\begin{pmatrix}
\mu_a B + \zeta\mathcal{M}_a - g_a/2 \\
\mu_a B - \zeta\mathcal{M}_a + g_a/2 \\
\mu_a B - \zeta\mathcal{M}_a + g_a/2 \\
\mu_a B + \zeta\mathcal{M}_a - g_a/2
\end{pmatrix},
$$
$$
v_a^{(\zeta)} = \frac{1}{2\sqrt{2\mathcal{M}_a(\mathcal{M}_a + \zeta g_a/2)}}
\begin{pmatrix}
\mathcal{M}_a - \zeta[\mu_a B - g_a/2] \\
\mathcal{M}_a + \zeta[\mu_a B + g_a/2] \\
-\mathcal{M}_a - \zeta[\mu_a B + g_a/2] \\
-\mathcal{M}_a + \zeta[\mu_a B - g_a/2]
\end{pmatrix}. \tag{70}
$$

It should be noted that the discrete quantum number $\zeta = \pm 1$ in Eqs. (68)-(70) does not correspond to the helicity quantum states.

Now our goal is to find the time dependent coefficients $a_a^{(\zeta)}(t)$ and $b_a^{(\zeta)}(t)$. On the basis of the general solution (68) of the wave equation (67) we obtain the ordinary differential equations for these functions which formally coincide with Eq. (31). However the mixing potential $\mathcal{V}$ is now defined in Eq. (67). To obtain the modified Eq. (31) we again use the orthonormality of the basis spinors (70). The initial condition for the functions $a_a^{(\zeta)}(t)$ and $b_a^{(\zeta)}(t)$ also coincide with Eq. (32), with $\nu_\beta^{(0)}$ corresponding to a definite helicity spinor.

Taking into account the fact that $\langle u_a^{(\zeta)}|\mathcal{V}|v_b^{(\zeta')}\rangle = 0$, we get that the equations for $a_a^{(\zeta)}(t)$ and $b_a^{(\zeta)}(t)$ decouple, i.e. the interaction $\mathcal{V}$ does not mix positive and negative energy eigenstates. In the following we will consider the evolution of only $a_a^{(\zeta)}(t)$ since the dynamics of $b_a^{(\zeta)}(t)$ is studied analogously.

Let us rewrite the modified Eq. (31) in the more conventional effective Hamiltonian form. For this purpose we introduce the "wave function" $\Psi'^{\mathrm{T}} = (a_1^-, a_2^-, a_1^+, a_2^+)$. Directly from the modified Eq. (31) for the functions $a_a^{(\zeta)}(t)$ we derive the equation for Ψ',

$$
\mathrm{i}\frac{\mathrm{d}\Psi'}{\mathrm{d}t} = H'\Psi', \quad H' =
\begin{pmatrix}
0 & h_- e^{\mathrm{i}\omega_- t} & 0 & H_- e^{\mathrm{i}\Omega_- t} \\
h_- e^{-\mathrm{i}\omega_- t} & 0 & H_+ e^{-\mathrm{i}\Omega_+ t} & 0 \\
0 & H_+ e^{\mathrm{i}\Omega_+ t} & 0 & h_+ e^{\mathrm{i}\omega_+ t} \\
H_- e^{-\mathrm{i}\Omega_- t} & 0 & h_+ e^{-\mathrm{i}\omega_+ t} & 0
\end{pmatrix}, \tag{71}
$$

where

$$
\begin{aligned}
h_\mp = & \langle u_a^\mp|\mathcal{V}|u_b^\mp\rangle = \frac{1}{8\sqrt{\mathcal{M}_a\mathcal{M}_b(\mathcal{M}_a \pm g_a/2)(\mathcal{M}_b \pm g_b/2)}} \\
& \times [2\mu B(g_a\mu_b B + g_b\mu_a B) \pm 4\mu B(\mu_a B\mathcal{M}_b + \mu_b B\mathcal{M}_a) \\
& \pm 2g(g_a\mathcal{M}_b + g_b\mathcal{M}_a) + 4g\mathcal{M}_a\mathcal{M}_b + gg_a g_b], \quad a \neq b, \\
H_\mp = & \langle u_1^\mp|\mathcal{V}|u_2^\pm\rangle = \langle u_2^\pm|\mathcal{V}|u_1^\mp\rangle = \frac{1}{8\sqrt{\mathcal{M}_1\mathcal{M}_2(\mathcal{M}_1 \pm g_1/2)(\mathcal{M}_2 \mp g_2/2)}} \\
& \times [2\mu B(g_1\mu_2 B + g_2\mu_1 B) \mp 4\mu B(\mu_1 B\mathcal{M}_2 - \mu_2 B\mathcal{M}_1) \\
& \mp 2g(g_1\mathcal{M}_2 - g_2\mathcal{M}_1) - 4g\mathcal{M}_1\mathcal{M}_2 + gg_1 g_2],
\end{aligned} \tag{72}
$$

as well as $\omega_\mp = E_1^\mp - E_2^\mp + (g_1 - g_2)/2$ and $\Omega_\mp = E_1^\mp - E_2^\pm + (g_1 - g_2)/2$.

Instead of Ψ' it is more convenient to use the transformed "wave function" Ψ defined by

$$\Psi' = \mathcal{U}\Psi, \quad \mathcal{U} = \mathrm{diag}\left\{e^{\mathrm{i}(\Omega+\omega_-)t/2}, e^{\mathrm{i}(\Omega-\omega_-)t/2}, e^{-\mathrm{i}(\Omega-\omega_+)t/2}, e^{-\mathrm{i}(\Omega+\omega_+)t/2}\right\}, \tag{73}$$

where $\Omega = (\Omega_- - \Omega_+)/2$, to exclude the explicit time dependence of the effective Hamiltonian H'. Using the property $\omega_+ + \omega_- = \Omega_+ + \Omega_-$, we arrive to the new Schrödinger equation for the "wave function" Ψ,

$$\mathrm{i}\frac{\mathrm{d}\Psi}{\mathrm{d}t} = H\Psi, \quad H = \mathcal{U}^\dagger H' \mathcal{U} - \mathrm{i}\mathcal{U}^\dagger \dot{\mathcal{U}} = \tag{74}$$
$$= \begin{pmatrix} (\Omega+\omega_-)/2 & h_- & 0 & H_- \\ h_- & (\Omega-\omega_-)/2 & H_+ & 0 \\ 0 & H_+ & -(\Omega-\omega_+)/2 & h_+ \\ H_- & 0 & h_+ & -(\Omega+\omega_+)/2 \end{pmatrix}.$$

Despite initially we used perturbation theory to account for the influence of the potential $\mathcal{V}$ on the dynamics of the system (67), the contribution of this potential is taken into account exactly in Eq. (74). It means that our method allows one to sum up all terms in the perturbation series.

As we mentioned above, the quantum number ζ does not correspond to a definite helicity eigenstate. Thus the initial condition, which we should add to Eq. (74), has to be derived from Eqs. (32) and (70) and also depend on the neutrino oscillations channel. For example, if we discuss $\nu_\beta^{\mathrm{L}} \to \nu_\alpha^{\mathrm{R}}$ neutrino oscillations, the proper initial condition for the "wave function" $\Psi(0) = \Psi_0$ is

$$\Psi_0^{\mathrm{T}} = \left(-\sin\theta\sqrt{\frac{\mathcal{M}_1 + g_1/2}{2\mathcal{M}_1}}, -\cos\theta\sqrt{\frac{\mathcal{M}_2 + g_2/2}{2\mathcal{M}_2}}, \right.$$
$$\left. \sin\theta\sqrt{\frac{\mathcal{M}_1 - g_1/2}{2\mathcal{M}_1}}, \cos\theta\sqrt{\frac{\mathcal{M}_2 - g_2/2}{2\mathcal{M}_2}} \right). \tag{75}$$

Suppose that one has found the solution of the system (74) and (75) as $\Psi^{\mathrm{T}}(t) = (\psi_1, \psi_2, \psi_3, \psi_4)$. Then the transition probability for $\nu_\beta^{\mathrm{L}} \to \nu_\alpha^{\mathrm{R}}$ oscillations channel can be found as

$$P_{\nu_\beta^{\mathrm{L}} \to \nu_\alpha^{\mathrm{R}}}(t) = \frac{1}{2}\left\{ \frac{\mu_1 B \cos\theta}{\sqrt{\mathcal{M}_1}} \left[\frac{\psi_1(t)}{\sqrt{\mathcal{M}_1 + g_1/2}} + \frac{\psi_3(t)}{\sqrt{\mathcal{M}_1 - g_1/2}} \right] \right.$$
$$\left. - \frac{\mu_2 B \sin\theta}{\sqrt{\mathcal{M}_2}} \left[\frac{\psi_2(t)}{\sqrt{\mathcal{M}_2 + g_2/2}} + \frac{\psi_4(t)}{\sqrt{\mathcal{M}_2 - g_2/2}} \right] \right\}^2. \tag{76}$$

To obtain Eq. (76) for simplicity we use the fact that initially we have rather broad (in space) wave packet, corresponding to the initial condition $\nu_\beta(\mathbf{r}, 0) = e^{\mathrm{i}kx}\nu_\beta^{(0)\mathrm{L}}$ [see Eq. (18)].

Now we demonstrate the consistency of the results of relativistic quantum mechanics approach to the description of neutrino spin flavor oscillations (see Eqs. (74)-(76), which look completely new) with the standard quantum mechanical method developed in Ref. [20]. We remind that the following effective Hamiltimian:

$$H'_{QM} = \begin{pmatrix} \Phi + g_1 & g & -\mu_1 B & -\mu B \\ g & -\Phi + g_2 & -\mu B & -\mu_2 B \\ -\mu_1 B & -\mu B & \Phi & 0 \\ -\mu B & -\mu_2 B & 0 & -\Phi \end{pmatrix}, \quad (77)$$

was proposed in Ref. [20] to describe the evolution of neutrino mass eigenstates in matter under the influence of an external magnetic field.

The effective Hamiltonian H'_{QM} acts in the space with the basis composed of helicity eigenstates of massive neutrinos. As we mentioned above, the helicity operator $(\mathbf{\Sigma}\mathbf{p})/|\mathbf{p}|$ does not commute with the Hamiltonian $\mathcal{H}_a$ in Eq. (67). Therefore the choice of the helicity eigenstates as the basis functions is justified only in the relatively weak external magnetic field case (see the detailed discussion in Ref. [40]) or in case of the small diagonal magnetic moments [39]. In our approach we use the basis spinors (70) which are the eigenfunctions of the Hamiltonian $\mathcal{H}_a$ and exactly take into account matter density and magnetic field strength. Thus these spinors are more appropriate basis functions for the description of spin flavor oscillations.

We have found that in frames of the relativistic quantum mechanics approach the dynamics of the neutrino system can be described by the Schrödinger like equation with the effective Hamiltonian (74). Let us decompose the energy levels (69) supposing that neutrinos are ultrarelativistic particles,

$$E_a^{(\zeta)} = k + \frac{g_a}{2} - \zeta\mathcal{M}_a + \frac{m_a^2}{2k} + \zeta\frac{m_a^2 g_a^2}{8k^2\mathcal{M}_a} + \cdots. \quad (78)$$

In Eq. (78) we keep the term $\sim m_a^2/k^2$ to examine the corrections to the conventional quantum mechanical approach.

Performing the similarity transformation of the effective Hamiltonian H in Eq. (74) and using the orthogonal matrix $\mathcal{R}$ $(\mathcal{R}^{\mathrm{T}}\mathcal{R} = I)$ of the following form:

$$\mathcal{R} = \begin{pmatrix} -\frac{\sqrt{\mathcal{M}_1 + g_1/2}}{\sqrt{2\mathcal{M}_1}} & 0 & \frac{\mu_1 B}{\sqrt{2\mathcal{M}_1(\mathcal{M}_1 + g_1/2)}} & 0 \\ 0 & -\frac{\sqrt{\mathcal{M}_2 + g_2/2}}{\sqrt{2\mathcal{M}_2}} & 0 & \frac{\mu_2 B}{\sqrt{2\mathcal{M}_2(\mathcal{M}_2 + g_2/2)}} \\ \frac{\sqrt{\mathcal{M}_1 - g_1/2}}{\sqrt{2\mathcal{M}_1}} & 0 & \frac{\mu_1 B}{\sqrt{2\mathcal{M}_1(\mathcal{M}_1 - g_1/2)}} & 0 \\ 0 & \frac{\sqrt{\mathcal{M}_2 - g_2/2}}{\sqrt{2\mathcal{M}_2}} & 0 & \frac{\mu_2 B}{\sqrt{2\mathcal{M}_2(\mathcal{M}_2 - g_2/2)}} \end{pmatrix}, \quad (79)$$

we can see that the Hamiltonian H transforms to $\mathcal{R}^{\mathrm{T}} H \mathcal{R} \approx H_{QM} + \delta H$, where

$$H_{QM} = \begin{pmatrix} \Phi + 3g_1/4 - g_2/4 & g & -\mu_1 B & -\mu B \\ g & -\Phi + 3g_2/4 - g_1/4 & -\mu B & -\mu_2 B \\ -\mu_1 B & -\mu B & \Phi - (g_1 + g_2)/4 & 0 \\ -\mu B & -\mu_2 B & 0 & -\Phi - (g_1 + g_2)/4 \end{pmatrix}, \tag{80}$$

and

$$\delta H = \frac{1}{16k^2} \mathrm{diag} \left(-m_1^2 \frac{g_1^3}{\mathcal{M}_1^2}, -m_2^2 \frac{g_2^3}{\mathcal{M}_2^2}, m_1^2 \frac{g_1^3}{\mathcal{M}_1^2}, m_2^2 \frac{g_2^3}{\mathcal{M}_2^2} \right), \tag{81}$$

is the correction to the standard effective Hamiltonian. It should be noted that the transformation matrix $\mathcal{R}$ in Eq. (79) depends on the magnetic field strength and the matter density.

The effective Hamiltonian H_{QM} is equivalent to H'_{QM} in Eq. (77) since $H_{QM} = H'_{QM} - \mathrm{tr}(H'_{QM})/4 \cdot I$, where I is the 4×4 unit matrix. It is known that the unit matrix does not change the particles dynamics. Thus the relativistic quantum mechanics approach is equivalent to the standard approach developed in Ref. [20].

Now let us discuss the correction δH [see Eq. (81)] to the quantum mechanical method. This correction results from the fact that we use the correct energy levels for a neutrino moving in dense matter and strong magnetic field. Note that in Eq. (81) we keep only the diagonal corrections $\sim m_a^2/k^2$ to the effective Hamiltonian (80). If we slightly change non-diagonal elements of the Hamiltonian, it will result in the small changes of the transition probability. However, if we add some small quantity to diagonal elements, it can produce the resonance enhancement of neutrino oscillations.

We should remind that the expressions for the basis spinors (70) were obtained in the approximation of neutrinos with small masses, whereas in Eq. (78) we expand the energy up to $\sim m_a^2/k^2$ terms. If we take into account $\sim m_a^2/k^2$ corrections to the basis spinors (70), we can expect that some non-diagonal entries in the effective Hamiltonian H (74) will also obtain $\sim m_a^2/k^2$ corrections: $h_\pm \to h_\pm + \delta h_\pm$ and $H_\pm \to H_\pm + \delta H_\pm$. However, using the explicit form of the effective Hamiltonian H (74) and the matrix $\mathcal{R}$ (79) we get that these additional contributions are washed out in diagonal entries in Eq. (81). We analyze the validity of the approximations made in the derivation of the correction (81) in Appendix B..

6. Spin Flavor Oscillations of Dirac Neutrinos in the Magnetized Envelope of a Supernova

In this section we study the application of the general formalism for the description of neutrino spin flavor oscillations, developed in Sec. 5., to the situation of neutrinos propagating in the expanding envelope formed after a supernova explosion [39]. We find an exact solution of the Schrödinger equation with the effective Hamiltonian (80) for the background matter profile present in an expanding envelope and in the supernova magnetic field. We also analyze the possibility of enhancement of neutrino oscillations.

To describe the dynamics of neutrino spin flavor oscillations one has to solve the evolution equation with the Hamiltonian (80). This problem, in its turn, requires to solve a secular equation which is the fourth-order algebraic equation in order to find the eigenvalues of the effective Hamiltonian. Although one can express the solution to such an equation in radicals, its actual form appears to be rather cumbersome for arbitrary parameters.

If we, however, consider the case of a neutrino propagating in the electrically neutral isoscalar matter, i.e. $n_e = n_p$ and $n_p = n_n$, a reasonable solution is possible to find. We will demonstrate later that it corresponds to a realistic physical situation. As one can infer from Eq. (36) for the case of the $\nu_e^{\mathrm{L}} \to \nu_\mu^{\mathrm{R}}$ oscillations channel, in a medium with this property one has the effective potentials $f_\alpha \equiv f_\mu = V_\mu = -G_{\mathrm{F}} n/\sqrt{2}$ and $f_\beta \equiv f_e = V_e = G_{\mathrm{F}} n/\sqrt{2}$, where $n \equiv n_e = n_p = n_n$. Using Eq. (37) we obtain that $g_1 = -g_2 = g_0$, where $g_0 = -V \cos 2\theta$, $g = V \sin 2\theta$, and $V = G_{\mathrm{F}} n/\sqrt{2}$.

Let us point out that background matter with these properties may well exist in some astrophysical environments. The matter profile of presupernovae is poorly known, and a variety of presupernova models with different profiles exist in the literature (see, e.g., Ref. [58]). Nevertheless, electrically neutral isoscalar matter may well exist in the inner parts of presupernovae consisting of elements heavier than hydrogen. Indeed, for example, the model W02Z in Ref. [58] predicts that in a $15 M_\odot$ presupernova one has $Y_e = n_e/(n_p + n_n) = 0.5$ in the O+Ne+Mg layer, between the Si+O and He layers, in the radius range $(0.007\text{–}0.2) R_\odot$.

We also discuss the model of neutrino magnetic moments in which the nondiagonal elements of the magnetic moments matrix (45) are much bigger than the diagonal magnetic moments. Such a magnetic moments matrix was previously discussed in our works [35, 37, 39] (see also Sec. 4.). Note that in case of negligible diagonal magnetic moments the helicity operator (43) commutes with the Hamiltonian $\mathcal{H}_a$ in Eq. (67) and hence the effective Hamiltonian (80) acts in the helicity eigenstates basis. In other words the effective Hamiltonian (77) proposed in the standard quantum mechanical approach [20] is justified in our case.

For neutrinos having such magnetic moments and propagating in isoscalar matter the effective Hamiltonian (80) is replaced by

$$H_{QM} \to \begin{pmatrix} \Phi + g_0 & g & 0 & -\mu B \\ g & -(\Phi + g_0) & -\mu B & 0 \\ 0 & -\mu B & \Phi & 0 \\ -\mu B & 0 & 0 & -\Phi \end{pmatrix}. \tag{82}$$

We now look for the stationary solutions of the Schrödinger equation with this Hamiltonian. After a straightforward calculation one finds

$$\Psi(t) = \sum_{\zeta = \pm 1} \left[\left(U_\zeta \otimes U_\zeta^\dagger \right) \exp\left(-\mathrm{i}\mathcal{E}_\zeta t\right) + \left(V_\zeta \otimes V_\zeta^\dagger \right) \exp\left(\mathrm{i}\mathcal{E}_\zeta t\right) \right] \Psi_0, \tag{83}$$

where we have denoted

$$\begin{aligned} \mathcal{E}_\pm &= \frac{1}{2}\sqrt{2V^2 + 4(\mu B)^2 + 4\Phi^2 - 4\Phi V \cos 2\theta \pm 2VR}, \\ R &= \sqrt{(V - 2\Phi \cos 2\theta)^2 + 4(\mu B)^2}. \end{aligned} \tag{84}$$

The vectors $U_\pm$ and $V_\pm$ are the eigenvectors corresponding to the energy eigenvalues $\mathcal{E}_\pm$ and $-\mathcal{E}_\pm$, respectively. They are given by ($\zeta = \pm$)

$$U_\zeta = \frac{1}{N_\zeta}\begin{pmatrix} Z_\zeta \\ \sin 2\theta(\mathcal{E}_\zeta - \Phi) \\ -\mu B \sin 2\theta \\ -\mu B Z_\zeta/(\mathcal{E}_\zeta + \Phi)\end{pmatrix}, \quad V_\zeta = \frac{1}{N_\zeta}\begin{pmatrix} -\sin 2\theta(\mathcal{E}_\zeta - \Phi) \\ Z_\zeta \\ \mu B Z_\zeta/(\mathcal{E}_\zeta + \Phi) \\ -\mu B \sin 2\theta \end{pmatrix}, \tag{85}$$

where

$$\begin{aligned} Z_\zeta =& \frac{V + \zeta R}{2} - \mathcal{E}_\zeta \cos 2\theta, \\ N_\zeta^2 =& Z_\zeta^2 \left[1 + \frac{(\mu B)^2}{(\mathcal{E}_\zeta + \Phi)^2}\right] + \sin^2(2\theta)\left[(\mu B)^2 + (\mathcal{E}_\zeta - \Phi)^2\right]. \end{aligned} \tag{86}$$

It should be noted that Eq. (83) is a general solution of the evolution equation with the effective Hamiltonian (82) satisfying the initial condition $\Psi(0) = \Psi_0$.

Note that we received the solution (83)-(86) under some assumptions on the external fields such as isoscalar matter with constant density and constant magnetic field. In Sec. 5. we showed that our method is equivalent to the quantum mechanical description of neutrino oscillations [20] which can be used for a more general case of coordinate dependent external fields. Nevertheless the assumption of constant matter density and magnetic field is quite realistic for certain astrophysical environments like a shock wave propagating inside an expanding envelope after a supernova explosion.

Consistently with Eqs. (2)-(4), and (18) with an initial spinor $\nu_\beta^{(0)}$ corresponding to a left polarized neutrino (43), we take the initial wave function $\Psi(0) \equiv \Psi_0$ in Eq. (83) as $\Psi_0^{\mathrm{T}} = (\psi_1^{\mathrm{L}}, \psi_2^{\mathrm{L}}, \psi_1^{\mathrm{R}}, \psi_2^{\mathrm{R}}) = (\sin\theta, \cos\theta, 0, 0)$. Using Eqs. (83)-(86) one finds the components of the quantum mechanical wave function corresponding to the right polarized neutrinos to be of the form

$$\begin{aligned} \psi_1^{\mathrm{R}}(t) =& \frac{\mu B}{N_+^2}\Bigg\{ \cos\theta\left[e^{\mathrm{i}\mathcal{E}_+ t}\frac{Z_+^2}{\mathcal{E}_+ + \Phi} - \sin^2(2\theta)(\mathcal{E}_+ - \Phi)e^{-\mathrm{i}\mathcal{E}_+ t}\right] \\ & - \sin\theta \sin 2\theta Z_+ \left[e^{-\mathrm{i}\mathcal{E}_+ t} + e^{\mathrm{i}\mathcal{E}_+ t}\frac{\mathcal{E}_+ - \Phi}{\mathcal{E}_+ + \Phi}\right]\Bigg\} + \{+ \to -\}, \\ \psi_2^{\mathrm{R}}(t) =& \frac{\mu B}{N_+^2}\Bigg\{ \sin\theta\left[\sin^2(2\theta)(\mathcal{E}_+ - \Phi)e^{\mathrm{i}\mathcal{E}_+ t} - e^{-\mathrm{i}\mathcal{E}_+ t}\frac{Z_+^2}{\mathcal{E}_+ + \Phi}\right] \\ & - \cos\theta \sin 2\theta Z_+ \left[e^{\mathrm{i}\mathcal{E}_+ t} + e^{-\mathrm{i}\mathcal{E}_+ t}\frac{\mathcal{E}_+ - \Phi}{\mathcal{E}_+ + \Phi}\right]\Bigg\} + \{+ \to -\}, \end{aligned} \tag{87}$$

where the $\{+ \to -\}$ stand for the terms similar to the terms preceding each of them but with all quantities with a subscript $+$ replaced with corresponding quantities with a subscript $-$. The wave function of the right-handed neutrino of the flavor "α", ν_α^{R}, can be written with help of Eqs. (2)-(4), and (87) as $\nu_\alpha^{\mathrm{R}}(t) = \cos\theta\psi_1^{\mathrm{R}}(t) - \sin\theta\psi_2^{\mathrm{R}}(t)$.

The probability for the transition $\nu_\beta^{\mathrm{L}} \to \nu_\alpha^{\mathrm{R}}$ is obtained as the square of the quantum mechanical wave function ν_α^{R}. One obtains

$$P_{\nu_\beta^{\mathrm{L}} \to \nu_\alpha^{\mathrm{R}}}(t) = \left|\nu_\alpha^{\mathrm{R}}\right|^2 = [C_+ \cos(\mathcal{E}_+ t) + C_- \cos(\mathcal{E}_- t)]^2 + [S_+ \sin(\mathcal{E}_+ t) + S_- \sin(\mathcal{E}_- t)]^2, \tag{88}$$

where ($\zeta = \pm$)

$$C_\zeta = \frac{\mu B}{N_\zeta^2}\left\{\frac{Z_\zeta^2}{\mathcal{E}_\zeta + \Phi} - \sin^2(2\theta)(\mathcal{E}_\zeta - \Phi)\right\},$$
$$S_\zeta = \frac{\mu B}{N_\zeta^2}\left\{\sin^2(2\theta)\frac{2\Phi Z_\zeta}{\mathcal{E}_\zeta + \Phi} + \cos 2\theta\left[\frac{Z_\zeta^2}{\mathcal{E}_\zeta + \Phi} + \sin^2(2\theta)(\mathcal{E}_\zeta - \Phi)\right]\right\}. \quad (89)$$

As a consistency check, one easily finds from Eq. (89) that $C_+ + C_- = 0$ as required for assuring $P(0) = 0$.

In the following we will limit our considerations to the case $\mathcal{E}_+ \approx \mathcal{E}_-$, corresponding to the situations where the effect of the interactions of neutrinos with matter (V) is small compared with that of the magnetic interactions (μB) or the vacuum contribution (Φ) or both [see Eq. (84)]. Note that in this case one can analyze the exact oscillation probability (88) analytically, which would be practically impossible in more general situations.

In the case $\mathcal{E}_+ \approx \mathcal{E}_-$, one can present the transition probability in Eq. (88) in the following form:

$$P(t) = P_0(t) + P_c(t)\cos(2\Omega t) + P_s(t)\sin(2\Omega t), \quad (90)$$

where

$$P_0(t) = \frac{1}{2}\left[S_+^2 + S_-^2 + 2S_+S_-\cos(2\delta\Omega t) - 4C_+C_-\sin^2(\delta\Omega t)\right],$$
$$P_c(t) = -\frac{1}{2}\left[(S_+^2 + S_-^2)\cos(2\delta\Omega t) + 2S_+S_- - 4C_+C_-\sin^2(\delta\Omega t)\right],$$
$$P_s(t) = \frac{1}{2}\left(S_+^2 - S_-^2\right)\sin(2\delta\Omega t), \quad (91)$$

and

$$\Omega = \frac{\mathcal{E}_+ + \mathcal{E}_-}{2}, \quad \delta\Omega = \frac{\mathcal{E}_+ - \mathcal{E}_-}{2}. \quad (92)$$

As one can infer from these expressions, the transition probability $P(t)$ is a rapidly oscillating function, with the frequency Ω, enveloped from up and down by the slowly varying functions $P_{u,d} = P_0 \pm \sqrt{P_c^2 + P_s^2}$, respectively.

The behavior of the transition probability for various matter densities ρ and the values of μB and for a fixed neutrino energy of $E = 10\,\mathrm{MeV}$ and squared mass difference of $\delta m^2 = 8\times 10^{-5}\,\mathrm{eV}^2$ is illustrated in Figs. 2-4.

As these plots show, at low matter densities the envelope functions give, at each propagation distance, the range of the possible values of the oscillation probability. At greater matter densities, where the probability oscillates less intensively, the envelope functions are not that useful in analyzing the physical situation.

One can find the maximum value of the upper envelope function, which is also the upper bound for the transition probability, given as

$$P_u^{(\max)} = \begin{cases} (S_+ - S_-)^2, & \text{if } B < B', \\ \dfrac{C_+C_-(S_+^2 - S_-^2)^2}{C_+C_-(S_+^2 + S_-^2) + (C_+C_-)^2 + (S_+S_-)^2}, & \text{if } B > B', \end{cases} \quad (93)$$

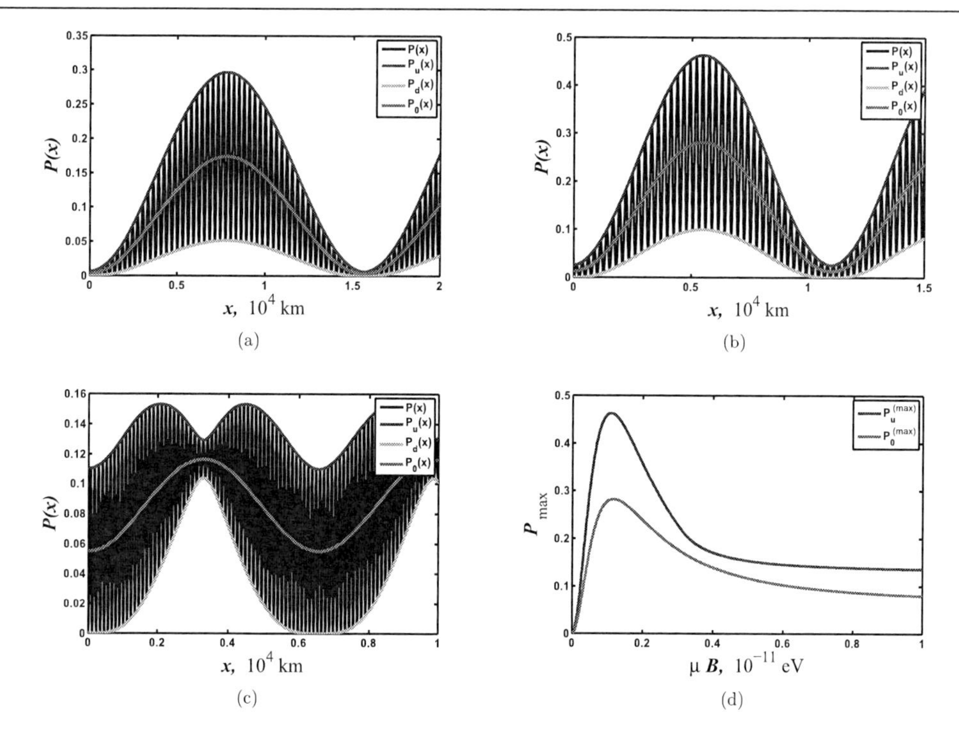

Figure 2. (color online) (a)-(c) The transition probability versus the distance passed by a neutrino beam in matter with the density $\rho = 10$ g/cc; (a) $\mu B = 5 \times 10^{-13}$ eV, (b) $\mu B = 1.1 \times 10^{-12}$ eV, (c) $\mu B = 5 \times 10^{-12}$ eV. We take that $E_\nu = 10$ MeV, $\delta m^2 = 8 \times 10^{-5}$ eV^2 and $\theta = 0.6$, which is quite close to the solar neutrinos' oscillations parameters. The black line is the function $P(x)$, the blue and green lines are the envelope functions $P_{u,d}(x)$, and the red line is the averaged transition probability $P_0(x)$. (d) The dependence of the maximal values of the functions $P(x)$ and $P_0(x)$, blue and red lines, respectively, on the magnetic energy μB for the given density. This figure is taken from Ref. [39].

where the value B' is the solution of the transcendent algebraic equation, $C_+ C_- = S_+ S_-$. The corresponding maximum values of the averaged transition probability $P_0(x)$ are given by

$$P_0^{(\max)} = \frac{1}{2}[(S_+ S_-)^2 - 4C_+ C_-], \tag{94}$$

for arbitrary values of B. The values of these maxima depend on the size of the quantity μB. These dependencies are plotted in Figs. 2(d)-4(d). In the case of rapid oscillations the physically relevant quantities, rather than the maxima, are the averaged values of the transition probability, which are also plotted in these figures.

As Figs. 2(d)-4(d) show, the interplay of the matter effect and the magnetic interaction can lead, for a given magnetic moment μ, to an enhanced spin flavor transition if the magnetic field B has a suitable strength relative to the density of matter ρ. In our numerical examples this occurs at $\mu B_{\max} = 1.1 \times 10^{-12}$ eV for $\rho = 10$ g/cc, at $\mu B_{\max} = 6.6 \times 10^{-13}$ eV for $\rho = 50$ g/cc, and at $\mu B_{\max} = 8 \times 10^{-13}$ eV for $\rho = 100$ g/cc. For these values of μB both the maxima and the average of the transition probability become considerably larger than for any other values of μB. Figs. 2(b)-4(b) correspond to the situation of maximal

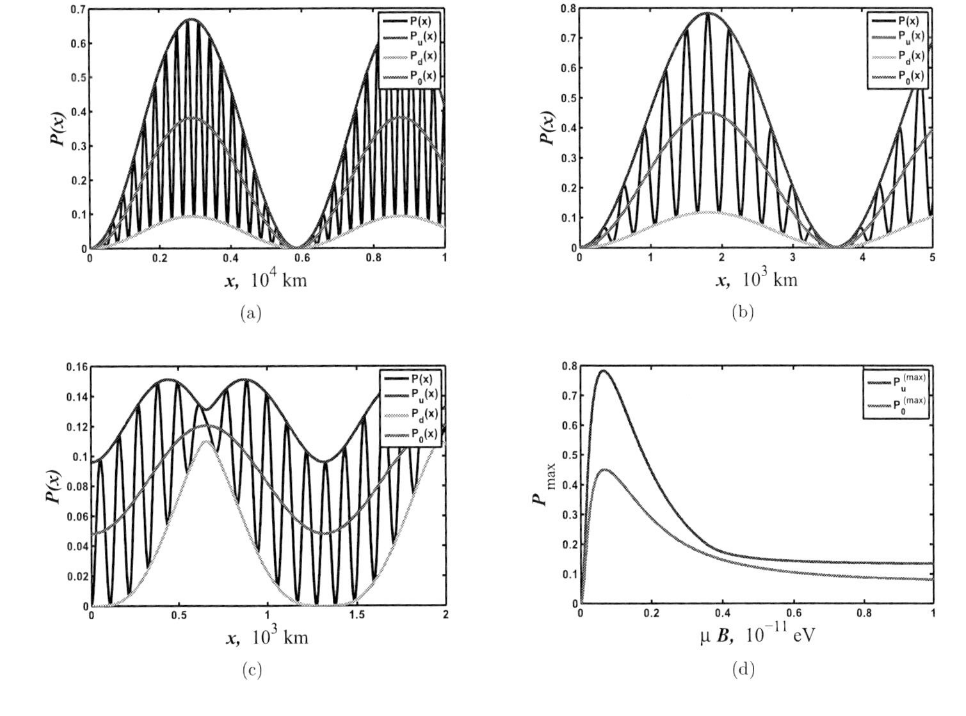

Figure 3. (color online) The same as in Fig. 2 for the density $\rho = 50\,\text{g/cc}$; (a) $\mu B = 3.5 \times 10^{-13}\,\text{eV}$, (b) $\mu B = 6.6 \times 10^{-13}\,\text{eV}$, (c) $\mu B = 5 \times 10^{-12}\,\text{eV}$. After Ref. [39].

enhancement, whereas Figs. 2(a)-4(a) and Figs. 2(c)-4(c) illustrate the situation above and below the optimal strength B_{max} of the magnetic field.

It is noteworthy that the enhanced transition probability is achieved towards the lower end of the μB region where substantial transitions all occur, that is, at relatively moderate magnetic fields. At larger values of μB the maximum of the transition probability approaches towards $\cos^2(2\theta)$. Indeed, if $\mu B \gg \max(\Phi, V)$, the transition probability can be written in the form $P(t) = \cos^2(2\theta)\sin^2(\mu B t)$ [see Eq. (64)]. It was found in Ref. [59] that neutrino spin flavor oscillations can be enhanced in a very strong magnetic field, with the transition probability being practically equal to unity. This phenomenon can be realized only for Dirac neutrinos with small nondiagonal magnetic moments and small mixing angle. As we can see from Figs. 2(d)-4(d) the situation is completely different for big nondiagonal magnetic moments.

One should notice that for long propagation distances consisting of several oscillation periods of the envelope functions, the enhancement effect would diminish considerably due to averaging. In the numerical examples presented in Figs. 2-4 the period of the envelope function is of the order of $10^3 - 10^4$ km, which is a typical size of a shock wave with the matter densities we have used in the plots (see, e.g., Ref. [60]). Thus the enhanced spin flavor transition could take place when neutrinos traverse a shock wave.

Let us recall that the above analysis was made by assuming neutrinos to be Dirac particles. We will see below (see Sec. 10.) that the corresponding results are quite different in the case of Majorana neutrinos.

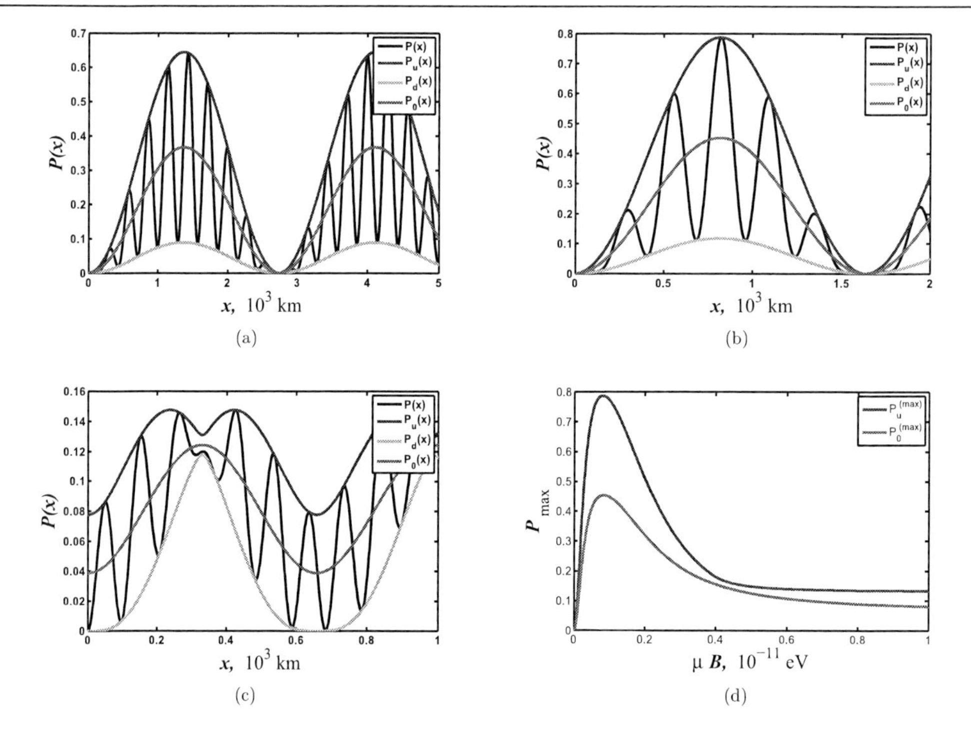

Figure 4. (color online) The same as in Fig. 2 for the density $\rho = 100\,\text{g/cc}$; (a) $\mu B = 4 \times 10^{-13}\,\text{eV}$, (b) $\mu B = 8 \times 10^{-13}\,\text{eV}$, (c) $\mu B = 5 \times 10^{-12}\,\text{eV}$. After Ref. [39].

7. Spin Flavor Oscillations between Electron and Sterile Astrophysical Neutrinos

In this section we continue with the studies of oscillations of supernova neutrinos on the basis of the effective Hamiltonian derived in Sec. 5.. In particular, we discuss the possibility of spin flavor oscillations between right polarized electron neutrino and additional sterile neutrino in an expanding envelope formed after a supernova explosion under the influence of a strong magnetic field. It is shown that the resonance enhancement of neutrino oscillations is possible if we take into account the correction to the effective Hamiltonian (81).

As we mentioned in Sec. 6., in general case the solution of the Schrödinger equation based on the effective Hamiltonians (80) and (81) is quite cumbersome and should be analyzed numerically. To study spin flavor oscillations analytically we consider, just for simplicity, the situation when the vacuum mixing angle between different neutrino eigenstates is small, $\theta \ll 1$. If $\theta \ll 1$, the matter interaction term and the the magnetic moments coincide in both flavor and mass eigenstates bases.

A resonance of neutrino oscillations can appear if the difference between two diagonal elements in the effective Hamiltonian is small [61]. In table 1, where we take into account the contributions of both Eqs. (80) and (81), we list the resonance conditions for various oscillations channels. It can be seen from this table that the corrections (81) to the effective Hamiltonian become important when we discuss $\nu^{\mathrm{R}}_{e,\mu,\tau} \leftrightarrow \nu_s^{\mathrm{L,R}}$ oscillations channel, where ν_s is the additional sterile neutrino.

Table 1. The resonance conditions for the $\nu_\alpha^{\mathrm{L,R}} \leftrightarrow \nu_\beta^{\mathrm{L,R}}$ oscillations channels which take into account the Hamiltonians (80) and (81).

No.	Oscillations channel	Resonance condition
1	$\nu_\beta^{\mathrm{L}} \leftrightarrow \nu_\alpha^{\mathrm{R}}$	$\Phi = \frac{f_\beta}{2} - \frac{1}{32k^2}\left(\frac{m_1^2 f_\alpha^3}{\mathcal{M}_1^2} + \frac{m_2^2 f_\beta^3}{\mathcal{M}_2^2}\right)$
2	$\nu_\beta^{\mathrm{R}} \leftrightarrow \nu_\alpha^{\mathrm{L}}$	$\Phi = -\frac{f_\alpha}{2} + \frac{1}{32k^2}\left(\frac{m_1^2 f_\alpha^3}{\mathcal{M}_1^2} + \frac{m_2^2 f_\beta^3}{\mathcal{M}_2^2}\right)$
3	$\nu_\beta^{\mathrm{L}} \leftrightarrow \nu_\alpha^{\mathrm{L}}$	$\Phi = \frac{f_\beta - f_\alpha}{2} + \frac{1}{32k^2}\left(\frac{m_1^2 f_\alpha^3}{\mathcal{M}_1^2} - \frac{m_2^2 f_\beta^3}{\mathcal{M}_2^2}\right)$
4	$\nu_\beta^{\mathrm{R}} \leftrightarrow \nu_\alpha^{\mathrm{R}}$	$\Phi = -\frac{1}{32k^2}\left(\frac{m_1^2 f_\alpha^3}{\mathcal{M}_1^2} - \frac{m_2^2 f_\beta^3}{\mathcal{M}_2^2}\right)$

As an example, we can study the case of $\nu_e^{\mathrm{R}} \leftrightarrow \nu_s^{\mathrm{L}}$ oscillations. Putting $\alpha \equiv s$ and $\beta \equiv e$ in table 1 and using Eq. (36), we obtain that $f_s = 0$, $f_e = \sqrt{2}G_{\mathrm{F}}(n_e - n_n/2)$ and $\mathcal{M}_2 \equiv \mathcal{M}_e = \sqrt{(\mu_{\nu_e}B)^2 + f_e^2}$. From table 1 we obtain the resonance condition for this oscillations channel as,

$$\delta m^2 = \frac{1}{4\sqrt{2}E_\nu} G_{\mathrm{F}} \left(n_e - \frac{n_n}{2}\right) m_{\nu_e}^2, \tag{95}$$

where we discuss the case of small diagonal magnetic moment of an electron neutrino: $\mu_{\nu_e}B \ll f_e$ and $E_\nu = k$ is the neutrino energy. In Appendix B. we discuss other small factors which can also contribute to the resonance condition (95).

As an example we consider $\nu_e^{\mathrm{R}} \leftrightarrow \nu_s^{\mathrm{L}}$ oscillations in an expanding envelope of a supernova explosion. It is known (see, e.g., Refs. [62, 63] for the most detailed analysis) that right polarized electron neutrinos can be created during the explosion of a core-collapse supernova. Indeed, if a neutrino is a Dirac particle, the spin flip can occur in the following reaction: $\nu_{\mathrm{L}} + (e^-, p, N) \rightarrow \nu_{\mathrm{R}} + (e^-, p, N)$, with electrons e^-, protons p, and nuclei N in the dense matter of a forming neutron star. This spin flip is due to the interaction of the neutrino diagonal magnetic moment with charged particles. Hence left polarized neutrinos are converted into right polarized ones with energy in the range $(100 - 200)$ MeV [64].

Besides the fact that the creation of right polarized neutrinos can provide additional supernova cooling [65], these particles can be observed in a terrestrial detector. When right polarized neutrinos propagate from the supernova explosion site towards the Earth, interacting with the galactic magnetic field, their helicity can change back and they become left polarized neutrinos. Although the flux of this kind of neutrinos is smaller than that of generic left polarized neutrinos [63, 66], potentially we can detect these particles. The analysis of the neutrino spin precession in the galactic magnetic field is given in Ref. [66].

Suppose that the flux of right polarized electron neutrinos is crossing an expanding envelope of a supernova. A shock wave can be formed in the envelope [67]. Approximately

1 s after the core collapse, the matter density in the shock wave region $L \sim 10^8$ cm can be up to 10^6 g/cm^3. We can also suppose that the matter density is approximately constant inside the shock wave (see also Sec. 6.).

For the electroneutral matter Eq. (95) reads

$$\delta m^2 \approx 5.0 \times 10^{-17}\,\mathrm{eV}^2 \times (3Y_e - 1)\left(\frac{\rho}{10^6\,\mathrm{g/cm}^3}\right)\left(\frac{E_\nu}{100\,\mathrm{MeV}}\right)^{-1}\left(\frac{m_{\nu_e}}{1\,\mathrm{eV}}\right)^2, \tag{96}$$

where $Y_e = n_e/(n_e + n_n)$ is the electrons fraction. From Eq. (96) we can see that for the matter with $Y_e > 1/3$ and $\rho \sim 10^6$ g/cm^3 and an electron neutrino with $E_\nu \sim 100$ MeV and $m_{\nu_e} \sim 1$ eV the mass squared difference should be $\delta m^2 \sim 10^{-17}$ eV^2.

Note that the possibility of existence of sterile neutrinos closely degenerate in mass with active neutrinos was recently discussed. In Ref. [68] it was examined how the fluxes of supernova neutrinos can be altered is presence of the almost degenerate neutrinos. The effect of spin flavor oscillations between active and sterile neutrinos with small δm^2 on the solar neutrino fluxes was discussed in Ref. [69]. The implications of the CP phases of the mixing between active and sterile neutrinos in the scenario with small δm^2 to the effective mass of an electron neutrinos were considered in Ref. [70]. In Ref. [71] it was examined the possibility to experimentally confirm, e.g., in the IceCube detector, the existence of almost degenerate in masses sterile neutrinos emitted with high energies from extragalactic sources. The range of δm^2 studied in Ref. [68] is $10^{-16}\,\mathrm{eV}^2 < \delta m^2 < 10^{-12}\,\mathrm{eV}^2$ which is quite close to the estimates obtained from Eq. (96). In Ref. [71] the sterile neutrinos with even smaller δm^2 were studied: $10^{-19}\,\mathrm{eV}^2 < \delta m^2 < 10^{-12}\,\mathrm{eV}^2$.

Besides the fulfillment of the resonance condition (95), to have the significant $\nu_e^{\mathrm{R}} \leftrightarrow \nu_s^{\mathrm{L}}$ transitions rate the strength of the magnetic field and distance traveled by the neutrinos beam, which we take to be equal to the shock wave size L, should satisfy the condition $\mu B L \approx \pi/2$. We can rewrite this condition as,

$$B \approx 5.3 \times 10^7\,\mathrm{G}\left(\frac{\mu}{10^{-12}\,\mu_{\mathrm{B}}}\right)^{-1}\left(\frac{L}{10^3\,\mathrm{km}}\right)^{-1}. \tag{97}$$

For the transition magnetic moment $\mu = 3 \times 10^{-12}\,\mu_{\mathrm{B}}$ [72] and $L \sim 10^8$ cm (see above), we get that $B \sim 10^7$ G. Supposing that the magnetic field of a neutron star depends on the distance as $B(r) = B_0 (R/r)^3$, where $R = 10$ km is the typical neutron star radius and $B_0 = 10^{13}$ G is the magnetic field on the surface of a young neutron star, we get that at $r = 10^8$ cm the magnetic field reaches 10^7 G, which is consistent with the estimates of Eq. (97). Note that magnetic fields in a supernova explosion can be even higher than 10^{13} G and can reach the values of $\sim 10^{16}$ G [73].

8. Majorana Neutrinos in Vacuum

In this section we apply the formalism, developed in Secs. 2.-5. for the description of the Dirac neutrinos evolution in various external fields, for the studies of oscillations of Majorana neutrinos. Here we study the evolution of mixed Majorana neutrinos in vacuum. We solve the initial condition problem for the two component neutrino wave functions and find the transition probability. However, besides flavor oscillations, for Majorana particles one can consider transitions between neutrinos and antineutrinos. We study this process also on the basis of relativistic quantum mechanics.

In general case the system of flavor neutrinos can be described with help of the appropriate number of left and right handed spinors [44]. We can however suggest that only left handed chirality components $\nu_\lambda^\mathrm{L} = (1/2)(1-\gamma^5)\nu_\lambda$ are present in our system. It is the case, for example, for neutrinos in frames of the standard model where the right-handed neutrinos are sterile. The general situation when one has both left and right handed particles will be studied in Sec. 11..

The general mass matrix, involving left-handed flavor neutrinos ν_λ^L, can be diagonalized with help of the matrix transformation [44],

$$\nu_\lambda^\mathrm{L} = \sum_a U_{\lambda a}\eta_a, \tag{98}$$

where $\lambda = \alpha, \beta$ is the flavor index and η_a, $a = 1, 2$, corresponds to a Majorana particle with a definite mass m_a. In the simplest case the mixing of the flavor states arises purely from Majorana mass terms between the left-handed neutrinos, and then the mixing matrix $U_{\lambda a}$ is a 2×2 and unitary matrix, i.e., $a = 1, 2$ and, assuming no CP violation, it can be parameterized in the same way as in Eq. (4).

We study the evolution of this system with the following initial condition [see also Eq. (18)]:

$$\nu_\alpha^\mathrm{L}(\mathbf{r}, 0) = 0, \quad \nu_\beta^\mathrm{L}(\mathbf{r}, 0) = \nu_\beta^{(0)} e^{\mathrm{i}\mathbf{kr}}, \tag{99}$$

where $\mathbf{k} = (0, 0, k)$ is the initial momentum and $\nu_\beta^{(0)\mathrm{T}} = (0, 1)$. The initial state is thus a left-handed neutrino of flavor "β" propagating along the z-axis to the positive direction.

As both the left-handed state ν_λ^L and Majorana state η_a have two degrees of freedom, we will describe them in the following by using two-component Weyl spinors. The Weyl spinor of a free Majorana particle obeys the wave equation (see, e.g., [74]),

$$\mathrm{i}\dot{\eta}_a + (\boldsymbol{\sigma}\mathbf{p})\eta_a + \mathrm{i}m_a\sigma_2\eta_a^* = 0. \tag{100}$$

Note that Eq. (100) can be formally derived from Eq. (7) if impose the Majorana condition $\psi_a = (\psi_a)^c$, where the index "c" means the charge conjugation, on the four component spinor ψ_a. The Majorana condition means that this spinor is represented as $\psi_a^\mathrm{T} = (\mathrm{i}\sigma_2\eta_a^*, \eta_a)$.

The general solution of Eq. (100) can be presented as [75]

$$\eta_a(\mathbf{r}, t) = \int \frac{\mathrm{d}^3\mathbf{p}}{(2\pi)^{3/2}} e^{\mathrm{i}\mathbf{pr}} \sum_{\zeta=\pm 1} \left[a_a^{(\zeta)}(\mathbf{p}) u_a^{(\zeta)}(\mathbf{p}) e^{-\mathrm{i}E_a t} + a_a^{(\zeta)*}(-\mathbf{p}) v_a^{(\zeta)}(-\mathbf{p}) e^{\mathrm{i}E_a t} \right], \tag{101}$$

where $E_a = \sqrt{m_a^2 + |\mathbf{p}|^2}$. As in Sec. 2. the coefficient $a_a^{(\zeta)}(\mathbf{p})$ is time independent and its value is determined by the initial condition (99).

The basis spinors $u_a^{(\zeta)}$ and $v_a^{(\zeta)}$ in Eq. (101) have the form

$$\begin{aligned} u_a^-(\mathbf{p}) &= \lambda_a w_-, \quad u_a^+(\mathbf{p}) = -\lambda_a \frac{m_a}{E_a + |\mathbf{p}|} w_+, \\ v_a^+(\mathbf{p}) &= \lambda_a w_-, \quad v_a^-(\mathbf{p}) = \lambda_a \frac{m_a}{E_a + |\mathbf{p}|} w_+, \end{aligned} \tag{102}$$

where $w_\pm$ are helicity amplitudes given by [76]

$$w_+ = \begin{pmatrix} e^{-i\phi/2}\cos(\vartheta/2) \\ e^{i\phi/2}\sin(\vartheta/2) \end{pmatrix}, \quad w_- = \begin{pmatrix} -e^{-i\phi/2}\sin(\vartheta/2) \\ e^{i\phi/2}\cos(\vartheta/2) \end{pmatrix}, \tag{103}$$

the angles ϕ and ϑ giving the direction of the momentum of the particle, $\mathbf{p} = |\mathbf{p}| \times (\sin\vartheta\cos\phi, \sin\vartheta\sin\phi, \cos\vartheta)$. The normalization factor λ_a in Eq. (102) can be chosen as

$$\lambda_a^{-2} = 1 - \frac{m_a^2}{(E_a + |\mathbf{p}|)^2}. \tag{104}$$

Let us mention the following properties of the helicity amplitudes $w_\pm$:

$$\begin{gathered} (\boldsymbol{\sigma}\mathbf{p})w_\pm = \pm|\mathbf{p}|w_\pm, \quad \mathrm{i}\sigma_2 w_\pm^* = \mp w_\mp, \quad w_\pm(-\mathbf{p}) = \mathrm{i}w_\mp(\mathbf{p}), \\ (w_+ \otimes w_-^{\mathrm{T}}) - (w_- \otimes w_+^{\mathrm{T}}) = \mathrm{i}\sigma_2, \quad \left(w_+ \otimes w_+^\dagger\right) + \left(w_- \otimes w_-^\dagger\right) = 1, \end{gathered} \tag{105}$$

which can be immediately obtained from Eq. (103) and which are useful in deriving the results given below.

The time-independent coefficients $a_a^\pm(\mathbf{p})$ in Eq. (101) have the following form [75]:

$$\begin{aligned} a_a^+(\mathbf{p}) =& \frac{1}{(2\pi)^{3/2}} \left[\eta_a^{(0)\dagger}(-\mathbf{p}) v_a^+(\mathbf{p}) + \frac{i m_a}{E_a + |\mathbf{p}|} v_a^{+\dagger}(-\mathbf{p}) \eta_a^{(0)}(\mathbf{p}) \right], \\ a_a^-(\mathbf{p}) =& \frac{1}{(2\pi)^{3/2}} \left[u_a^{-\dagger}(\mathbf{p}) \eta_a^{(0)}(\mathbf{p}) - \frac{i m_a}{E_a + |\mathbf{p}|} \eta_a^{(0)\dagger}(-\mathbf{p}) u_a^-(-\mathbf{p}) \right], \end{aligned} \tag{106}$$

where $\eta_a^{(0)}(\mathbf{p})$ is the Fourier transform of the initial wave function η_a,

$$\eta_a^{(0)}(\mathbf{p}) = \int \mathrm{d}^3\mathbf{p} e^{-\mathrm{i}\mathbf{p}\mathbf{r}} \eta_a^{(0)}(\mathbf{r}).$$

Using Eqs. (101)-(106) we then obtain the following expression for the wave function for the neutrino mass eigenstate:

$$\begin{aligned} \eta_a(\mathbf{r}, t) = \int \frac{\mathrm{d}^3\mathbf{p}}{(2\pi)^3} e^{\mathrm{i}\mathbf{p}\mathbf{r}} \lambda_a^2 \Bigg[\Bigg\{ & \left(e^{-\mathrm{i}E_a t} - \left[\frac{m_a}{E_a + |\mathbf{p}|} \right]^2 e^{\mathrm{i}E_a t} \right) \left(w_- \otimes w_-^\dagger \right) \\ & + \left(e^{\mathrm{i}E_a t} - \left[\frac{m_a}{E_a + |\mathbf{p}|} \right]^2 e^{-\mathrm{i}E_a t} \right) \left(w_+ \otimes w_+^\dagger \right) \Bigg\} \eta_a^{(0)}(\mathbf{p}) \\ & - 2 \frac{m_a}{E_a + |\mathbf{p}|} \sin(E_a t) \sigma_2 \eta_a^{(0)*}(-\mathbf{p}) \Bigg]. \end{aligned} \tag{107}$$

From Eqs. (105) and (107) it follows that a mass eigenstate particle initially in the left polarized state $\eta_a^{(0)}(\mathbf{r}) \sim w_-(\mathbf{k}) e^{\mathrm{i}\mathbf{k}\mathbf{r}}$ is described at later times by

$$\begin{aligned} \eta_a(\mathbf{r}, t) \sim \lambda_a^2 \Bigg\{ & \left(e^{-\mathrm{i}E_a t} - \left[\frac{m_a}{E_a + |\mathbf{k}|} \right]^2 e^{\mathrm{i}E_a t} \right) e^{\mathrm{i}\mathbf{k}\mathbf{r}} w_-(\mathbf{k}) \\ & - 2\mathrm{i} \frac{m_a}{E_a + |\mathbf{k}|} \sin(E_a t) e^{-\mathrm{i}\mathbf{k}\mathbf{r}} w_+(\mathbf{k}) \Bigg\}. \end{aligned} \tag{108}$$

Let us notice that the second term in Eq. (108) describes an antineutrino state. Indeed the spinor $w_+(\mathbf{k})$ satisfies the relation, $(\boldsymbol{\sigma}\mathbf{k})w_+(\mathbf{k}) = |\mathbf{k}|w_+(\mathbf{k})$, see Eq. (105). Therefore it corresponds to an antiparticle [see Ref. [77]]. This term is responsible for the neutrino-to-antineutrino flavor state transition $\nu_\beta^{\mathrm{L}} \leftrightarrow (\nu_\alpha^{\mathrm{L}})^c$.

According to Eqs. (4) and (98), the wave function of the left-handed neutrino of the flavor "α" is $\nu_\alpha^{\mathrm{L}} = \cos\theta\eta_1^{\mathrm{L}} - \sin\theta\eta_2^{\mathrm{L}}$. From Eqs. (98) and (108) it then follows that the probability of the transition $\nu_\beta^{\mathrm{L}} \to \nu_\alpha^{\mathrm{L}}$ in vacuum is given by

$$\begin{aligned} P_{\nu_\beta^{\mathrm{L}} \to \nu_\alpha^{\mathrm{L}}}(t) = &|\nu_\alpha^{\mathrm{L}}|^2 = \sin^2(2\theta)\bigg\{ \sin^2(\Phi t) \\ &+ \frac{1}{2|\mathbf{k}|^2}\cos(\sigma t)\sin(\Phi t)[m_1^2\sin(E_1 t) - m_2^2\sin(E_2 t)]\bigg\} \\ &+ \mathcal{O}\left(\frac{m_a^4}{|\mathbf{k}|^4}\right), \end{aligned} \tag{109}$$

where σ and Φ are introduced in Eq. (21).

We can compare Eq. (109) with analogous expression for Dirac particles (20). The leading terms in both equations coincide, whereas the next-to-leading terms have different signs. Therefore one can reveal the nature of neutrinos, i.e. say if neutrino mass eigenstates are Dirac or Majorana particles, studying the corrections to the transition probability formulas.

Analogously we can calculate the transition probability for the process $\nu_\beta^{\mathrm{L}} \to (\nu_\alpha^{\mathrm{L}})^c$ using the second term in Eq. (108),

$$P_{\nu_\beta^{\mathrm{L}} \to (\nu_\alpha^{\mathrm{L}})^c}(t) = |(\nu_\alpha^{\mathrm{L}})^c|^2 = \frac{\sin^2(2\theta)}{4|\mathbf{k}|^2}[m_1\sin(E_1 t) - m_2\sin(E_2 t)]^2 + \mathcal{O}\left(\frac{m_a^4}{|\mathbf{k}|^4}\right). \tag{110}$$

Note that the next-to-leading term in Eq. (109) and leading term in Eq. (110) have the same order of magnitude $\sim m_a^2/|\mathbf{k}|^2$.

Before moving to consider Majorana neutrinos in magnetic fields in Sec. 9., we make a general comment concerning the validity of our approach based on relativistic field theory involving (classical) first quantized Majorana fields. It has been stated [78] that the dynamics of massive Majorana fields cannot be described within the classical field theory approach due to the fact that the mass term of the Lagrangian, $\eta^{\mathrm{T}}\mathrm{i}\sigma_2\eta$, vanishes when η is represented as a c-number function (see, e.g., Eq. (154) below). Note that Eq. (100) is a direct consequence of the Dirac equation if we suggest that the four-component wave function satisfies the Majorana condition. Therefore a solution of Eq. (100), i.e., wave functions and energy levels, in principle does not depend on the existence of a Lagrangian resulting in this equation. The wave equations describing elementary particles should follow from the quantum field theory principles. However quite often these quantum equations allow classical solutions (see Ref. [79] for many interesting examples). We demonstrate in Secs. 2.-5. (see also Refs. [33, 34, 35, 37, 40, 36, 39]) that oscillations of Dirac neutrinos in vacuum and various external fields can be described in the framework of the classical field theory. The main result of this section was to show that the quantum Eq. (100) for massive Majorana particles can be solved [see Eq. (108)] in the framework of the classical field theory as well.

9. Majorana Neutrinos in Matter and Transversal Magnetic Field

In this sections we apply the formalism developed in Sec. 8. to the more general case of Majorana neutrinos propagating in background matter and interacting with an external magnetic field. We show that the initial condition problem of the mass eigenstates can be expressed in terms of the Schrödinger equation with an effective Hamiltonian which coincides with previously proposed one [20].

For describing the evolution of two Majorana mass eigenstates in matter under the influence of an external magnetic field, the wave equation (100) is to be modified to the following form:

$$\mathrm{i}\dot{\eta}_a + \left(\boldsymbol{\sigma}\mathbf{p} - \frac{g_a}{2}\right)\eta_a + \mathrm{i}m_a\sigma_2\eta_a^* - \frac{g}{2}\eta_b - \mathrm{i}\mu(\boldsymbol{\sigma}\mathbf{B})\sigma_2\epsilon_{ab}\eta_b^* = 0, \quad a \neq b, \tag{111}$$

where $\epsilon_{ab} = \mathrm{i}(\sigma_2)_{ab}$, and g_a and g were defined in Eq. (27). Note that Eq. (111) can be formally derived from Eq. (67) if one neglects vector current interactions, i.e., replace $(1-\gamma^5)/2$ with $-\gamma^5/2$, and takes into account the fact that the magnetic moment matrix of Majorana neutrinos is antisymmetric (see, e.g., Ref. [80]). We will apply the same initial condition (99) as in the vacuum case. It should be mentioned that the evolution of Majorana neutrinos in matter and in a magnetic field has been previously discussed in Ref. [81].

The general solution of Eq. (111) can be expressed in the following form:

$$\begin{aligned}\eta_a(\mathbf{r},t) = \int \frac{\mathrm{d}^3\mathbf{p}}{(2\pi)^{3/2}} e^{\mathrm{i}\mathbf{p}\mathbf{r}} \sum_{\zeta=\pm1} \big[& a_a^{(\zeta)}(\mathbf{p},t) u_a^{(\zeta)}(\mathbf{p}) \exp(-\mathrm{i}E_a^{(\zeta)}t) \\ & + a_a^{(\zeta)*}(-\mathbf{p},t) v_a^{(\zeta)}(-\mathbf{p}) \exp(\mathrm{i}E_a^{(\zeta)}t)\big],\end{aligned} \tag{112}$$

where the energy levels are given in Eq. (29) (see Ref. [51]). The basis spinors in Eq. (112) can be chosen as

$$\begin{aligned} u_a^-(\mathbf{p}) &= \lambda_a^- w_-, \quad u_a^+(\mathbf{p}) = -\lambda_a^+ \frac{m_a}{E_a^+ + (|\mathbf{p}| - g_a/2)} w_+, \\ v_a^+(\mathbf{p}) &= \lambda_a^+ w_-, \quad v_a^-(\mathbf{p}) = \lambda_a^- \frac{m_a}{E_a^- + (|\mathbf{p}| + g_a/2)} w_+, \end{aligned} \tag{113}$$

where the normalization factors $\lambda_a^{(\zeta)}$, $\zeta = \pm$, are given by

$$(\lambda_a^{(\zeta)})^{-2} = 1 - \frac{m_a^2}{[E_a + (|\mathbf{p}| - \zeta g_a/2)]^2}. \tag{114}$$

Let us consider the propagation of Majorana neutrinos in the transversal magnetic field. Using a similar technique as in the Dirac case in Sec. 5. and assuming $k \gg m_a$, we end up with the following ordinary differential equations for the coefficients $a_a^{(\zeta)}$,

$$\mathrm{i}\frac{\mathrm{d}\Psi'}{\mathrm{d}t} = H'\Psi', \quad H' = \begin{pmatrix} 0 & ge^{\mathrm{i}\omega_- t}/2 & 0 & \mu Be^{\mathrm{i}\Omega_- t} \\ ge^{-\mathrm{i}\omega_- t}/2 & 0 & -\mu Be^{-\mathrm{i}\Omega_+ t} & 0 \\ 0 & -\mu Be^{\mathrm{i}\Omega_+ t} & 0 & -ge^{\mathrm{i}\omega_+ t}/2 \\ \mu Be^{-\mathrm{i}\Omega_- t} & 0 & -ge^{-\mathrm{i}\omega_+ t}/2 & 0 \end{pmatrix}, \tag{115}$$

where $\Psi'^{\mathrm{T}} = (a_1^-, a_2^-, a_1^+, a_2^+)$ and

$$\omega_\pm = E_1^\pm - E_2^\pm \approx 2\Phi \mp \frac{g_1 - g_2}{2}, \quad \Omega_\mp = E_1^\mp - E_2^\pm \approx 2\Phi \pm \frac{g_1 + g_2}{2}, \tag{116}$$

which should be compared with Eq. (71).

By making the matrix transformation

$$\Psi' = \mathcal{U}\Psi, \quad \mathcal{U} = \mathrm{diag}\left\{e^{\mathrm{i}(\Phi+g_1/2)t}, e^{-\mathrm{i}(\Phi-g_2/2)t}, e^{\mathrm{i}(\Phi-g_1/2)t}, e^{-\mathrm{i}(\Phi+g_2/2)t}\right\}, \tag{117}$$

we can recast Eq. (115) into the form

$$\begin{aligned} \mathrm{i}\frac{\mathrm{d}\Psi}{\mathrm{d}t} =& H\Psi, \quad H = \mathcal{U}^\dagger H' \mathcal{U} - \mathrm{i}\mathcal{U}^\dagger \dot{\mathcal{U}} \\ =& \begin{pmatrix} \Phi + g_1/2 & g/2 & 0 & \mu B \\ g/2 & -\Phi + g_2/2 & -\mu B & 0 \\ 0 & -\mu B & \Phi - g_1/2 & -g/2 \\ \mu B & 0 & -g/2 & -\Phi - g_2/2 \end{pmatrix}. \end{aligned} \tag{118}$$

Let us note that the analogous effective Hamiltonian has been used in describing the spin flavor oscillations of Majorana neutrinos within the quantum mechanical approach (see, e.g., Ref. [20]) if we use the basis $\Psi_{QM}^{\mathrm{T}} = (\psi_1^{\mathrm{L}}, \psi_2^{\mathrm{L}}, [\psi_1^{\mathrm{L}}]^c, [\psi_2^{\mathrm{L}}]^c)$.

Note that the consistent derivation of the master Eq. (111) should be done in the framework of the quantum field theory (see, e.g., Ref. [78]), supposing that the spinors η_a are expressed via anticommuting operators. This quantum field theory treatment is important to explain the asymmetry of the magnetic moment matrix. However, it is possible to see that the main Eq. (111) can also be reduced to the standard Schrödinger evolution Eq. (118) for neutrino spin flavor oscillations if we suppose that the wave functions η_a are c-number objects. That is why one can again conclude that classical and quantum field theory methods for studying Majorana neutrinos' propagation in external fields are equivalent.

10. Spin Flavor Oscillations of Majorana Neutrinos in the Expanding Envelope of a Supernova

In this section we discuss the application of the formalism developed in Sec. 9. to the description of spin flavor oscillations of Majorana neutrinos in an expanding envelope of a supernova and compare the results with the case of Dirac neutrinos studied in Sec. 6..

The dynamics of the system of two Majorana neutrinos in matter under the influence of an external magnetic field is governed by the Schrödinger equation (118). However, as in Sec. 6., the explicit analytical solution of this equation is quite cumbersome. That is why we again consider the situation when $n_e = n_p = n_n = n$, which results in $g_1 = -g_2$ (see Eqs. (36) and (37) as well as Sec. 6.). In this case the eigenvalues of the Hamiltonian (118) $\lambda = \pm\mathcal{E}_\pm$ are given by

$$\mathcal{E}_\pm = \frac{1}{2}\sqrt{V^2 + 4(\mu B)^2 + 4\Phi^2 \pm 4VR}, \quad R = \sqrt{(\Phi\cos 2\theta)^2 + (\mu B)^2}, \tag{119}$$

where $V = G_{\mathrm{F}} n/\sqrt{2}$ as in Sec 6.. The time evolution of the wave function is described by the formula,

$$\Psi(t) = \sum_{\zeta=\pm 1} \left[\left(U_\zeta \otimes U_\zeta^\dagger \right) \exp\left(-\mathrm{i}\mathcal{E}_\zeta t\right) + \left(V_\zeta \otimes V_\zeta^\dagger \right) \exp\left(\mathrm{i}\mathcal{E}_\zeta t\right) \right] \Psi_0, \tag{120}$$

where U_ζ and V_ζ are the eigenvectors of the Hamiltonian (118), given as

$$U_\zeta = \frac{1}{N_\zeta} \begin{pmatrix} -x_\zeta \\ -y_\zeta \\ 1 \\ -z_\zeta \end{pmatrix}, \quad V_\zeta = \frac{1}{N_\zeta} \begin{pmatrix} -y_\zeta \\ x_\zeta \\ z_\zeta \\ 1 \end{pmatrix}, \tag{121}$$

where

$$\begin{aligned} x_\zeta = & \frac{\mu B (\mathcal{E}_\zeta + \Phi)}{\Sigma_\zeta} V \sin 2\theta, \quad y_\zeta = \frac{\mu B}{\mathcal{E}_\zeta + \Phi - V\cos 2\theta/2} \left[1 + \frac{(\mathcal{E}_\zeta + \Phi)}{2\Sigma_\zeta} V^2 \sin^2(2\theta) \right], \\ z_\zeta = & \frac{V \sin 2\theta}{2(\mathcal{E}_\zeta + \Phi + V\cos 2\theta/2)} \left[1 + \frac{2(\mu B)^2 (\mathcal{E}_\zeta + \Phi)}{\Sigma_\zeta} \right], \\ \Sigma_\zeta = & \frac{V}{2} [2\mathcal{E}_\zeta (\mathcal{E}_\zeta + \Phi) - V^2/2 + \Phi V \cos 2\theta] \cos 2\theta + \zeta R V (\mathcal{E}_\zeta + \Phi - V\cos 2\theta/2). \end{aligned} \tag{122}$$

The normalization coefficient N_ζ in Eq. (121) is given by $N_\zeta = \sqrt{1 + x_\zeta^2 + y_\zeta^2 + z_\zeta^2}$.

Proceeding along the same lines as in Sec. 6., we obtain from Eqs. (98) and (120)-(122) the probability of the process $\nu_\beta^{\mathrm{L}} \to \nu_\alpha^{\mathrm{R}}$ as,

$$P_{\nu_\beta^{\mathrm{L}} \to \nu_\alpha^{\mathrm{R}}}(t) = [C_+ \cos(\mathcal{E}_+ t) + C_- \cos(\mathcal{E}_- t)]^2 + [S_+ \sin(\mathcal{E}_+ t) + S_- \sin(\mathcal{E}_- t)]^2, \tag{123}$$

where

$$C_\zeta = -\frac{1}{N_\zeta^2} [\sin 2\theta (x_\zeta + y_\zeta z_\zeta) + \cos 2\theta (y_\zeta - x_\zeta z_\zeta)], \quad S_\zeta = \frac{1}{N_\zeta^2} (y_\zeta + x_\zeta z_\zeta). \tag{124}$$

Consistently with Eq. (99), we have taken the initial wave function as

$$\Psi_0^{\mathrm{T}} = (\sin\theta, \cos\theta, 0, 0). \tag{125}$$

With help of Eqs. (122) and (124) it is easy to check that $C_+ + C_- = 0$ guaranteeing $P(0) = 0$.

Note that formally Eq. (123) corresponds to the transitions $\nu_\beta^{\mathrm{L}} \to \nu_\alpha^{\mathrm{R}}$. However, virtually it describes oscillations between active neutrinos $\nu_\beta^{\mathrm{L}} \leftrightarrow (\nu_\alpha^{\mathrm{L}})^c$ since $\nu_\alpha^{\mathrm{R}} = (\nu_\alpha^{\mathrm{L}})^c$ for Majorana particles.

As in the previous case of Eq. (88), Eq. (123) can be treated analytically for relatively small values of the effective potential V. The ensuing envelope functions $P_{u,d} = P_0 \pm \sqrt{P_c^2 + P_s^2}$ depend on the coefficients C_ζ and S_ζ in the same way as in Eq. (91). The transition probabilities at various values of the matter density and the magnetic field are presented in Fig. 5.

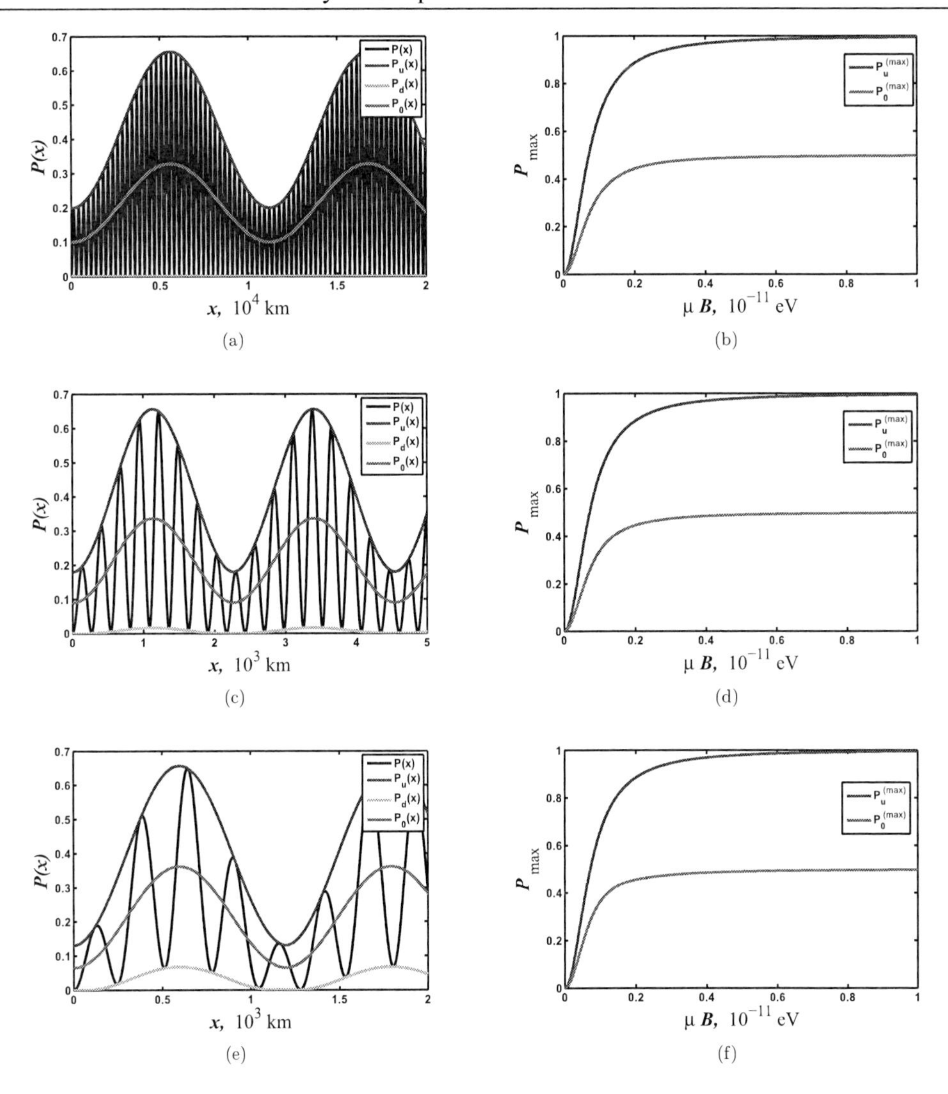

Figure 5. (color online) (a), (c) and (e) The transition probability versus the distance passed by a neutrino beam. The neutrino parameters have the same values as in Figs. 2-4, $E_\nu = 10\,\text{MeV}$, $\delta m^2 = 8 \times 10^{-5}\,\text{eV}^2$ and $\theta = 0.6$. The magnetic energy is equal to $\mu B = 10^{-12}\,\text{eV}$. The black line is the function $P(x)$, the blue and green lines are the envelope functions $P_{u,d}(x)$ and the red line is the averaged transition probability $P_0(x)$. (b), (d) and (f) The dependence of the maximal values of the functions $P(x)$ and $P_0(x)$, blue and red lines respectively, on the magnetic energy for the given density. Panels (a) and (b) correspond to the matter density $\rho = 10\,\text{g/cc}$, (c) and (d) to $\rho = 50\,\text{g/cc}$ and (e) and (f) to $\rho = 100\,\text{g/cc}$. This figure is taken from Ref. [39].

Despite the formal similarity between Dirac and Majorana transition probabilities [see Eqs. (88) and (123)] the actual dynamics is quite different in these two cases, as one can see by comparing Figs. 2(d)-4(d) and Fig. 5, panels (b), (d) and (f). In particular, in the Majorana case $P_u^{(\max)} = 4|C_+C_-|$ for arbitrary B, to be compared with Eq. (93), while the function $P_0^{(\max)}$ has the same form as in the Dirac case given in Eq. (94). In contrast with the

Dirac case, the averaged transition probability does not achieve its maximal value at some moderate magnetic field $B_{\max}$ value, but both $P_u^{\max}$ and $P_0^{\max}$ are monotonically increasing functions of the strength of the magnetic field with 1 and 1/2 as their asymptotic values, respectively. We can understand this behavior when we recall that, at $\mu B \gg \max(\Phi, V)$, the effective Hamiltonian Eq. (118) becomes

$$H_\infty = \mathrm{i}\mu B\gamma^2, \quad \mathrm{i}\gamma^2 = \begin{pmatrix} 0 & 0 & 0 & 1 \\ 0 & 0 & -1 & 0 \\ 0 & -1 & 0 & 0 \\ 1 & 0 & 0 & 0 \end{pmatrix}. \tag{126}$$

The Schrödinger equation with the effective Hamiltonian (126) has the formal solution

$$\Psi(t) = \exp(-\mathrm{i}H_\infty t)\Psi(0) = [\cos(\mu Bt) + \gamma^2 \sin(\mu Bt)]\Psi(0). \tag{127}$$

Using Eqs. (98), (125) and (127) we then immediately arrive to the following expression for the transition probability, $P(t) = |\nu_\alpha^{\mathrm{R}}|^2 = \sin^2(\mu Bt)$, which explains the behavior of the function $P_u^{(\max)}$ at strong magnetic fields. Note that the analogous result was also obtained in Ref. [59].

Finally, it is worth of noticing that in contrast to the Dirac case, the behavior of the transition probability in the Majorana case is qualitatively similar for different matter densities and different magnetic fields [see Fig. 5, panels (a), (c), and (e)].

The problem of Majorana neutrinos' spin flavor oscillations was studied in Refs. [82, 83] with help of numerical codes. For example, in Ref. [82] the evolution equation for three neutrino flavors propagating inside a presupernova star with zero metallicity, e.g., corresponding to W02Z model [58], was solved for the realistic matter and magnetic field profiles. Although our analytical transition probability formula (123) is valid only for the constant matter density and magnetic field strength, it is interesting to compare our results with the numerical simulations of Ref. [82]. In those calculations the authors used magnetic fields $B \sim 10^{10}\,\mathrm{G}$ and magnetic moments $\sim 10^{-12}\,\mu_{\mathrm{B}}$ that give us the magnetic energy $\mu B \sim 10^{-11}\,\mathrm{eV}$. This value is the maximal magnetic energy used in our work.

It was found in Ref. [82] that spin flavor conversion is practically adiabatic for low-energy neutrinos corresponding to $E_\nu \sim 5\,\mathrm{MeV}$ inside the region where $Y_e \approx 0.5$ and the averaged transition probability for the channel $\nu_\mu \to \bar{\nu}_e$ is close 0.5. This big transition probability is due to the RSF-H and RSF-L resonances at the distance $\approx 0.01R_\odot$. Even so that we study two neutrino oscillations scheme, we obtained the analogous behavior of $P_0^{(\max)}$ (see Fig. 5). However, in our case this big transition probability is due to the presence of the strong magnetic field (see Refs. [14, 59]). We cannot compare our transition probability formula (123) with the results of Ref. [82] for higher energies, $E_\nu > 25\,\mathrm{MeV}$, since spin flavor oscillations become strongly nonadiabatic for these kinds of energies and one has to take into account the coordinate dependence of the matter density which should decrease with radius as $1/r^3$ [84].

11. Evolution of Neutrinos Emitted by Classical Sources

In this section we present an alternative approach to the description of neutrino flavor oscillations which involves the studies of the evolution of mixed massive neutrinos in vacuum under the influence of external classical fields [38]. These fields are supposed to be localized in space and treated as "sources" which emit flavor neutrinos.

As in Secs. 2.-5. we consider a number of flavor neutrinos ν_λ, which, in principle, can be arbitrary: $\lambda = e, \mu, \tau, \dots$. Note that we will formulate the dynamics of the system in terms of the left ν_λ^{L} and right ν_λ^{R} handed chiral components of the neutrino spinors.

The external sources are denoted as J_λ^μ. It should be noted that, if one describes the emission of mixed massive neutrinos in any process within the quantum field theory, there is a nonzero probability of emission of neutrinos of different flavors [85]. Although a source of one particular flavor can prevail, one should always consider the situation when the sources of all flavors are present.

The Lorentz invariant Lagrangian describing the evolution of this system has the form,

$$\begin{aligned}\mathcal{L} = & \sum_\lambda \left(\overline{\nu_\lambda^{\mathrm{L}}} \mathrm{i}\gamma^\mu \partial_\mu \nu_\lambda^{\mathrm{L}} + \overline{\nu_\lambda^{\mathrm{R}}} \mathrm{i}\gamma^\mu \partial_\mu \nu_\lambda^{\mathrm{R}} \right) \\ & - \sum_{\lambda\lambda'} \left(m_{\lambda\lambda'}^{\mathrm{D}} \overline{\nu_\lambda^{\mathrm{L}}} \nu_{\lambda'}^{\mathrm{R}} + m_{\lambda\lambda'}^{\mathrm{L}} \left(\nu_\lambda^{\mathrm{L}}\right)^{\mathrm{T}} C \nu_{\lambda'}^{\mathrm{L}} + m_{\lambda\lambda'}^{\mathrm{R}} \left(\nu_\lambda^{\mathrm{R}}\right)^{\mathrm{T}} C \nu_{\lambda'}^{\mathrm{R}} + \mathrm{h.c.} \right) \\ & + \sum_\lambda \left(\overline{\nu_\lambda^{\mathrm{L}}} \gamma_\mu \ell_\lambda^{\mathrm{L}} J_\lambda^\mu + \mathrm{h.c.} \right), \end{aligned} \tag{128}$$

where $(m_{\lambda\lambda'}^{\mathrm{D}})$, $(m_{\lambda\lambda'}^{\mathrm{L}})$, $(m_{\lambda\lambda'}^{\mathrm{R}})$ are the Dirac as well as (left and right) Majorana mass matrices and $C = \mathrm{i}\gamma^2\gamma^0$ is the charge conjugation matrix. These mass matrices should satisfy certain requirements which are discussed in Ref. [44]. The Lagrangian (128) is CPT-invariant. In this section we adopt the notations for Dirac matrices as in Ref. [86]. The analogous mass terms are generated in theoretical models based on the type-II seesaw mechanism (see, e.g., Ref. [45]).

The external fields $J_\lambda^\mu = J_\lambda^\mu(\mathbf{r}, t)$ can be arbitrary functions. If we suppose that neutrinos interact with external sources in frames of the electroweak model, the spinor ℓ_λ in Eq. (128) is the charged $\mathrm{SU}(2)$ isodoublet partner of ν_λ. For example, we can study the neutrino emission in a process like the inverse β-decay: $p + e^- \to \nu_e + n$. In this case the external fields in Eq. (128) are (see, e.g., Ref. [87])

$$J_{\nu_e}^\mu = -\sqrt{2} G_{\mathrm{F}} \bar{\Psi}_n \gamma^\mu (1 - \alpha\gamma^5) \Psi_p, \quad J_{\nu_\mu}^\mu = 0, \quad J_{\nu_\tau}^\mu = 0, \tag{129}$$

where Ψ_p and Ψ_n are the wave functions of a proton and a neutron, $\alpha \approx 1.25$. In Eq. (129) we assume that a source consists of electrons, protons and neutrons.

To proceed in the analysis of the dynamics of the system (128) we should diagonalize the mass term of the Lagrangian. The diagonalized Lagrangian is expressed in terms of the Majorana fields with different masses [44]. In general case the number of these Majorana fields is double than the number of flavors in the initial Lagrangian. In the following we discuss the case when only Dirac mass term is presented. Then we study the general situation.

11.1. Dirac Mass Term

In this section we suppose that only Dirac mass matrix is present in Eq. (128). Analogously to the discussion in Sec. 2. we introduce the new set of spinor fields ψ_a. In our situation the mixing matrix is a square unitary matrix $(U_{\lambda a})$,

$$\nu_\lambda = \sum_a U_{\lambda a}\psi_a. \tag{130}$$

By definition the mass eigenstates ψ_a are Dirac particles.

When we transform the Lagrangian (128) using Eq. (130), it is rewritten in the following way:

$$\mathcal{L} = \sum_a \bar{\psi}_a(\mathrm{i}\gamma^\mu\partial_\mu - m_a)\psi_a + \sum_a(\bar{\psi}_a\xi_a + \text{h.c.}), \tag{131}$$

where ξ_a is the external source for the fermion ψ_a,

$$\xi_a = \sum_\lambda U^\dagger_{a\lambda}\gamma^{\mathrm{L}}_\mu \ell_\lambda f^\mu_\lambda. \tag{132}$$

Using Eqs. (131) and (132) we receive the inhomogeneous Dirac equation for the fermion ψ_a,

$$(\mathrm{i}\gamma^\mu\partial_\mu - m_a)\psi_a = -\xi_a. \tag{133}$$

As in Sec. 2. the masses m_a are the eigenvalues of the matrix $(m^{\mathrm{D}}_{\lambda\lambda'})$.

The solution to Eq. (132) for the arbitrary spinor ξ_a is expressed with help of the retarded Green function for a spinor field (see Ref. [88]),

$$\psi_a(\mathbf{r}, t) = \int \mathrm{d}^3\mathbf{r}'\mathrm{d}t' S^{\mathrm{ret}}_a(\mathbf{r}-\mathbf{r}', t-t')\xi_a(\mathbf{r}', t'). \tag{134}$$

The explicit form of $S^{\mathrm{ret}}_a(\mathbf{r}, t)$ can be also found in Ref. [88],

$$S^{\mathrm{ret}}_a(\mathbf{r}, t) = (\mathrm{i}\gamma^\mu\partial_\mu + m_a)D^{\mathrm{ret}}_a(\mathbf{r}, t). \tag{135}$$

In Eq. (135) $D^{\mathrm{ret}}_a(\mathbf{r}, t)$ is the retarded Green function for a scalar field [88],

$$D^{\mathrm{ret}}_a(\mathbf{r}, t) = \int \frac{\mathrm{d}^4 p}{(2\pi)^4}\frac{e^{\mathrm{i}px}}{m_a^2 - p^2 + \mathrm{i}\epsilon p^0} = \frac{1}{2\pi}\theta(t)\left\{\delta(s^2) - \theta(s^2)\frac{m_a}{2s}J_1(m_a s)\right\}, \tag{136}$$

where $\theta(t)$ is the Heaviside step function, $J_1(z)$ is the first order Bessel function, $s = \sqrt{t^2 - r^2}$ and $r = |\mathbf{r}|$.

To proceed in further calculations it is necessary to define the behavior of the external sources. We assume that the sources are localized in space and emit harmonic radiation,

$$\ell_\lambda(\mathbf{r}, t)J^\mu_\lambda(\mathbf{r}, t) = \theta(t)l_\lambda J^{(0)\mu}_\lambda e^{-\mathrm{i}Et}\delta^3(\mathbf{r}), \tag{137}$$

where $l_\lambda = \ell^{(0)}_\lambda$ is the time independent component of the spinor ℓ_λ, $J^{(0)\mu}_\lambda$ is the amplitude of the function $J^\mu_\lambda(\mathbf{r}, t)$, and E is the frequency of the source. Using Eqs. (132) and (137) we obtain for ξ_a,

$$\xi_a(\mathbf{r}, t) = \theta(t)\xi^{(0)}_a e^{-\mathrm{i}Et}\delta^3(\mathbf{r}), \quad \xi^{(0)}_a = \sum_\lambda U^\dagger_{a\lambda}\gamma^{\mathrm{L}}_\mu l_\lambda J^{(0)\mu}_\lambda. \tag{138}$$

With help of Eqs. (135)-(138) we can rewrite Eq. (134) in the following way:

$$\psi_a(\mathbf{r},t) = \left(\mathrm{i}\gamma^\mu \frac{\partial}{\partial x^\mu} + m_a\right) \left\{ e^{-\mathrm{i}Et} \int_0^t \mathrm{d}\tau D_a^{\mathrm{ret}}(\mathbf{r},\tau) e^{\mathrm{i}E\tau} \right\} \xi_a^{(0)}. \tag{139}$$

One can see that two major terms appear in Eq. (139) [see also Eq. (136)] while integrating over τ

$$\int_0^t \mathrm{d}\tau D_a^{\mathrm{ret}}(\mathbf{r},\tau) e^{\mathrm{i}E\tau} = \mathcal{I}_1 + \mathcal{I}_2. \tag{140}$$

They are

$$\mathcal{I}_1 = \frac{1}{2\pi} \int_0^t \mathrm{d}\tau \delta(s^2) e^{\mathrm{i}E\tau} = \frac{\theta(t-r)}{4\pi r} e^{\mathrm{i}Er}, \tag{141}$$

$$\mathcal{I}_2 = -\frac{m_a}{4\pi} \int_0^t \mathrm{d}\tau \theta(s^2) \frac{J_1(m_a s)}{s} e^{\mathrm{i}E\tau} = -\frac{m_a}{4\pi} \theta(t-r) \int_0^{x_m} \mathrm{d}x \frac{J_1(m_a x)}{\sqrt{r^2+x^2}} e^{\mathrm{i}E\sqrt{r^2+x^2}}, \tag{142}$$

where $x_m = \sqrt{t^2 - r^2}$ and $s = \sqrt{\tau^2 - r^2}$. It is interesting to notice that both $\mathcal{I}_1$ and $\mathcal{I}_2$ in Eqs. (141) and (142) are equal to zero if $t < r$. It means that the initial perturbation from a source placed at the point $\mathbf{r} = 0$ reaches a detector placed at the point $\mathbf{r}$ only after the time $t > r$.

Despite the integral $\mathcal{I}_1$ in Eq. (141) is expressed in terms of the elementary functions, the integral $\mathcal{I}_2$ in Eq. (142) cannot be computed analytically for arbitrary r and t. However some reasonable assumptions can be made to simplify the considered expression. Suppose that an observer is at the fixed distance from a source. As we have already mentioned one detects a signal starting from $t > r$. It is obvious that a nonstationary rapidly oscillating signal is detected when the wave front just arrives to a detector, i.e. when $t \gtrsim r$. The situation is analogous to waves propagating on the water surface. Therefore, if we suppose that one starts observing particles when the non-stationary signal attenuates, i.e. $t \gg r$ or $x_m \to \infty$, we can avoid relaxation phenomena. In this case the integral $\mathcal{I}_2$ can be computed analytically,

$$\mathcal{I}_2 = -\frac{1}{4\pi r}(e^{\mathrm{i}Er} - e^{\mathrm{i}p_a r}), \tag{143}$$

where $p_a = \sqrt{E^2 - m_a^2}$ is the analog of the particle momentum.

To obtain Eq. (143) we use the known values of the integrals,

$$\begin{aligned}
\int_0^\infty \mathrm{d}x \frac{J_\nu(mx)}{\sqrt{r^2+x^2}} \sin(E\sqrt{r^2+x^2}) =& \frac{\pi}{2} J_{\nu/2}\left[\frac{r}{2}\left(E - \sqrt{E^2-m^2}\right)\right] \\
& \times J_{-\nu/2}\left[\frac{r}{2}\left(E + \sqrt{E^2-m^2}\right)\right], \\
\int_0^\infty \mathrm{d}x \frac{J_\nu(mx)}{\sqrt{r^2+x^2}} \cos(E\sqrt{r^2+x^2}) =& -\frac{\pi}{2} J_{\nu/2}\left[\frac{r}{2}\left(E - \sqrt{E^2-m^2}\right)\right] \\
& \times N_{-\nu/2}\left[\frac{r}{2}\left(E + \sqrt{E^2-m^2}\right)\right],
\end{aligned} \tag{144}$$

and the fact that the Bessel and Neumann functions of $\pm 1/2$ order

$$J_{1/2}(z) = N_{-1/2}(z) = \sqrt{\frac{2}{\pi z}} \sin z, \quad J_{-1/2}(z) = \sqrt{\frac{2}{\pi z}} \cos z, \tag{145}$$

are expressed in terms of the elementary functions. The approximations made in derivation of Eq. (143) are analysed in Appendix C..

Using Eqs. (139)-(145) we obtain the field distribution of the fermion ψ_a in the following form:

$$\psi_a(\mathbf{r},t) = e^{-\mathrm{i}Et+\mathrm{i}p_a r} O_a \frac{\xi_a^{(0)}}{4\pi r}, \tag{146}$$

where $O_a = \gamma^0 E - (\boldsymbol{\gamma}\mathbf{n})p_a + m_a$ and $\mathbf{n}$ is the unit vector towards a detector. It should be noted that in deriving Eq. (146) we differentiate only exponential rather than the factor $1/r$ because the derivative of $1/r$ is proportional to $1/r^2$. Such a term is negligible at large distances from a source. We also remind that Eq. (146) is valid for $t \gg r$.

Let us turn to the description of the evolution of the fields ν_λ. Using Eqs. (130) and (146), we obtain the corresponding wave function,

$$\nu_\lambda(\mathbf{r},t) = e^{-\mathrm{i}Et} \sum_{a\lambda'} U_{\lambda a} U_{a\lambda'}^\dagger e^{\mathrm{i}p_a r} O_a \gamma_\mu^{\mathrm{L}} l_{\lambda'} \frac{J_{\lambda'}^{(0)\mu}}{4\pi r}. \tag{147}$$

As in Secs. 2.-5. we study the evolution of only two fermions. The mixing matrix is given in Eq. (4). We also choose the amplitudes of the sources in the following way: $J_\alpha^{(0)\mu} = 0$, $J_\beta^{(0)\mu} \equiv J^\mu \neq 0$ and $l_\beta \equiv l$. This choice of the sources corresponds to the emission of neutrinos of the flavor "β" and detection of the neutrinos belonging to the flavor "α" at the distance r from a source. Using Eqs. (4) and (147) we obtain the wave functions of each of the particles $\nu_{\alpha,\beta}$,

$$\begin{aligned}
\nu_\alpha(\mathbf{r},t) =& \sin\theta\cos\theta e^{-\mathrm{i}Et} \frac{J^\mu}{4\pi r}(e^{\mathrm{i}p_1 r}O_1 - e^{\mathrm{i}p_2 r}O_2)\gamma_\mu^{\mathrm{L}} l, \\
\nu_\beta(\mathbf{r},t) =& e^{-\mathrm{i}Et} \frac{J^\mu}{4\pi r}(\sin^2\theta e^{\mathrm{i}p_1 r}O_1 + \cos^2\theta e^{\mathrm{i}p_2 r}O_2)\gamma_\mu^{\mathrm{L}} l.
\end{aligned} \tag{148}$$

In the following we discuss the high frequency approximation, $E \gg m_{1,2}$, which corresponds to the emission of ultrarelativistic neutrinos.

The probability to detect a neutrino of the flavor "λ" can be calculated as $P_\lambda(\mathbf{r},t) = |\nu_\lambda(\mathbf{r},t)|^2$. Finally using Eq. (148) we for the probabilities,

$$\begin{aligned}
P_\alpha(r) =& -2E^2 \frac{f^{\dagger\mu} f^\nu}{(4\pi r)^2} \langle T_{\mu\nu} \rangle \left\{ \sin^2(2\theta)\sin^2\left(\frac{\Delta m^2}{4E} r\right) + \mathcal{O}\left(\frac{m_a^2}{E^2}\right) \right\}, \\
P_\beta(r) =& -2E^2 \frac{f^{\dagger\mu} f^\nu}{(4\pi r)^2} \langle T_{\mu\nu} \rangle \left\{ 1 - \sin^2(2\theta)\sin^2\left(\frac{\Delta m^2}{4E} r\right) + \mathcal{O}\left(\frac{m_a^2}{E^2}\right) \right\},
\end{aligned} \tag{149}$$

where $\langle T_{\mu\nu} \rangle = l^\dagger \gamma_\mu^\dagger (\boldsymbol{\alpha}\mathbf{n}) \gamma_\nu l$. It is possible to calculate the components of the tensor $\langle T_{\mu\nu} \rangle$. They depend on the properties of the fermion $\ell_\beta \equiv \ell$,

$$\begin{aligned}
\langle T_{00} \rangle = -(\mathbf{vn}), \quad \langle T_{0i} \rangle = \langle T_{i0} \rangle^* = n_i - \mathrm{i}[\mathbf{n} \times \boldsymbol{\zeta}]_i, \\
\langle T_{ij} \rangle = \delta_{ij}(\mathbf{vn}) - (v_i n_j + v_j n_i) + \mathrm{i}\varepsilon_{ijk} n_k,
\end{aligned} \tag{150}$$

where $\mathbf{v} = \langle \boldsymbol{\alpha} \rangle$ is the velocity of fermion ℓ and $\boldsymbol{\zeta} = \langle \boldsymbol{\Sigma} \rangle$ is its spin.

Let us discuss the simplified case when the spatial components of the four-vector J^{μ} are equal to zero: $\mathbf{J} = 0$. It corresponds to the neutrino emission by a nonmoving and unpolarized source. Using Eqs. (149) and (150) we can find the transition and survival probabilities in the following way:

$$P_{\beta\to\alpha}(r) \sim \sin^2(2\theta)\sin^2\left(\frac{\Delta m^2}{4E}r\right), \tag{151}$$

$$P_{\beta\to\beta}(r) \sim 1 - \sin^2(2\theta)\sin^2\left(\frac{\Delta m^2}{4E}r\right). \tag{152}$$

where we drop the common factor $2E^2(\mathbf{vn})|J^0|^2/(4\pi r)^2$ to have the probabilities normalized to unity. It should be noted that Eqs. (151) and (152) are the same as the common formulae for the description of neutrino flavor oscillations in vacuum [44, 47].

11.2. General Mass Term

When both Dirac and Majorana mass matrices are present in Eq. (128), left and right handed chiral components should be transformed independently (see also Ref. [44]),

$$\nu_\lambda^{\mathrm{L}} = \sum_a U_{\lambda a}\Psi_a^{\mathrm{L}}, \quad \nu_\lambda^{\mathrm{R}} = \sum_a V_{\lambda a}\Psi_a^{\mathrm{R}}. \tag{153}$$

with help of the matrices $(U_{\lambda a})$ and $V_{\lambda a}$ to diagonalize the general mass term. Note that these matrices are rectangular and nonhermitian. The modern parameterization for these matrices is given in Ref. [89].

However one cannot say whether particles $\Psi_a^{\mathrm{L,R}}$ are Majorana or Dirac although they correspond to definite mass eigenstates. That is why we introduce a new field $\psi_a = \Psi_a^{\mathrm{L}} + (\Psi_a^{\mathrm{L}})^c$ which is Majorana by definition.

The Lagrangian for the particles ψ_a takes the form,

$$\mathcal{L} = \sum_a \left(\overline{\psi_a^{\mathrm{L}}}\mathrm{i}\gamma^\mu\partial_\mu\psi_a^{\mathrm{L}} - m_a\overline{\psi_a^{\mathrm{L}}}\psi_a^{\mathrm{R}} + \overline{\psi_a^{\mathrm{L}}}\xi_a^{\mathrm{R}} + \mathrm{h.c.}\right), \tag{154}$$

where the external sources $\xi_a \equiv \xi_a^{\mathrm{R}}$ have the same form as in Eq. (132).

As we mentioned in Sec. 8., Majorana spinors are equivalent to two component Weil spinors [75]. Hence we can rewrite the spinors $\psi_a^{\mathrm{L,R}}$ and ξ_a as

$$\psi_a^{\mathrm{L}} = \begin{pmatrix} 0 \\ \eta_a \end{pmatrix}, \quad \psi_a^{\mathrm{R}} = \begin{pmatrix} \mathrm{i}\sigma_2\eta_a^* \\ 0 \end{pmatrix}, \quad \xi_a = \begin{pmatrix} \phi_a \\ 0 \end{pmatrix}, \tag{155}$$

where

$$\phi_a = \sum_\lambda U_{a\lambda}^\dagger \chi_\lambda J_\lambda^0. \tag{156}$$

To obtain Eq. (156) we suppose that, as in Sec. 11.1., the vectors J_λ^μ have only time component J_λ^0. We also assume that $\left(\ell_\lambda^{\mathrm{L}}\right)^{\mathrm{T}} = (0, \chi_\lambda)$. To derive Eq. (156) it is crucial that only left handed currents interactions are presented in Eq. (128).

It is useful to rewrite the Lagrangian (154) in terms of the two component spinors η_a and ϕ_a [74],

$$\mathcal{L} = \sum_a \mathrm{i}\eta_a^\dagger(\partial_t - \boldsymbol{\sigma}\boldsymbol{\nabla})\eta_a + \sum_a \left(\frac{\mathrm{i}}{2} m_a \eta_a^\dagger \sigma_2 \eta_a^* + \eta_a^\dagger \phi_a + \text{h.c.} \right). \qquad (157)$$

Using Eq. (157) we can receive the wave equation for two-component spinors,

$$\left(\frac{\partial}{\partial t} - \boldsymbol{\sigma}\boldsymbol{\nabla} \right) \eta_a + m_a \sigma_2 \eta_a^* = \mathrm{i}\phi_a. \qquad (158)$$

The solutions to Eq. (158) have the form (see, e.g., Ref. [74]),

$$\eta_a(\mathbf{r},t) = \int \mathrm{d}^3\mathbf{r}'\mathrm{d}t' \mathcal{S}_a^{\text{ret}}(\mathbf{r}-\mathbf{r}',t-t')\phi_a(\mathbf{r}',t'), \qquad (159)$$

$$\eta_a^*(\mathbf{r},t) = \int \mathrm{d}^3\mathbf{r}'\mathrm{d}t' \mathcal{R}_a^{\text{ret}}(\mathbf{r}-\mathbf{r}',t-t')\phi_a(\mathbf{r}',t'), \qquad (160)$$

where the retarded Green functions are expressed as (see also Ref. [74]),

$$\mathcal{S}_a^{\text{ret}}(\mathbf{r},t) = \mathrm{i}\tilde{\sigma}^\mu \partial_\mu D_a^{\text{ret}}(\mathbf{r},t), \quad \mathcal{R}_a^{\text{ret}}(\mathbf{r},t) = \mathrm{i}m_a \sigma_2 D_a^{\text{ret}}(\mathbf{r},t). \qquad (161)$$

Here $\partial_\mu = (\partial_t, \boldsymbol{\nabla})$, $\tilde{\sigma}^\mu = (\sigma^0, \boldsymbol{\sigma})$, σ^0 is the 2×2 unit matrix, $\boldsymbol{\sigma}$ are Pauli matrices, and $D_a^{\text{ret}}(\mathbf{r},t)$ is given in Eq. (136). One can check that Eq. (159) and (160), along with the definition of the retarded Green functions (161), represent the solutions to Eq. (158) by means of direct substituting.

Let us assume that the sources ϕ_a depend on time and spatial coordinates as in Sec. 11.1.,

$$\phi_a(\mathbf{r},t) = \theta(t)\phi_a^{(0)} e^{-\mathrm{i}Et}\delta^3(\mathbf{r}), \quad \phi_a^{(0)} = \sum_\lambda U_{a\lambda}^\dagger J_\lambda \chi_\lambda^{(0)}, \qquad (162)$$

where $\chi_\lambda^{(0)}$ is the time independent component of χ_λ. In deriving of Eq. (162) from Eq. (156) we introduce the new quantities $J_\lambda \equiv J_\lambda^{(0)0}$, where $J_\lambda^{(0)0}$ is the time independent part of J_λ^0, to simplify the notations.

On the basis of Eqs. (159)-(162) and using the technique developed in Sec. 11.1. we get the particles wave functions,

$$\eta_a(\mathbf{r},t) = \mathrm{i}\tilde{\sigma}^\mu \partial_\mu [e^{-\mathrm{i}Et+\mathrm{i}p_a r}] \frac{\phi_a^{(0)}}{4\pi r}, \qquad (163)$$

$$\eta_a^*(\mathbf{r},t) = \mathrm{i}m_a \sigma_2 e^{-\mathrm{i}Et+\mathrm{i}p_a r} \frac{\phi_a^{(0)}}{4\pi r}. \qquad (164)$$

To derive Eqs. (163) and (164) we suppose that $t \gg r$. The derivatives in Eq. (163) are applied on the exponent only because of the same reasons as in Sec. 11.1..

Using Eqs. (153), (163) and (164) as well as the following identity:

$$(\nu_\lambda^{\mathrm{L}})^c = (\nu_\lambda^c)^{\mathrm{R}} = \sum_a U_{\lambda a}^* \psi_a^{\mathrm{R}}, \quad (\nu_\lambda^{\mathrm{L}})^c = C \left(\overline{\nu_\lambda^{\mathrm{L}}} \right)^{\mathrm{T}}, \qquad (165)$$

we get the wave functions of ν_λ^{L} and $\left(\nu_\lambda^{\mathrm{L}}\right)^c$ as

$$\nu_\lambda^{\mathrm{L}}(\mathbf{r},t) = \frac{2E}{4\pi r} e^{-\mathrm{i}Et} \sum_{a\lambda'} e^{\mathrm{i}p_a r} U_{\lambda a} U_{a\lambda'}^\dagger J_{\lambda'} \begin{pmatrix} 0 \\ \tilde{\chi}_{\lambda'}^{(0)} \end{pmatrix}, \tag{166}$$

$$\left(\nu_\lambda^{\mathrm{L}}\right)^c(\mathbf{r},t) = -\frac{2E}{4\pi r} e^{-\mathrm{i}Et} \sum_{a\lambda'} \frac{m_a}{2E} e^{\mathrm{i}p_a r} U_{\lambda a}^* U_{a\lambda'}^\dagger J_{\lambda'} \begin{pmatrix} \chi_{\lambda'}^{(0)} \\ 0 \end{pmatrix}, \tag{167}$$

where $\tilde{\chi}_{\lambda'}^{(0)} = (1/2)[1 - (\boldsymbol{\sigma}\mathbf{n})]\chi_{\lambda'}^{(0)}$. In deriving of Eq. (166) we suppose that

$$[E - p_a(\boldsymbol{\sigma}\mathbf{n})] \approx E[1 - (\boldsymbol{\sigma}\mathbf{n})], \tag{168}$$

that is valid for relativistic neutrinos.

It is possible to construct two four component Majorana spinors from two-component Weil spinors,

$$\psi_a^{(1)} = \begin{pmatrix} \mathrm{i}\sigma_2(\eta_a)^* \\ \eta_a \end{pmatrix}, \quad \psi_a^{(2)} = \begin{pmatrix} \mathrm{i}\sigma_2\eta_a^* \\ (\eta_a^*)^* \end{pmatrix}, \tag{169}$$

where η_a and η_a^* are defined in Eqs. (163) and (164). One can see that these spinors satisfy the Majorana condition $\left[\psi_a^{(1,2)}\right]^c = \psi_a^{(1,2)}$. We use the spinor $\psi_a^{(1)}$ to receive Eq. (166) and $\psi_a^{(2)}$ for Eq. (167). In this case we get that ν_λ^{L} and $\left(\nu_\lambda^{\mathrm{L}}\right)^c$ are obtained as a result of the evolution of particles emitted from the same source.

It should be mentioned that both ν_λ^{L} and $\left(\nu_\lambda^{\mathrm{L}}\right)^c$ in Eqs. (166) and (167) propagate forward in time. To explain this fact let us discuss the complex conjugated equation (159). Performing the same computations one arrives to the analog of Eq. (167) which would depend on time as $e^{\mathrm{i}Et}$. However this wave function describes a particle emitted by the source different from that discussed here. Indeed, if we studied the complex conjugated Eq. (159), the integrand there would be $\mathcal{S}_a^{\mathrm{ret}*}\phi_a^*$. It would mean that the source in Eq. (162) would be proportional to $\phi_a^{(0)*}e^{\mathrm{i}Et}$, that, in its turn, would signify that $\left(\nu_\lambda^{\mathrm{L}}\right)^c$ would be emitted in a process involving a lepton which is a charge conjugated counterpart to that discussed here.

Let us illustrate this problem on a more physical example. Suppose that we study a *neutrino* emission in a process like the inverse β-decay,

$$p + \ell_\lambda^- \to n + \nu_\lambda \Rightarrow n + \sum_a U_{\lambda a}\psi_a^{\mathrm{L}}, \tag{170}$$

where p, n and ℓ_λ^- stand for a proton, a neutron and for a negatively charged lepton. The complex conjugated Eq. (159) would correspond to a process,

$$n + \ell_\lambda^+ \to p + \tilde{\nu}_\lambda \Rightarrow p + \sum_a U_{\lambda a}^* \left(\psi_a^{\mathrm{L}}\right)^c \Rightarrow p + \sum_a U_{\lambda a}^* \psi_a^{\mathrm{R}}, \tag{171}$$

where $\tilde{\nu}_\lambda$ and ℓ_λ^+ denote an *antineutrino* and a positively charged lepton, which is the charge conjugated counterpart to ℓ_λ^-. In Eq. (171) we use the facts that only left handed interactions exist in nature and the fields ψ_a describe Majorana particles. As one can see, Eqs. (170) and (171) represent two different processes. Hence, if we used $(\eta_a)^*$ to obtain $\left(\nu_\lambda^{\mathrm{L}}\right)^c(\mathbf{r},t)$, i.e. after a beam of neutrinos passes some distance r, it would correspond to the initial reaction (171) rather than (170). Note that the same result also follows from Eq. (169) if we replace $\psi_a^{(1)} \leftrightarrow \psi_a^{(2)}$ there.

Let us suppose for simplicity that the momentum of the fermion ℓ_λ is parallel to the neutrino momentum. It takes place if a relativistic incoming lepton is studied. We also assume that this fermion is in a state with the definite helicity,

$$\frac{1}{2}[1 - (\boldsymbol{\sigma}\mathbf{n})]\chi_\lambda^{(0)} = \chi_\lambda^{(0)}. \tag{172}$$

This expression is again natural for the relativistic fermion ℓ_λ. One can notice that the Lagrangian (128) is written in terms of the left handed chiral projections of ℓ_λ. Therefore, if we study a relativistic lepton, it will have its spin directed oppositely to the particle momentum as one can see from Eq. (172).

Using Eqs. (166), (167) and (172) as well as the orthonormality of the two-component spinors $\chi_\lambda^{(0)}$, $\left(\chi_\lambda^{(0)\dagger}\chi_{\lambda'}^{(0)}\right) = \delta_{\lambda\lambda'}$, we get the probabilities to detect ν_λ^{L} and $\left(\nu_\lambda^{\mathrm{L}}\right)^c$ as

$$P_{\nu_\lambda^{\mathrm{L}}}(r) \sim \sum_{ab,\lambda'} e^{\mathrm{i}(p_a - p_b)r} U_{\lambda a} U_{a\lambda'}^\dagger U_{b\lambda}^\dagger U_{\lambda' b} |J_{\lambda'}|^2, \tag{173}$$

$$P_{(\nu_\lambda^{\mathrm{L}})^c}(r) \sim \sum_{ab,\lambda'} \frac{m_a m_b}{(2E)^2} e^{\mathrm{i}(p_a - p_b)r} U_{\lambda a}^* U_{a\lambda'}^\dagger U_{b\lambda}^{\mathrm{T}} U_{\lambda' b} |J_{\lambda'}|^2. \tag{174}$$

In Eqs. (173) and (174) we drop the factor $(2E)^2/(4\pi r)^2$.

We can see that Eqs. (173) and (174) contain the oscillating exponent. This our result reproduces the usual formulae for neutrino oscillations in vacuum [44]. It should be also noticed that the expressions for the probabilities depend on r rather than on t in contrast to our previous works [33, 34, 36, 35, 40, 37, 39]. Note that the problem whether neutrino oscillations happen in space or in time was also discussed in Ref. [90]. The similar coordinate dependence of the probabilities was obtained in Refs. [26, 27] where the problem of neutrino oscillations in vacuum was studied.

It was mentioned in Ref. [26] that oscillations between active and sterile neutrinos are possible in case of the nonunitary matrix $(U_{\lambda a})$, i.e. the presence of only Majorana mass terms is not sufficient for the existence of this kind of transitions. The situation is analogous to that considered in the pioneering work [91] where oscillations between neutrinos and antineutrinos were studied. In Eq. (174) we obtain that the probability to detect $\left(\nu_\lambda^{\mathrm{L}}\right)^c$ is suppressed by the factor $m_a m_b/E^2$. It is in agreement with the results of Refs. [26, 92] as well as with Eq. (110).

12. Quantum Field Theory Description of Neutrino Oscillations in Background Matter

In the present section we study neutrino oscillations in background matter (see Sec. 3.) in frames of the quantum field theory approach. In particular we will be interested to examine the influence of background matter on the transitions between neutrinos and antineutrinos discussed in Secs. 8. and 11.2..

Despite we could receive satisfactory results for the description of neutrino-to-antineutrino transitions in vacuum in frames of the relativistic quantum mechanics approach [see Eqs. (110) as well as (174)] the accurate treatment of this process has to be done within the quantum field theory because of the following reason. For the case of Majorana neutrinos, particles are identical to their antiparticles and only in frames of the quantum field theory one has the most accurate description of antiparticle states. Moreover, as we have seen in Sec. 8., "antineutrino" states appear along with the small factor m_a/E_ν. Typically various approaches to the description of neutrino oscillations give contradictory results at this order of accuracy [94]. That is why we will use the quantum field theory treatment to capture this tiny effect.

As in Sec. 11., we will study the system of flavor neutrinos ν_λ, $\lambda = e, \mu, \tau, \ldots$, propagating in dense background matter between two spatial points, $\mathbf{x}_1$ and $\mathbf{x}_2$. We suggest that emission and absorption of neutrinos is due to the interaction with leptons $l^\pm_\lambda$ and heavy nucleons N. i.e these interactions are localized in two spatial regions: a "source" and a "detector". On the contrary, the neutrino interaction with background matter is uniformly distributed along the total neutrino propagation distance $\mathbf{L} = \mathbf{x}_2 - \mathbf{x}_1$.

In general case the dynamics of the system of mixed massive neutrinos should be formulated in frames of the left ν^{L}_λ and right ν^{R}_λ handed projections of flavor neutrinos [see also Eq. (128)],

$$\begin{aligned}
\mathcal{L} = & \sum_\lambda \left(\overline{\nu^{\mathrm{L}}_\lambda} \mathrm{i}\gamma^\mu \partial_\mu \nu^{\mathrm{L}}_\lambda + \overline{\nu^{\mathrm{R}}_\lambda} \mathrm{i}\gamma^\mu \partial_\mu \nu^{\mathrm{R}}_\lambda \right) \\
& - \sum_{\lambda\lambda'} \left(m^{\mathrm{D}}_{\lambda\lambda'} \overline{\nu^{\mathrm{L}}_\lambda} \nu^{\mathrm{R}}_{\lambda'} + m^{\mathrm{L}}_{\lambda\lambda'} \left(\nu^{\mathrm{L}}_\lambda\right)^{\mathrm{T}} C \nu^{\mathrm{L}}_{\lambda'} + m^{\mathrm{R}}_{\lambda\lambda'} \left(\nu^{\mathrm{R}}_\lambda\right)^{\mathrm{T}} C \nu^{\mathrm{R}}_{\lambda'} + \mathrm{h.c.} \right) \\
& - \sum_\lambda \overline{\nu^{\mathrm{L}}_\lambda} \gamma_\mu \nu^{\mathrm{L}}_{\lambda'} f^\mu_{\lambda\lambda'} - \sqrt{2} G_{\mathrm{F}} \left(j^\mu J_\mu + \mathrm{h.c.} \right),
\end{aligned} \tag{175}$$

where $m^{\mathrm{D}}_{\lambda\lambda'}$ and $m^{\mathrm{L,R}}_{\lambda\lambda'}$ are Dirac and Majorana mass matrices defined in Sec. 11., and

$$j^\mu = \sum_\lambda \overline{\nu^{\mathrm{L}}_\lambda} \gamma^\mu l^{\mathrm{L}}_\lambda, \tag{176}$$

is the neutrino-lepton current. The effective potential of the neutrino interaction with background matter $f^\mu_{\lambda\lambda'}$ [see Eq. (23)] is supposed to be coordinate independent, whereas the nuclear current J_μ is localized in space. We will take into account the interaction with background matter exactly. As in Sec. 3. here we study the general case and consider the matrix $(f^\mu_{\lambda\lambda'})$ to be nondiagonal. Note that in Eq. (175) right handed neutrinos do not participate in interactions with other particles, i.e. they are sterile.

As we will see below, the transitions between neutrinos and antineutrinos manifest in the process like $(N_1, N_2) + l^-_\beta \to (\text{neutrinos}) \to (N'_1, N'_2) + l^+_\alpha$. It should be noted that in this process massive neutrinos appear as virtual particles rather than show up explicitly [26, 27]. The S-matrix element for this kind of reaction has the form [26],

$$S = -\frac{1}{2!} \left(\sqrt{2} G_{\mathrm{F}} \right)^2 \int \mathrm{d}^4 x \mathrm{d}^4 y \langle l^+_\alpha; N'_1 N'_2 | T\{ j^\mu(x) J_\mu(x) j^\nu(y) J_\nu(y) \} | l^-_\beta; N_1 N_2 \rangle, \tag{177}$$

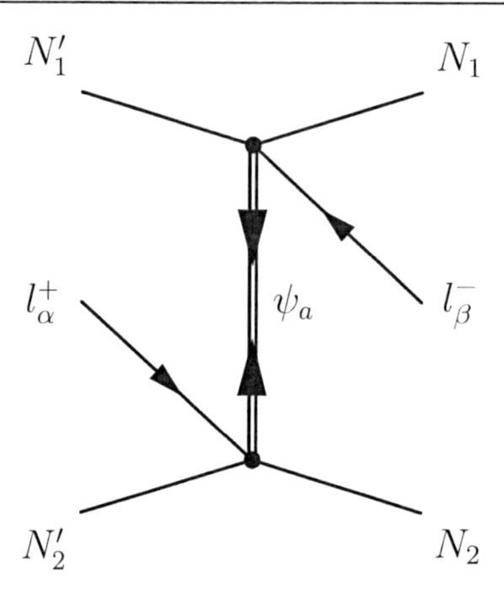

Figure 6. The Feynmann diagram corresponding to the lepton-nuclei scattering, $(N_1, N_2) + l_\beta^- \to \text{(neutrinos)} \to (N_1', N_2') + l_\alpha^+$, with the S-matrix element shown in Eq. (177).

where $N_{1,2}$ and $N_{1,2}'$ are states of the initial and final nucleons. This process is schematically illustrated in Fig 6. Note that the propagation of internal neutrinos is shown by a broad line which means that it accounts for the interaction with background matter exactly.

Now we have to diagonalize the mass matrices $m_{\lambda\lambda'}^{\mathrm{D}}$ and $m_{\lambda\lambda'}^{\mathrm{L,R}}$ in Eq. (175) by means of the matrix transformations (153), with the rectangular matrices $(U_{\lambda a})$ and $(V_{\lambda a})$ being nonunitary and independent. The amplitude of the process shown in Fig. 6 is nonzero if virtual neutrinos correspond to Majorana fields. That is why we introduce a new Majorana field $\psi_a = \Psi_a^{\mathrm{L}} + (\Psi_a^{\mathrm{L}})^c$ as in Sec. 11.2..

Let us rewrite the Lagrangian (175) in terms of the Majorana fields ψ_a,

$$\mathcal{L} = \sum_a \bar{\psi}_a (\mathrm{i}\gamma^\mu \partial_\mu - m_a)\psi_a - \sum_{ab} g_{ab}^\mu \overline{\psi_a^{\mathrm{L}}} \gamma_\mu \psi_b^{\mathrm{L}} - \text{(sources)}, \tag{178}$$

where (g_{ab}^μ) is the matrix of the effective potentials of the neutrino interaction with matter in the mass eigenstates basis which is defined in Eq. (25). In Eq. (178) we do not present the expression for the sources in the explicit form since they are supposed to be localized in space.

With help of Eq. (153) the T-product of the neutrino-lepton currents in Eq. (177) takes the form,

$$\begin{aligned}\langle l_\alpha^+ | T\{j^\mu(x) j^\nu(y)\} | l_\beta^- \rangle = & \frac{e^{\mathrm{i}p_\alpha x - \mathrm{i}p_\beta y}}{2\mathcal{V}\sqrt{E_\alpha E_\beta}} \sum_{ab} U_{\alpha a}^* U_{\beta b}^* \\ & \times u^{\mathrm{T}}(-p_\alpha) \left(\gamma_\mu^{\mathrm{L}}\right)^{\mathrm{T}} \langle 0 | T\{ [\bar{\psi}_a(x)]^{\mathrm{T}} \bar{\psi}_b(y)\} | 0 \rangle \gamma_\nu^{\mathrm{L}} u(p_\beta),\end{aligned} \tag{179}$$

where $u(p_{\alpha,\beta})$ are the outgoing and incoming spinors with four momenta $p_{\alpha,\beta} = (E_{\alpha,\beta}, \mathbf{p}_{\alpha,\beta})$ and $\mathcal{V}$ is the normalization volume. In Eq. (179) we use the representation of lepton states in the form of a plane wave from Ref. [76].

Since the mass of the nucleons is much bigger that the energies of leptons, we can suggest that they are at rest, i.e. we replace the nucleon currents

$$J_\mu(x) = \delta_{0\mu}\delta(\mathbf{x} - \mathbf{x}_2), \quad J_\mu(x) = \delta_{0\mu}\delta(\mathbf{y} - \mathbf{x}_1), \tag{180}$$

in Eq. (177).

The T-product of Majorana neutrino fields in Eq. (179) can be rewritten as (see, e.g., Ref. [74])

$$\langle 0|T\{[\bar{\psi}_a(x)]^{\mathrm{T}}\bar{\psi}_b(y)\}|0\rangle = -CS_{ab}(x-y), \tag{181}$$

where C is the charge conjugation matrix and $S_{ab}(x-y) = \langle 0|T\{\psi_a(x)\bar{\psi}_b(y)\}|0\rangle$ is the usual Feynmann propagator corresponding to a Dirac particle.

To simplify the calculations we suggest that the matrix (g^{μ}_{ab}) in Eq. (178) is close to diagonal. Under this assumption the propagator in Eq. (181) has the diagonal form: $S_{ab}(x) = \delta_{ab}S_a(x)$. The explicit form of the function $S_a(x)$ is given in Appendix D.. Moreover, as in Secs. 3. and 5., we will study the situation of nonmoving and unpolarized matter, with matrix (g^{μ}_{ab}) having only the zeroth component.

Using the results of Appendix D. we can represent the S-matrix element in Sec. (177) as

$$\begin{aligned} S =& 2\pi\delta(E_\alpha - E_\beta)\frac{G_{\mathrm{F}}^2}{4\mathcal{V}\sqrt{E_\alpha E_\beta}}M_{\beta\alpha}, \\ M_{\beta\alpha} =& -\frac{e^{\mathrm{i}\mathbf{p}_\beta\mathbf{x}_1 - \mathrm{i}\mathbf{p}_\alpha\mathbf{x}_2}}{2\pi L}\sum_a m_a U^*_{\alpha a}U^*_{\beta a}\bar{u}(p_\alpha)F_a u(p_\beta), \end{aligned} \tag{182}$$

where

$$\begin{aligned} F_a =& \frac{1}{2|g_a|p_a}\left[e^{\mathrm{i}k_1 L}\left(\frac{g_a^2}{2} + |g_a|p_a - g_a k_1(\boldsymbol{\alpha}\mathbf{n})\right)\right. \\ & \left. - e^{\mathrm{i}k_2 L}\left(\frac{g_a^2}{2} - |g_a|p_a - g_a k_2(\boldsymbol{\alpha}\mathbf{n})\right)\right](1-\gamma^5), \end{aligned} \tag{183}$$

and $k_{1,2}^2 = (p_a \pm |g_a|/2)^2$, $p_a = \sqrt{E^2 - m_a^2}$ is the neutrino "momentum", $E = E_\alpha = E_\beta$ is the energy of leptons which is conserved since nucleons are supposed to be at rest.

Let us discuss the case of high energy leptons with $E_\lambda \gg m_\lambda$, where m_λ is the lepton mass. In this case the biggest contribution to the matrix element of the total process arises from the channel in which the helicity of leptons changes from $-1/2$ to $1/2$. Thus the matrix element in Eq. (182) takes the form,

$$M(l^-_\beta \to l^+_\alpha) \approx -\frac{2e^{\mathrm{i}\mathbf{p}_\beta\mathbf{x}_1 - \mathrm{i}\mathbf{p}_\alpha\mathbf{x}_2}}{\pi L}\sum_a E m_a U^*_{\alpha a}U^*_{\beta a}\langle F_a\rangle e^{-\mathrm{i}\varphi/2}\sin\frac{\vartheta}{2} \tag{184}$$

where the spherical angles φ and ϑ fix the direction of the outgoing lepton momentum with respect to the incoming lepton momentum, which, for simplicity, is chosen to be directed along the vector $\mathbf{L}$.

Using Eq. (183) we can calculate the function F_a in Eq. (184) in the high energy leptons approximation,

$$\langle F_a\rangle = \bar{u}(p_\alpha)F_a u(p_\beta) = \begin{cases} \left(1 - \dfrac{|g_a|}{2p_a}\right)e^{\mathrm{i}(p_a - |g_a|/2)L}, & \text{if } p_a > \dfrac{|g_a|}{2}, \\ 0, & \text{if } p_a < \dfrac{|g_a|}{2}. \end{cases} \tag{185}$$

The details of the averaging over the lepton states in Eqs. (184) and (185) can be found in Ref. [26].

Finally the total cross section for the leptons scattering $(N_1, N_2) + l_\beta^- \to (\text{neutrinos}) \to (N_1', N_2') + l_\alpha^+$ can be presented in the following form:

$$\sigma(l_\beta^- \to l_\alpha^+) \sim \frac{G_F^4 E^2}{L^2} \sum_{ab} m_a m_b U_{\alpha a}^* U_{\beta a}^* U_{\alpha b} U_{\beta b} \langle F_a \rangle \langle F_b^* \rangle \frac{v_\alpha}{v_\beta} \tag{186}$$

where $v_{\alpha,\beta}$ are the velocities of outgoing and incoming leptons. If we study the most general Lagrangian (175), which contains both Dirac and Majorana mass matrices, Eqs. (185) and (186) involve transitions between neutrinos and antineutrinos studied in Refs. [26, 92] (see also Secs. 8. and 11.).

Using Eqs. (185) and (186) one can conclude that for light neutrinos with $m_a \ll E$ the matter contribution is negligible since $p_a \approx E - m_a^2/2E \gg |g_a|$. However the existence of heavy neutrinos is not excluded in some theoretical models [95]. If we study the neutrino propagation in the dense nuclear matter with $|g_a| \sim$ eV, the "momentum" of such heavy neutrinos can be comparable with the effective potential of matter interaction at the appropriate choice of lepton energy. Nevertheless we can see in Eq. (185) that the contribution of such heavy mass eigenstates to neutrino-to-antineutrino transitions will be strongly suppressed by interaction with background matter. Note that such a behavior of the transition probability in presence of background matter is in agreement with the results of Ref. [93] where oscillations between neutrinos and antineutrinos were studied in frames of the standard quantum mechanical approach.

In the derivation of the main results (185) and (186) we neglected the contribution of the nondiagonal elements of the matrix (g_{ab}^μ). One can expect that this contribution can be responsible for the enhancement of the probability of neutrino-to-antineutrino transitions as it can for neutrino flavor oscillation in matter (see Sec. 3.). If one takes into account these nondiagonal elements, the neutrino propagator in Eq. (181) also acquires nondiagonal entries. It is rather difficult to analyze this kind of "nondiagonal" propagators analytically. Thus, the approach for the description of neutrino oscillations based on the quantum field theory does not seem to be applicable for the situation of big nondiagonal elements of the matrix (g_{ab}^μ).

13. Conclusion

In conclusion we mention that in the present work we summarized our resent achievements in the theoretical description of neutrino oscillations in vacuum and external fields. Basically we studied three approaches: relativistic quantum mechanics approach (Secs. 2.-5., 8., and 9.), classical external sources method (Sec. 11.), and quantum field theory treatment of neutrino oscillations (Sec. 12.).

The formulation of the initial condition problem (2) for the system of flavor neutrinos is the main feature of the relativistic quantum mechanics method. Then one looks for wave functions of flavor neutrinos at subsequent moments of time. In vacuum the Cauchy problem for neutrino mass eigenstates (3) can be solved exactly [see Eq. (12)] giving us neutrino wave functions for arbitrary initial condition. When we study the time evolution of flavor neutrinos in presence of external fields within this formalism, we can reduce the dynamics of the system to a Schrödinger like equation [see, e.g., Eqs. (74) and (118)].

Note that, using this formalism, one solves not only the problem of the structure of the effective Hamiltonian but also we clearly point out what kind of space of wave functions this effective Hamiltonian acts in. In frames of the standard quantum mechanical approach [47, 20] one studies the evolution of a quantum mechanical neutrino "wave function" $\nu^{\mathrm{T}} = (\nu_\alpha, \nu_\beta, \dots)$, but the origin of the components of this wave function is disguised. Using the relativistic quantum mechanics method one immediately gets the answer to this questions. If we take the coefficients $a_a^{(\zeta)}$ of the decomposition of a four component neutrino wave function over the basis spinors [see, e.g., Eqs. (68) and (112)] and then make a certain transformation, like in Eqs. (73) and (117), of these coefficients, we arrive to the quantum mechanical neutrino "wave function".

Note that in our approach the coefficients $a_a^{(\zeta)}$ are c-numbers rather than operator as in quantum field theory. It means that this method may be called as a classical field theory treatment of neutrino oscillations since we do not use second quantized neutrino wave functions.

Besides resolving the conceptual problem of neutrino oscillations, the relativistic quantum mechanics method can be applicable for the description of neutrino oscillations not only in vacuum (Sec. 2.), but also in background matter (Sec. 3.), spin flavor oscillations in an external magnetic field (Sec. 4.), and in the combination of background matter and magnetic field (Sec. 5.). Note that both Dirac and Majorana neutrinos can be treated within this formalism (see Secs. 8. and 9.).

When we study the evolution of neutrinos in an external field, we use exact solutions of wave equations in this external field. That is why our treatment is valid for arbitrary strength of external fields. Moreover, the dynamics of mixed neutrinos is usually described in the mass eigenstates basis. In general case, external fields, like interaction with background matter or with an external electromagnetic field, are not diagonal in this basis [see Eqs. (25) and (45)], i.e. the neutrino mass eigenstates are not independent in presence of an external field [see, e.g., Eq. (67)]. Nevertheless in frames of the relativistic quantum mechanics method we can treat this kind of coupled mass eigenstates and receive results which are consistent with the standard quantum mechanical approach.

As it was mentioned above, the relativistic quantum mechanics method is a field theory counterpart of the usual quantum mechanical description of neutrino oscillation in a sense that it allows one to describe the evolution of mixed massive neutrinos from the first principles and thus it throws daylight upon some unclear issues of the quantum mechanical approach. However, trying to follow quantum mechanical approach as close as possible, we also adopted some of its weaknesses. For example, as in the quantum mechanical description, in frames of our method neutrino "wave functions" evolve in time rather than in space. It is, however, a contradiction with the majority of the experiments where one measures neutrino oscillations in space rather than in time.

Among other disadvantages of the present method we mention its limited applicability to description of neutrinos with small initial momentum in the case of nondiagonal external fields in the mass eigenstates basis. In this situation the ordinary differential equations for $a_a^{(\zeta)}(t)$ and $b_a^{(\zeta)}(t)$ evolution [see, e.g., Eq. (31)] start to entangle, making possible the transitions between "positive" and "negative" energy states. Thus, for example, instead of

4×4 effective Schrödinger equation for spin flavor oscillations one gets 8×8 differential equation, which is quite difficult to analyze.

We should also mention the fact that to get the consistency with the quantum mechanical approach, one should consider rather broad initial wave packet [see Eq. (18)] which is rather difficult to implement in practice. In Sec 2. we also demonstrated, for the case of neutrino evolution in vacuum, that spatially limited initial wave packets do not reveal flavor oscillations since they propagate as massless particles.

As we mentioned above, the relativistic quantum mechanics approach predicts neutrino oscillations in time rather than in space as they are observed in experiments. To resolve this contradiction in Sec. 11. we develop a new approach to the description of neutrino oscillations in vacuum which is based on the consideration of flavor neutrinos emitted by classical sources. Within this method one could study neutrino oscillations in space of both Dirac [Eqs. (151) and (152)] and Majorana [Eqs. (173)] neutrinos. For the case of Majorana neutrinos we could also describe the transitions between neutrinos and antineutrinos [see Eq. (174)]. This our result is also consistent with the previous studies [26, 92]. In frames of this formalism we exactly take into account neutrino masses as well as examine what kind of neutrino sources could give us a stable oscillations picture (see also Appendix C.).

However, the external classical source approach is not free from disadvantages. This method is based on the use of the retarded Green functions [Eqs. (135) and (161)] in the neutrino mass eigenstates basis. Thus, if one tries to apply this formalism for the description of neutrino oscillations in an external field, which, as we mentioned above, is typically nondiagonal in the mass eigenstate basis, one encounters a problem to find a nondiagonal retarded Green function. It is quite difficult to calculate this kind of function analytically. Thus this approach is unlikely to be applicable to the description of neutrino oscillations in an external field, in contrast to the relativistic quantum mechanics method with gives satisfactory results in presence of external fields at least for ultrarelativistic neutrinos, unless we study a special case of an external field which is diagonal in the neutrino mass eigenstates basis.

We made an attempt to describe transitions between neutrinos and antineutrinos using the relativistic quantum mechanics method [Eq. (110)] and the external classical source approach [Eq. (174)] and obtained the corresponding transition probability formulae which resemble the previously derived ones [26, 92]. However, in case of Majorana neutrinos, particles are identical to their antiparticles. The concept of antiparticles is an inherent component of quantum field theory. Thus the approaches based on classical physics does not seem to be an appropriate tool for the description of oscillations between neutrinos and antineutrinos. That is why in Sec. 12. we use the quantum field theory method to study particles and antiparticles in background matter in case of Majorana neutrinos. Note that the analogous method for the description of neutrino oscillations in vacuum was previously used in Refs. [26, 27, 94].

It should be noted that oscillations between neutrinos and antineutinos manifest in ($0\nu 2\beta$)-decay [17]. During this process two nucleons inside a nucleus exchange a virtual Majorana neutrino. That is why it is important to examine the influence of dense nuclear matter on the process of Majorana neutrinos propagation.

In our analysis we considered the situation when the interaction with matter is diagonal in the mass eigenstates basis since the case of a nondiagonal neutrino propagator is rather difficult to study analytically. This difficulty is analogous to that in the external classical source approach. Nevertheless we found that this type of matter interaction suppresses transitions between neutrinos and antineutrinos. Moreover, in case of hypothetical very heavy Majorana neutrinos the transition probability vanishes [see Eqs. (185) and (186)]. It means that one cannot explore the presence of this kind of heavy neutrinos studying $(0\nu 2\beta)$-decay.

In Secs. 6., 7., and 10. we discussed several applications of the results obtained in frames of the relativistic quantum mechanics method to the studies of astrophysical neutrinos emitted during the core collapse of a supernova. First we studied spin flavor oscillations of Dirac (Sec. 6.) and Majorana (Sec. 10.) neutrinos in an expanding envelope under the influence of the magnetic field of a supernova. It was found that for Dirac neutrinos the amplification of neutrino oscillations can be achieved in a moderate magnetic field [see Figs. 2(d)-4(d)]. On the contrary, in the Majorana neutrinos case, neutrino oscillations can be enhanced only in a strong magnetic field [see Fig. 5(b), (d) and (f)].

In Sec. 7. another channel of oscillations of astrophysical neutrinos was considered. We studied the possibility of spin flavor oscillations between electron and additional quasi-degenerate in mass sterile neutrinos. We obtained that the correction (81) to the effective Hamiltonian (80), found in frames of the relativistic quantum mechanics method, results in the appearance of the new resonance in this oscillations channel.

Finally in Appendix A. we present the general solution of the ordinary differential equations for the coefficients $a_a^{(\zeta)}$ which one encounters in the relativistic quantum mechanics method. In Appendix B. we discuss the validity of the relativistic quantum mechanics approach to the description of neutrino spin flavor oscillations. In particular, we analyze various factors which influence oscillations between electron and sterile neutrinos with very small δm^2 (see also Sec. 7.). In Appendix C. we study how neutrino wave functions converge to the values corresponding to the stable oscillations picture in the external classical source approach (Sec. 11.). In Appendix D. we present the details of the calculation of the S-matrix element involving transitions between neutrinos and antineutrinos in frames of the quantum field theory description of neutrino oscillations (Sec. 12.).

Summarizing we can say that a theoretical approach for the description of neutrino oscillations should satisfy the following requirements: the evolution equation governing the dynamics of mixed neutrinos should be derived from Lorentz invariant Lagrangian of the system and thus it should account for the relativistic invariance, at least implicitly; and such a method should be applicable in presence of various external fields. Among various theoretical approaches [25, 96] the relativistic quantum mechanics method is likely to be most appropriate formalism for the description of neutrino oscillations.

Acknowledgments

This work has been supported by Conicyt (Chile), Programa Bicentenario PSD-91-2006. The author is thankful to S. G. Kovalenko, T. Morozumi, and J. Maalampi for helpful discussions.

A. Solution to the Ordinary Differential Equations for the Functions $a_a^{(\zeta)}$

In this Appendix we describe the formalism for the analysis of ordinary differential equations for the functions $a_a^{(\zeta)}$ which one encounters in Secs. 3. and 4..

Let us study the time evolution of the two component spinor $\mathbf{Z}^{\mathrm{T}} = (Z_1, Z_2)$ which is governed by the Schrödinger equation of the form,

$$\mathrm{i}\dot{\mathbf{Z}} = H\mathbf{Z}, \tag{187}$$

where the Hamiltonian has the following form:

$$H = g \begin{pmatrix} 0 & e^{\mathrm{i}\omega t} \\ e^{-\mathrm{i}\omega t} & 0 \end{pmatrix}. \tag{188}$$

Here g and ω are real parameters. Eq. (187) should be supplied with the initial condition $\mathbf{Z}(0)$. To find the solution of Eqs. (187) and (188) we introduce the new spinor $\mathbf{Z}'$ by the relation, $\mathbf{Z} = \mathcal{U}\mathbf{Z}'$, where the unitary matrix $\mathcal{U}$ reads

$$\mathcal{U} = \begin{pmatrix} e^{\mathrm{i}\omega t/2} & 0 \\ 0 & e^{-\mathrm{i}\omega t/2} \end{pmatrix}. \tag{189}$$

Now Eq. (187) is rewritten in the following way:

$$\mathrm{i}\dot{\mathbf{Z}}' = H'\mathbf{Z}', \tag{190}$$

with the new Hamiltonian H' which is obtained with help of Eqs. (187)-(189),

$$H' = \mathcal{U}^{\dagger} H \mathcal{U} - \mathrm{i}\mathcal{U}^{\dagger}\dot{\mathcal{U}} = \begin{pmatrix} \omega/2 & g \\ g & -\omega/2 \end{pmatrix} \tag{191}$$

Note that the initial condition for the spinor $\mathbf{Z}'(0)$ is the same as for $\mathbf{Z}(0)$, $\mathbf{Z}'(0) = \mathbf{Z}(0)$, due to the special form of the matrix $\mathcal{U}$ in Eq. (189).

Supposing that the Hamiltonian H' in Eqs. (190) and (191) does not depend on time we get the solution to Eq. (190) as

$$\mathbf{Z}'(t) = \exp\left(-\mathrm{i}H't\right)\mathbf{Z}'(0) = \left(\cos\Omega t - \mathrm{i}(\boldsymbol{\sigma}\mathbf{n})\sin\Omega t\right)\mathbf{Z}'(0), \tag{192}$$

where $\mathbf{n} = (g, 0, \omega/2)/\Omega$ is the unit vector and $\Omega = \sqrt{g^2 + (\omega/2)^2}$. Using Eqs. (189) and (192) we arrive to the expressions for the components of $\mathbf{Z}$ written in terms of the initial condition $\mathbf{Z}(0)$:

$$\begin{aligned} Z_1(t) &= \left(\cos\Omega t - \mathrm{i}\frac{\omega}{2\Omega}\sin\Omega t\right) e^{\mathrm{i}\omega t/2} Z_1(0) - \mathrm{i}\frac{g}{\Omega}\sin(\Omega t) e^{\mathrm{i}\omega t/2} Z_2(0), \\ Z_2(t) &= \left(\cos\Omega t + \mathrm{i}\frac{\omega}{2\Omega}\sin\Omega t\right) e^{-\mathrm{i}\omega t/2} Z_2(0) - \mathrm{i}\frac{g}{\Omega}\sin(\Omega t) e^{-\mathrm{i}\omega t/2} Z_1(0), \end{aligned} \tag{193}$$

which can be directly applied for the analysis of ordinary differential equations from Secs. 3. and 4..

To get the solution of Eq. (33) we identify the components of the spinor $\mathbf{Z}$ with $a_{1,2}^{-}$ and the parameter ω with ω_{-} (see Sec. 3.). Finally we arrive to Eq. (34). We can also apply Eq. (193) to obtain the solution of Eq. (53). For this purpose one considers two cases:

- For $\mathbf{Z}^{\mathrm{T}} = (a_1^+, a_2^+)$, $g = -\mu B$, $\omega = \omega_+$ and $\Omega = \Omega_+$;
- For $\mathbf{Z}^{\mathrm{T}} = (a_1^-, a_2^-)$, $g = \mu B$, $\omega = \omega_-$ and $\Omega = \Omega_-$.

Using these formulae together with Eqs. (193) we readily arrive to Eqs. (54)-(56). Note that the dynamics of the system (187) and (188) is analogous to the quantum mechanical description of neutrino spin flavor oscillation in a twisting magnetic field studied in Ref. [97].

B. Analysis of Approximations Made in the Derivation of the Effective Hamiltonian

In this Appendix we analyze the validity of the relativistic quantum mechanics approach for the description of neutrino spin flavor oscillations used in Sec. 5.. The correction to the effective Hamiltonian (81) is a rather small quantity. Therefore we should evaluate other factors which can also give the contributions, comparable with Eq. (81), to the effective Hamiltonian. In this section we analyze the contributions to the effective Hamiltonian from longitudinal magnetic field, matter polarization and possible corrections from the new interactions.

First we should remind that we use the relativistic quantum mechanics approach, with the external fields being independent of spatial coordinates. If external fields depend on the spatial coordinates, Dirac wave packets theory reveals various additional phenomena such as particles creation by the external field inhomogeneity [98]. For the approximation of the spatially constant external fields to be valid, the typical length scale of the external field variation L_{ext} should be much greater than the Compton length of a neutrino [98]: $L_{\text{ext}} \gg_C = \hbar/m_\nu c$ [49]. For a neutrino with $m_\nu \sim 1\,\text{eV}$ this condition reads $L_{\text{ext}} \gg 10^{-5}$ cm, that is fulfilled for almost all realistic external fields.

A general remark on the "perturbative" approach used in Sec. 5. should be made. In the modified Eq. (31), which includes the interaction with the magnetic field, the terms containing $a_a^\pm$ and $a_a^\mp$ are coupled and the coupling terms are proportional to g and μB (72). As we demonstrate in Sec. 2., mass eigenstates decouple in vacuum are their values depend on the initial condition only. While solving the modified Eq. (31), we could just take a term linear in g and μB (see, e.g., Eq. (61) as well as Refs. [34, 35]). However we take into account these terms exactly in the further analysis [see Eqs. (71) and (74)]. It is equivalent to the summation of all terms in the perturbation series.

While deriving the effective Hamiltonian (74) in Sec. 5. we supposed that magnetic field is transverse with respect to the neutrino motion. The effect of the longitudinal magnetic field on neutrino oscillations was studied in Ref. [55]. It was found there that diagonal entries of the effective Hamiltonian receive additional small contributions $\mu_a B_\parallel (m_a/k)$. In order to neglect the longitudinal magnetic field contribution in comparison with our corrections (81), its strength should satisfy the condition,

$$\frac{B_\parallel}{B_\perp} \ll \frac{1}{16kB_\perp|\mu_a m_a - \mu_b m_b|} \left| \frac{m_a^2 g_a^3}{\mathcal{M}_a^2} - \frac{m_b^2 g_b^3}{\mathcal{M}_b^2} \right|, \tag{194}$$

where $B_\perp$ is the transverse component of the magnetic field.

A regular magnetic field of a supernova typically has both poloidal and toriodal components. Of course, an irregular turbulent magnetic field can be also present but its length scale appears to be quite small. A toroidal magnetic field can be $\sim 10^{16}\,\text{G}$ and is concentrated near the equator of the star at the distance $\sim 10\,\text{km}$ from the star center [99]. Thus in our case a toroidal magnetic field is unlikely to significantly contribute to the dynamics of neutrino oscillations.

Let us evaluate the fraction of neutrinos for which the new correction (81) to the effective Hamiltonian gives bigger contribution to the resonance enhancement of oscillations compared to that of the longitudinal component of the poloidal magnetic field. Using Eq. (194) we find that these neutrinos should be emitted inside the solid angle near the equatorial plane with the spread 2ϑ, where $\vartheta \sim B_{\parallel}/B_{\perp}$. Assuming the radially symmetric neutrino emission we find that about 10% of the total neutrino flux is affected by the new resonance (95), i.e. the influence of the longitudinal magnetic field is negligible for oscillations of such particles.

The next important approximation made in the deviation of Eq. (74) was the assumption of negligible polarization of matter which can be not true if we study rather strong magnetic fields. The effect of matter polarization on neutrino oscillations was previously discussed in Refs. [49, 100, 50]. Matter polarization produces the following contributions to the diagonal entries of the effective Hamiltonian [49, 50]: $g_a(\boldsymbol{\lambda}_f\boldsymbol{\beta}_\nu)$ (left polarized neutrinos) and $g_a(\boldsymbol{\lambda}_f\boldsymbol{\beta}_\nu)(m_a/k)$ (right polarized neutrinos). Here $\boldsymbol{\beta}_\nu$ is the neutrino velocity and $\boldsymbol{\lambda}_f$ is the mean polarization vector of background fermions and we keep only the leading order in m_a/k.

First we estimate the contribution to the right polarized neutrinos effective potential. It is clear that one should take into account only the polarization of electrons since nucleons are much heavier. For the weakly degenerate electrons we should discuss the case of weak field limit (see Ref. [100]), since $2eB/m_e^2 \sim 10^{-8} \ll 1$, where m_e is the electron's mass and $B \sim 10^7\,\text{G}$. Since the temperature inside the shock wave region can be about several MeV [101] the electrons are relativistic. Hence their mean polarization can be estimated as, $|\boldsymbol{\lambda}_f| \sim \mu_{\text{B}} B m_e/3T_e^2$, where T_e is the temperature of electrons. Therefore we get that the new correction to the effective Hamiltonian (81) becomes bigger than the contribution of matter polarization to the effective potential of right polarized neutrinos at $T_e > 4\,\text{MeV}$.

To evaluate the contribution $g_a(\boldsymbol{\lambda}_f\boldsymbol{\beta}_\nu)$ to the effective Hamiltonian, which corresponds to the left polarized neutrinos effective potential, we notice that the vector $\boldsymbol{\lambda}_f$ should be directed along the external magnetic field. For this term to be much less than the new correction to the effective Hamiltonian (81), the angle ϑ, defined above, should be very small: $\vartheta \ll 10^{-8}$, for $T_e \sim 4\,\text{MeV}$. It means that we can neglect the polarization effects only if neutrinos are emitted very close the equator of a star. We should, however, remind that this kind of matter polarization term contributes only to $(H_{QM})_{22}$ in Eq. (80) since only this entry corresponds to ν_e^{L}. Thus the presence of the term $g_a(\boldsymbol{\lambda}_f\boldsymbol{\beta}_\nu)$ does not directly affect our results since we study $\nu_e^{\text{R}} \leftrightarrow \nu_s^{\text{L}}$ oscillations channel (see also Table 1).

The presence of big Dirac neutrino magnetic moments implies the existence of new interactions, beyond the standard model, which electromagnetically couple left and right polarized neutrinos. It is probable that these new interactions also contribute to the effective potential of the right polarized neutrino interaction with background matter. Despite this additional effective potential is likely to be small, one should evaluate it and compare with the correction (81).

The most generic $\mathrm{SU}(2)_L \times \mathrm{U}(1)_Y$ gauge invariant and remormalizable interaction which produces Dirac neutrino magnetic moment was discussed in Ref. [102]. It was found that neutrino magnetic moments arise from the effective Lagrangian involving the dimension $n = 6$ operators $\mathcal{O}_j$,

$$\mathcal{L}_{\text{eff}} = \sum_j \frac{C_j}{\Lambda^2}\mathcal{O}_j + \text{h.c.}, \tag{195}$$

where $\Lambda \sim 1\,\text{TeV}$ is a scale of the new physics and C_j are the effective operator coupling constants. The sum in Eq. (195) spans all the operators of the given dimension.

One of the operators $\mathcal{O}_j$ also contributes to the effective potential of a right-handed neutrino in matter,

$$\mathcal{O} = \kappa \bar{L}\tau_a \tilde{\phi}\sigma^{\mu\nu}\nu_{\text{R}} W^a_{\mu\nu}, \tag{196}$$

where κ is the coupling constant, τ_a are Pauli matrices, $L^{\text{T}} = (\nu_{\text{L}}, e_{\text{L}})$ is the $\mathrm{SU}(2)_L$ isodoublet, $\tilde{\phi} = \mathrm{i}\tau_2\phi^*$, with ϕ being a Higgs field, and $W^a_{\mu\nu} = \partial_\mu W^a_\nu - \partial_\nu W^a_\mu - \kappa\epsilon_{abc}W^b_\mu W^c_\nu$ is the $\mathrm{SU}(2)_L$ field strength tensor.

Assuming the spontaneous symmetry breaking at the electroweak scale, $\phi^{\text{T}} \to (0, v/\sqrt{2})$, we can rewrite the Lagrangian in the form,

$$\mathcal{L}_{\text{eff}} = \frac{C\kappa v}{\sqrt{2}}\bar{e}_{\text{L}}\sigma^{\mu\nu}\nu_{\text{R}}(W^1_{\mu\nu} - \mathrm{i}W^2_{\mu\nu}) + \text{h.c.}, \tag{197}$$

which implies that a process $e^- + \nu_{\text{R}} \to e^- + \nu_{\text{L}}$ should happen in background matter.

Using the results of Ref. [102] we can evaluate the contribution of the Lagrangian (197) to the effective Hamiltonian (80) as

$$\delta V_{\text{R}} \sim V_{sm}\left(\frac{\mu_\nu}{\mu_{\text{B}}}\right)^2\left(\frac{E_\nu}{m_e}\right)^2\frac{|\kappa|^2}{G_{\text{F}}M_W^2}, \tag{198}$$

where $V_{sm} \sim G_{\text{F}}n_e$ is the standard model effective potential and M_W is the W boson mass. Taking $\mu_\nu \sim 10^{-12}\mu_{\text{B}}$, $E_\nu \sim 100\,\text{MeV}$ and $m_\nu \sim 1\,\text{eV}$ (see Sec. 7.) we can get that the ratio of the correction to the effective potential (198) and new correction (81) is $\sim 10^{-4}$. It means that the influence of new interactions, which generate neutrino magnetic moments, are not important for neutrino spin flavor oscillations.

Now let us estimate the influence of the diagonal magnetic moments μ_a on the dynamics of spin flavor oscillations. It was found in Ref. [66] that to get the significant ν_e^{R} luminosity $\sim 10^{50}$ erg/s the diagonal magnetic moment should be $\mu_{\nu_e} = 10^{-13}\,\mu_{\text{B}}$. Eq. (95) was obtained under the assumption $\mu_{\nu_e} \ll \mu$. In Eq. (97) we use $\mu = 3 \times 10^{-12}\,\mu_{\text{B}}$, i.e. the condition of the Eq. (95) validity is satisfied. In Fig. 7 we present the numerical solution of the Schrödinger equation (74). We remind that our simplified model with $\mu_{\nu_e} = 0$ gives the transition probability $P(x) = \sin^2(\mu B x)$ if the resonance condition (95) is fulfilled. As one can see on Fig. 7 there is almost no difference in the dynamics of spin flavor oscillations in our simplified model and more realistic situation which involves non-zero diagonal magnetic moment of an electron neutrino.

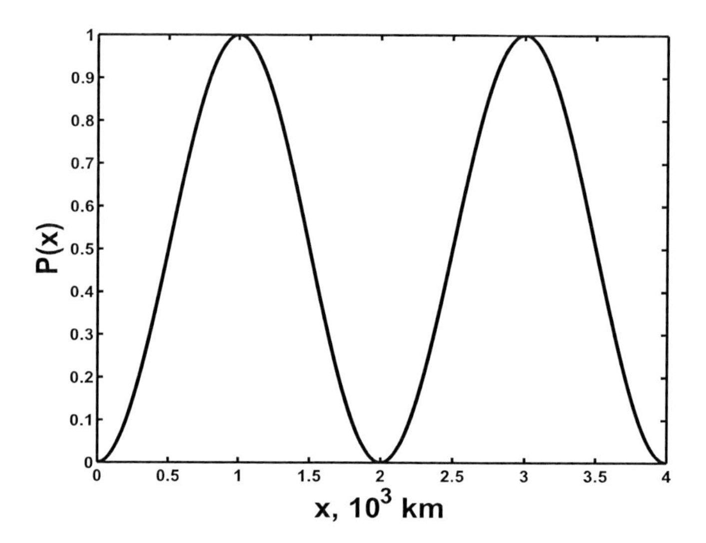

Figure 7. The numerical transition probability for neutrinos with $\mu = 3 \times 10^{-12}\,\mu_{\text{B}}$, $\mu_{\nu_e} = 10^{-13}\,\mu_{\text{B}}$, $E = 100\,\text{MeV}$, $m_{\nu_e} = 1\,\text{eV}$ and $\delta m^2 = 5 \times 10^{-17}\,\text{eV}^2$ moving in background matter with $\rho = 10^6\,\text{g}/\text{cm}^3$ and interacting with the magnetic field $B = 10^7\,\text{G}$. After Ref. [40].

C. Evaluation of Integrals

In this Appendix we calculate the integrals which one encounters while studying neutrino oscillations in the model with classical sources in Sec. 11.. It is interesting to evaluate the inexactitude which is made when we approach to the limit $x_m \to \infty$ in Eq. (142). Let us discuss two functions

$$F(r,t) = \int_0^{x_m} \mathrm{d}x \frac{J_1(mx)}{\sqrt{r^2+x^2}} e^{\mathrm{i}E\sqrt{r^2+x^2}} = \int_0^{y_m} \mathrm{d}y \frac{J_1(\rho y)}{\sqrt{1+y^2}} e^{\mathrm{i}\mathcal{E}\sqrt{1+y^2}}, \tag{199}$$

$$F_0(r) = \frac{1}{\rho}(e^{\mathrm{i}\mathcal{E}} - e^{\mathrm{i}P}). \tag{200}$$

These functions are proportional to $\mathcal{I}_2$ in Eqs. (142) and (143) respectively. In Eqs. (199) and (200) we use dimensionless parameters $\rho = mr$, $\mathcal{E} = Er = \gamma\rho$, $\gamma = E/m$, $y_m = \sqrt{(t/r)^2 - 1}$ and $P = \sqrt{\mathcal{E}^2 - \rho^2}$.

In case we study neutrinos, we get that $\mathcal{E} \gg \rho \gg 1$ in almost all realistic situations. For example, suppose we study a neutrino emitted in a supernova explosion is our Galaxy. The typical distance is $r \sim 10\,\text{kpc}$. Taking $m \sim 1\,\text{eV}$ and $E \sim 10\,\text{MeV}$, we receive that $\mathcal{E} \sim 10^{35}$ and $\rho \sim 10^{28}$.

Basing on the analysis of Sec. 11.1. we can rewrite Eq. (199) as

$$F(r,t) = F_0(r) - \delta F, \quad \delta F = \int_{y_m}^{+\infty} \mathrm{d}y \frac{J_1(\rho y)}{\sqrt{1+y^2}} e^{\mathrm{i}\mathcal{E}\sqrt{1+y^2}}. \tag{201}$$

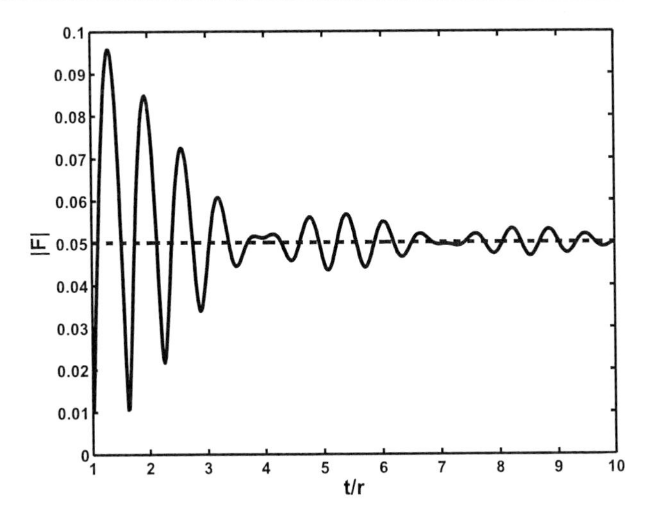

Figure 8. The absolute values of the functions $F(r,t)$ and $F_0(r)$ versus t. This figure is taken from Ref. [38].

Using the fact that $\rho \gg 1$, $y_m \gg 1$ and the representation for the Bessel function,

$$J_1(z) \approx \sqrt{\frac{2}{\pi z}} \cos\left(z - \frac{3\pi}{4}\right) \quad \text{at} \quad z \to +\infty, \tag{202}$$

we obtain for the function δF in Eq. (201) the following expression:

$$\delta F \approx -\frac{1}{\sqrt{2\pi\rho}}\{\text{ci}([\gamma+1]\rho y_m) + \text{ci}([\gamma-1]\rho y_m) + \text{i}[\text{si}([\gamma+1]\rho y_m) + \text{si}([\gamma-1]\rho y_m)]\}, \tag{203}$$

where $\text{ci}(z)$ and $\text{si}(z)$ are cosine and sine integrals. Using the asymptotic expression,

$$|\text{ci}(z)| \sim |\text{si}(z)| \sim \frac{1}{z} \quad \text{at} \quad z \to +\infty, \tag{204}$$

we obtain that the function δF approaches to zero as $1/(y_m \rho^{3/2})$ at great values of y_m and ρ. Note that this result remains valid for a particle with an arbitrary γ factor, i.e. rapid oscillations of the function δF will attenuate even for slow particles. This analysis substantiates the approximations made in Sec. 11.1..

Finally let us illustrate the behavior of the functions $F(r,t)$ and $F_0(r)$. On Fig. 8 we present the absolute values of these functions versus t. This figure is plotted for $\rho = 1$ and $\mathcal{E} = 10$. The solid line is the absolute value of the function $F(r,t)$ and the dashed line of the function $F_0(r)$. As we mention in Sec. 11.1. the relaxation phenomena occur when $t \gtrsim r$. It can be also seen on Fig. 8. It is possible to notice that $|F(r,t)| \to |F_0(r)|$ at great values of t as it is predicted in Sec. 11.1..

D. Calculation of the S-matrix Element

In this Appendix we present the detailed calculation of the S-matrix element in Sec. 12.. Note that analogous calculation for the case of virtual neutrinos propagating in vacuum, i.e. when the matrix $(f^{\mu}_{\lambda\lambda'})$ is absent in Eq. (175), is presented in Ref. [26].

Using Eqs. (179)-(181) as well as after the spatial integration and the elimination of the combinatorial factor we can cast Eq. (177) in the form,

$$S = \frac{G_F^2}{\mathcal{V}\sqrt{E_\alpha E_\beta}} \int \mathrm{d}x_0 \mathrm{d}y_0 \bar{u}(p_\alpha)\gamma^0 P_R S_a(\mathbf{L}, x_0 - y_0) P_R \gamma^0 u(p_\beta) e^{\mathrm{i}\mathbf{p}_\beta \mathbf{x}_1 - \mathrm{i}\mathbf{p}_\alpha \mathbf{x}_2} e^{\mathrm{i}E_\alpha x_0 - \mathrm{i}E_\beta y_0}, \tag{205}$$

where we use the fact that $u^T(-p_\alpha)C = \bar{u}(p_\alpha)$. The integration with respect to x_0 and y_0 can be performed with help of the new variables: $T = (x_0 + y_0)/2$ and $t = x_0 - y_0$. After this integration one gets the energy conservation δ-function in Eq. (182).

The Fourier transform of the neutrino propagator $S_a(x)$ in Eq. (205) was found in Ref. [103] and has the form,

$$S_a(k) = \frac{(k^2 - m_a^2 - g_a^2/4 - \mathrm{i}\sigma_{\mu\nu}\gamma^5 g_a^\mu k^\nu)(\gamma^\mu k_\mu + m + \gamma_\mu \gamma^5 g_a^\mu/2)}{(k^2 - m^2 - g_a^2/4)^2 - (g_a k)^2 + k^2 g_a^2}, \tag{206}$$

where $g_a^\mu \equiv g_{aa}^\mu$ are the diagonal elements of the matrix (g_{ab}^μ) [see Eq. (25)].

In case of the nonmoving and unpolarized matter the momentum integration in Eq. (205) can be performed using the calculus of residues. Finally we arrive to the following result:

$$\int e^{\mathrm{i}\mathbf{k}\mathbf{L}} \frac{\mathrm{d}^3\mathbf{k}}{(2\pi)^3} \frac{p_a^2 - g_a^2/4 - \mathbf{k}^2 - g_a(\boldsymbol{\alpha}\mathbf{k})}{(\mathbf{k}^2 - k_1^2)(\mathbf{k}^2 - k_2^2)} (1 - \gamma^5) \approx -\frac{F_a}{4\pi L}. \tag{207}$$

where F_a is defined in Eq. (183). In Eq. (207) we neglect several small terms since $k_{1,2}L \gg 1$.

It is convenient to perform momentum integration in Eq. (207) using cylindrical coordinates pointing $\mathbf{L}$ along the z-axis. Note that the integration over k_ϕ is trivial and gives 2π since the terms in the integrand [see also Eq. (206)] containing k_ϕ just vanish. The result of the integration over $\mathbf{k}$ can be presented in the form,

$$\int \frac{\mathrm{d}^2\mathbf{k}}{(2\pi)^2} \frac{F(k_\rho, k_z)}{(\mathbf{k}^2 - k_1^2)(\mathbf{k}^2 - k_2^2)} = J_a + J_b + J_c, \tag{208}$$

where $\mathrm{d}^2\mathbf{k} = k_\rho \mathrm{d}k_\rho \mathrm{d}k_z$ and

$$\begin{aligned} J_a &= \int_0^{k_2} \frac{k_\rho \mathrm{d}k_\rho}{(2\pi)^2} \int_{-\infty}^{+\infty} \mathrm{d}k_z \frac{F(k_\rho, k_z)}{(k_z^2 - k_{z1}'^2)(k_z^2 - k_{z2}'^2)}, \\ J_b &= \int_{k_2}^{k_1} \frac{k_\rho \mathrm{d}k_\rho}{(2\pi)^2} \int_{-\infty}^{+\infty} \mathrm{d}k_z \frac{F(k_\rho, k_z)}{(k_z^2 - k_{z1}'^2)(k_z^2 + k_{z2}''^2)} \\ J_c &= \int_{k_1}^{+\infty} \frac{k_\rho \mathrm{d}k_\rho}{(2\pi)^2} \int_{-\infty}^{+\infty} \mathrm{d}k_z \frac{F(k_\rho, k_z)}{(k_z^2 + k_{z1}''^2)(k_z^2 + k_{z2}''^2)}. \end{aligned} \tag{209}$$

Here we use the notations $k'_{z1,2} = \sqrt{k_{1,2}^2 - k_\rho^2}$ and $k''_{z1,2} = \sqrt{k_\rho^2 - k_{1,2}^2}$. The contours of the integration for each of the integrals in Eq. (209) are shown in Fig 9.

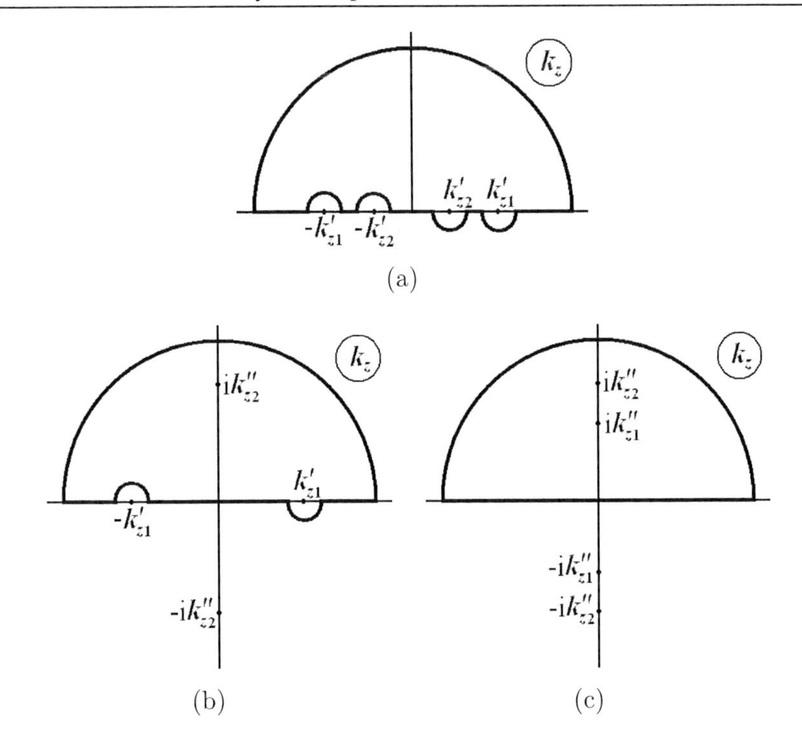

Figure 9. Integration contours for each of the integrals in Eq. (209). The panel (a) corresponds to J_a, (b) to J_b, and (c) to J_c.

References

[1] K. Abe, *et al.* (Super-Kamiokande Collaboration), 1010.0118 [hep-ex]; B. Aharmim, *et al.* (SNO Collaboration), *Phys. Rev. C* **81**, 055504 (2010), 0910.2984 [nucl-ex]; G. Bellini, *et al.* (Borexino Collaboration), *Phys. Rev. D* **82**, 033006 (2010), 0808.2868 [astro-ph]; J. N. Abdurashitov, *et al.* (SAGE Collaboration), *Phys. Rev. C* **80**, 015807 (2009), 0901.2200 [nucl-ex].

[2] J. N. Bahcall and M. H. Pinsonneault, *Phys. Rev. Lett.* **92**, 121301 (2004), astro-ph/0402114.

[3] R. Wendell, *et al.* (Super-Kamiokande Collaboration), *Phys. Rev. D* **81**, 092004 (2010), 1002.3471 [hep-ex]. B. Aharmim, *et al.* (SNO Collaboration), *Phys. Rev. D* **80**, 012001 (2009), 0902.2776 [hep-ex].

[4] R. Abbasi, *et al.* (IceCube Collaboration), *Astropart. Phys.* **34**, 48 (2010), 1004.2357 [astro-ph.HE].

[5] K. Hirata, *et al.* (Kamiokande II Collaboration), *Phys. Rev. Lett.* **58**, 1490 (1987); R. M. Bionta, *et al.* (IMB Collaboration), *Phys. Rev. Lett.* **58**, 1494 (1987); E. N. Alekseev *et al.*, *JETP Lett.* **45**, 589 (1987); V. L. Dadykin, *et al.*, *JETP Lett.* **45**, 593 (1987).

[6] A. Gando, *et al.* (KamLAND Collaboration), 1009.4771 [hep-ex]; P. Adamson, *et al.* (MINOS Collaboration), *Phys. Rev. D* **82**, 051102(R) (2010), 1006.0996 [hep-ex]; N. Agafonova, *et al.* (OPERA Collaboration), *Phys. Lett. B* **691**, 138 (2010), 1006.1623 [hep-ex]; A. A. Aguilar-Arevalo, *et al.* (MiniBooNE Collaboration), *Phys. Rev. Lett.* **102**, 101802 (2009), 0812.2243 [hep-ex].

[7] Y. Kurimoto, *et al.* (SciBooNE Collaboration), *Phys. Rev. D* **81**, 111102(R) (2010), 1005.0059 [hep-ex].

[8] R. N. Mohapatra and A. Yu. Smirnov, *Ann. Rev. Nucl. Part. Sci.* **56**, 569 (2006), hep-ph/0603118.

[9] L. Wolfenstein, Phys. Rev. D **17**, 2369 (1978); S. P. Mikheev and A. Yu. Smirnov, *Sov. J. Nucl. Phys.* **42**, 913 (1985).

[10] A. Cisneros, *Astrophys. Space Sci.* **10**, 87 (1971); M. B. Voloshin, M. I. Vysotskiĭ, and L. B. Okun', *Sov. Phys. JETP* **64**, 446 (1986).

[11] P. C. de Holanda and A. Yu. Smirnov, *JCAP* **0302**, 001 (2003), hep-ph/0212270.

[12] C. Biggio, M. Blennow, and E. Fernandez-Martinez, *JHEP* **0908**, 090 (2009), 0907.0097 [hep-ph].

[13] A. Yu. Ignatiev and G. C. Joshi, *Phys. Rev. D* **51**, 2411 (1995), hep-ph/9407346.

[14] C. Giunti and A. Studenikin, *Phys. Atom. Nucl.* **72**, 2089 (2009), 0812.3646 [hep-ph].

[15] M. Fukugita and T. Yanagida, *Physics of Neutrinos and Applications to Astrophysics* (Springer, Berlin, 2003), pp. 461–479.

[16] F. T. Avignone, III, S. R. Elliott, and J. Engel, *Rev. Mod. Phys.* **80**, 481 (2008), 0708.1033 [nucl-ex].

[17] S. M. Bilenky, *Lect. Notes Phys.* **817**, 139 (2010), 1001.1946 [hep-ph].

[18] J. Argyriades, *et al.* (NEMO Collaboration), *Phys. Rev. C* **80**, 032501(R) (2009), 0810.0248 [hep-ex]; C. Arnaboldi, *et al.* (CUORICINO Collaboration), *Phys. Rev. C* **78**, 035502 (2008), 0802.3439 [hep-ex].

[19] S. M. Bilenky and B. Pontecorvo, *Phys. Rept.* **41**, 225 (1978).

[20] C.-S. Lim and W. J. Marciano, *Phys. Rev. D* **37**, 1368 (1988).

[21] B. Kayser, *Phys. Rev. D* **24**, 110 (1981).

[22] C. Guinti, *JHEP* **0211**, 017 (2002), hep-ph/0205014.

[23] A. D. Dolgov, *et al.*, *Eur. Phys. J. C* **44**, 431 (2005), hep-ph/0506203.

[24] L. Stodolsky, *Phys. Rev. D* **58**, 036006 (1998), hep-ph/9802387.

[25] A. E. Bernardini, M. M. Guzzo, and C. C. Nishi, 1004.0734 [hep-ph].

[26] I. Yu. Kobzarev, *et al.*, *Sov. J. Nucl. Phys.* **35**, 708 (1982).

[27] C. Giunti, *et al.*, Phys. Rev. D **48**, 4310 (1993), hep-ph/9305276; W. Grimus and P. Stockinger, *Phys. Rev. D* **54**, 3414 (1996), hep-ph/9603430.

[28] A. Ioannisian and A. Pilaftsis, *Phys. Rev. D* **59**, 053003 (1999), hep-ph/9809503.

[29] D. V. Naumov and V. A. Naumov, *J. Phys. G* **37**, 105014 (2010), 1008.0306 [hep-ph].

[30] M. Blasone and G. Vitiello, Ann. Phys. **244**, 283 (1995), hep-ph/9501263; Erratum *ibid.* **249**, 363 (1996).

[31] J. Wu, *et al.*, *Ann. Phys. Rev. D* **82**, 013006 (2010), 1005.3260 [hep-ph].

[32] C. Anastopoulos and N. Savvidou, 1005.4307 [quant-ph].

[33] M. Dvornikov, *Phys. Lett B* **610**, 262 (2005), hep-ph/0411101; in *Proceedings of the IPM School and Conference on Lepton and Hadron Physics*, ed. by Y. Farzan, eConf C0605151 (2007), hep-ph/0609139.

[34] M. Dvornikov, *Eur. Phys. J. C* **47**, 437 (2006), hep-ph/0601156.

[35] M. Dvornikov and J. Maalampi, *Phys. Lett. B* **657**, 217 (2007), hep-ph/0701209.

[36] M. Dvornikov, *J. Phys. Conf. Ser.* **110**, 082005 (2008), 0708.2975 [hep-ph].

[37] M. Dvornikov, *J. Phys. G* **35**, 025003 (2008), 0708.2328 [hep-ph].

[38] M. S. Dvornikov, *Phys. Atom. Nucl.* **72**, 116 (2009), hep-ph/0610047.

[39] M. Dvornikov and J. Maalampi, *Phys. Rev. D* **79**, 113015 (2009), 0809.0963 [hep-ph].

[40] M. Dvornikov, 1008.3115 [hep-ph].

[41] A. E. Bernardini and S. De Leo, *Phys. Rev. D* **70**, 053010 (2004), hep-ph/0411134; *Eur. Phys. J. C* **37**, 471 (2004), hep-ph/0411153; *Phys. Rev. D* **71**, 076008 (2005), hep-ph/0504239; C. C. Nishi, *Phys. Rev. D* **73**, 053013 (2006), hep-ph/0506109.

[42] K. Kiers and N. Weiss, *Phys. Rev. D* **57**, 3091 (1998), hep-ph/9710289.

[43] R. N. Mohapatra, *New J. Phys.* **6**, 82 (2004), hep-ph/0411131.

[44] I. Yu. Kobzarev, *et al.*, *Sov. J. Nucl. Phys.* **32**, 823 (1980).

[45] J. Schechter and J. W. F. Valle, *Phys. Rev. D* **22**, 2227 (1980).

[46] N. N. Bogoliubov, D. D. Shirkov, *Introduction to the Theory of Quantized Fields*, (Wiley, New York, 1980), third ed., p. 607.

[47] V. Gribov and B. Pontecorvo, *Phys. Lett B* **28**, 493 (1969).

[48] P. M. Morse and H. Feshbach, *Methods of Theoretical Physics*, (McGraw-Hill, New York, 1953), vol. 1, Chapter 7, pp. 854–857.

[49] M. Dvornikov and A. Studenikin, *JHEP* **0209**, 016 (2002), hep-ph/0202113.

[50] A. E. Lobanov, and A. I. Studenikin, *Phys. Lett. B* **515**, 94 (2001), hep-ph/0106101. A. Grigiriev, A. Lobanov, and A. Studenikin, *Phys. Lett. B* **535**, 187 (2002), hep-ph/0202276.

[51] A. Studenikin and A. Ternov, *Phys. Lett B* **608**, 107 (2005), hep-ph/0412408. A. E. Lobanov, *Phys. Lett B* **619**, 136 (2005), hep-ph/0506007.

[52] M. Deniz, *et al.* (TEXONO Collaboration), *Phys. Rev. D* **82**, 033004 (2010), 1006.1947 [hep-ph].

[53] T. Kikuchi, H. Minakata, and S. Uchinami, *JHEP* **0903**, 114 (2009), 0809.3312 [hep-ph].

[54] A. G. Beda, *et al.*, *Phys. Atom. Nucl.* **70**, 1873 (2007), 0705.4576 [hep-ex].

[55] E. Kh. Akhmedov and M. Yu. Khlopov, *Sov. J. Nucl. Phys.* **47**, 689 (1988).

[56] A. M. Egorov, A. E. Lobanov, and A. I. Studenikin, *Phys. Lett. B* **491**, 137 (2000), hep-ph/9910476.

[57] I. M. Ternov, V. G. Bagrov, and A. M. Khapaev, *JETP* **21**, 613 (1965).

[58] S. E. Woosley, A. Heger, and T. A. Weaver, *Rev. Mod. Phys.* **74**, 1015 (2002).

[59] G. G. Likhachev and A. I. Studenikin, *JETP* **81**, 419 (1995).

[60] S. Kawagoe, *et al.*, *J. Phys. Conf. Ser.* **39**, 294 (2006).

[61] E. Kh. Akhmedov, A. Lanza, and D. W. Sciama, *Phys. Rev. D* **56**, 6117 (1997), hep-ph/9702436.

[62] D. Nötzold, *Phys. Rev. D* **38**, 1658 (1988).

[63] R. Barbieri and R. N. Mohapatra, *Phys. Rev. Lett.* **61**, 27 (1988).

[64] A. Ayala, J. C. D'Olivo, and M. Torres, *Nucl. Phys. B* **564**, 204 (2000), hep-ph/9907398; A. V. Kuznetsov and N. V. Mikheev, *JCAP* **0711**, 204 (2000), 0709.0110 [hep-ph].

[65] A. V. Kuznetsov, N. V. Mikheev, and A. A. Okrugin, *Int. J. Mod. Phys. A* **24**, 5977 (2009), 0907.2905 [hep-ph].

[66] O. Lychkovskiy and S. Blinnikov, *Phys. Atom. Nucl.* **73**, 614 (2010), 0905.3658 [hep-ph].

[67] R. Tòmas, *et al.*, *JCAP* **0409**, 015 (2004), astro-ph/0407132.

[68] P. Keränen, *et al.*, *Phys. Lett. B* **574**, 162 (2003), hep-ph/0307041; *ibid.* **597**, 374 (2004), hep-ph/0401082.

[69] V. A. Kutvitskiĭ, V. B. Semikoz, and D. D. Sokoloff, *Astron. Rep.* **53**, 166 (2009), 0809.3172 [astro-ph]; C. R. Das, J. Pulido, and M. Picariello, *Phys. Rev. D* **79**, 073010 (2010), 0902.1310 [hep-ph].

[70] J. Maalampi and J. Riittinen, *Phys. Rev. D* **81**, 037301 (2010), 0912.4628 [hep-ph].

[71] A. Esmaili, *Phys. Rev. D* **81**, 013006 (2010), 0909.5410 [hep-ph].

[72] G. G. Raffelt, *Phys. Rev. Lett.* **64**, 2856 (1990).

[73] S. Akiyama, *et al.*, *Astrophys. J* **584**, 954 (2003), astro-ph/0208128.

[74] See pp. 292–296 in Ref. [15].

[75] K. M. Case, *Phys. Rev.* **107**, 307 (1957).

[76] V. B. Berestetskiĭ, E. M. Lifschitz, and L. P. Pitaevskiĭ, *Quantum Electrodynamics* (Moscow, Nauka, 1989), pp. 108–112.

[77] See pp. 137–141 in Ref. [76].

[78] J. Schechter and J. W. F. Valle, *Phys. Rev. D* **24**, 1883 (1981).

[79] D. Giulini, *et al.*, *Decoherence and the Appearence of a Classical World in Quantum Theory* (Springer-Verlag, Berlin, 1996).

[80] See pp. 477–478 in Ref. [15].

[81] S. Pastor, *Master's Thesis* (University of Valencia, 1996).

[82] S. Ando and K. Sato, *Phys. Rev. D* **68**, 023003 (2003), hep-ph/0305052.

[83] S. Ando and K. Sato, *JCAP* **0310**, 001 (2003), hep-ph/0309060.

[84] M. Kachelrieß, *et al.*, *Phys. Rev. D* **65**, 073016 (2002), hep-ph/0108100.

[85] R. E. Shrock, *Phys. Lett. B* **96**, 159 (1980).

[86] C. Itzykson and J.-B. Zuber, *Quantum Field Theory* (Moscow, Mir, 1984), vol. 2, pp. 385–388.

[87] L. B. Okun', *Leptons and Quarks* (Moscow, Nauka, 1990), 2nd ed., pp. 36–41.

[88] See pp. 140–141 and 602–607 in Ref. [46].

[89] Z.-Z. Xing, *Phys. Lett. B* **660**, 515 (2008), 0709.2220 [hep-ph].

[90] E. Kh. Akhmedov and A. Yu. Smirnov, *Phys. Atom. Nucl.* **72**, 1363 (2009), 0905.1903 [hep-ph].

[91] B. Pontecorvo, *JETP* **7**, 172 (1958); for the contemporary description see pp. 286–287 in Ref. [15].

[92] J. Schechter and J. W. F. Valle, *Phys. Rev. D* **23**, 1666 (1981); A. de Gouvêa, B. Kayser, and R. N. Mohapatra, *Phys. Rev. D* **67**, 053004 (2003), hep-ph/0211394.

[93] S. Esposito and N. Tancredi, *Mod. Phys. Lett. A* **12**, 1829 (1997), hep-ph/9705351.

[94] M. Beuthe, *Phys. Rev. D* **66**, 013003 (2002), hep-ph/0202068.

[95] A. Atre, *et al.*, *JHEP* **0905**, 030 (2009), 0901.3589 [hep-ph].

[96] M. Beuthe, *Phys. Rep.* **375**, 105 (2003), hep-ph/0109119.

[97] A. Yu. Smirnov, *Phys. Lett B* **260**, 161 (1991); E. Kh. Akhmedov, P. I. Krastev, and A. Yu. Smirnov, *Z. Phys. C* **48**, 701 (1991).

[98] C. Itzykson and J.-B. Zuber, *Quantum Field Theory* (Moscow, Mir, 1984), vol. 1, pp. 80–84.

[99] N. V. Ardeljan, S. G. Bisnovatyi-Kogan, and S. G. Moiseenko, *MNRAS* **359**, 333 (2005), astro-ph/0410234.

[100] H. Nunokawa, *et al.*, *Nucl. Phys. B* **501**, 17 (1997), hep-ph/9701420.

[101] K. Sumiyoshi, *et al.*, *Astrophys. J.* **629**, 922 (2005), astro-ph/0506620.

[102] N. F. Bell, *et al.*, *Phys. Rev. Lett.* **95**, 151802 (2005), hep-ph/0504134.

[103] I. Pivovarov and A. Studenikin, *PoS* **HEP2005**, 191 (2006), hep-ph/0512031.

In: Neutrinos: Properties, Sources and Detection
Editor: Joshua P. Greene, pp. 91-118
ISBN 978-1-61209-650-6

Chapter 3

NEUTRINO-NUCLEUS INTERACTIONS AT LOW AND INTERMEDIATE NEUTRINO ENERGIES

V.Ch. Chasioti *
Department of Informatics and Computer Technology,
T.E.I. of Western Macedonia, Kastoria, GR 52100 Greece

Abstract

Neutrino studies have been paid considerable attention in nuclear physics, astrophysics and cosmology. It has been pointed out that neutrinos are sensitive probes for investigating astrophysical processes and stellar evolution. The study of solar neutrinos provides important information on nuclear processes inside the sun as well as on matter densities. Moreover, Supernova neutrinos provide sensitive probes for study supernova explosions, neutrino properties and stellar collapse mechanisms. In the present work we have developed a formalism describing (anti)-neutrino-nucleus cross-sections in neutral and charged-current processes, taking into account vector and axial vector contributions of the hadronic current. The calculation of the single-particle transition matrix elements, based on the multipole expansion treatment of the relevant hadronic currents, was improved ending to a more compact form. The differential and integrated neutrino-nucleus cross sections were evaluated for low and intermediate neutrino energies ($0\ MeV \leq \epsilon_\nu \leq 100\ MeV$) by examining the dependence of the cross sections on the scattering angle and initial neutrino-energy of Fermi and Gamow-Teller type transition rates. The required many-body nuclear wave-functions were calculated by utilizing the quasi-particle random phase approximation. The results presented refer to the neutral current scattering of electron neutrino on the various nuclear isotopes such as Mo, Fe, O, and Cd which play a significant role in astrophysical neutrino studies. Finally we have explored the nuclear response of some of these isotopes as a supernova neutrino detector assuming Fermi-Dirac distribution for supernova neutrino spectra. In the present work inelastic neutrino-nucleus reaction cross sections at low and intermediate neutrino energies are studied. The required many-body nuclear wave-functions are calculated in the context of quasi-particle random phase approximation (QRPA) that uses realistic two-body forces. The results presented here refer to

*E-mail address: v.hasioti@kastoria.teikoz.gr, vhassioti@yahoo.gr

the differential, integrated and total cross sections of neutral-current induced reactions of even-even isotopes used recently at low-energy astrophysical neutrino searches.

PACS 23.40.Bw , 25.30.Pt, 21.60.Jz, 26.30.+k.

Keywords: Semi-leptonic electroweak interactions, Neutrino-nucleus reactions, Inelastic cross sections,Quasi-particle random phase approximation

1. Introduction

It is well known that, neutrinos and their interactions with nuclei have attracted a great deal of attention, since they play a fundamental role to nuclear physics, cosmology and to various astrophysical processes, especially in the dynamics of core-collapse supernova-nucleosynthesis [1, 2, 3, 4, 5, 6]. Moreover, neutrinos proved to be interesting tools for testing weak interaction properties, by examining nuclear structure, and for exploring the limits of the standard model [7]. In spite of the important role the neutrinos play in many phenomena in nature, numerous questions concerning their properties, oscillation-characteristics, their role in star evolutions and in the dark matter of the universe, remain still unanswered. The main goal of experimental [8, 9, 10] and theoretical studies [11, 12, 13, 14, 15, 16, 17, 18, 19] is to shed light on the above open problems to which neutrinos are absolutely crucial.

Among the probes which involve neutrinos, the neutrino-nucleus interaction possess a prominent position. Thus, the study of neutrino scattering with nuclei [20, 21, 22, 23, 24, 25, 26] is a good way to detect or distinguish neutrinos of different flavor and explore the basic structure of the weak interactions. Also, specific neutrino-induced transitions between discrete nuclear states with good quantum numbers of spin, isospin and parity allows us to study the structure of the weak hadronic currents. Furthermore, terrestrial experiments performed to detect astrophysical neutrinos, as well as neutrino induced nucleosynthesis interpreted through several neutrino-nucleus interaction theories, constitute good sources of explanation for neutrino properties. There are four categories of neutrino-nucleus processes: the two types of charged-current (CC) reactions of neutrinos and antineutrinos and the two types of neutral-current (NC) ones. In the charged-current reactions a neutrino ν_l (antineutrino $\bar{\nu}_l$) with $l = e, \mu, \tau$, transforms one neutron (proton) of a nucleus to a proton (neutron), and a charged lepton l^- (anti-lepton l^+) is emitted as

$$\begin{aligned} \nu_l + (A, Z) &\longrightarrow l^- + (A, Z+1)^* \\ \bar{\nu}_l + (A, Z) &\longrightarrow l^+ + (A, Z-1)^* \end{aligned} \tag{1}$$

These reactions are also called neutrino (anti-neutrino) capture, since they can be considered as the reverse processes of lepton-capture. They are mediated by exchange of heavy $W^\pm$ bosons according to the (lowest order) Feynman diagram shown in Fig.1(a). In neutral-current reactions (neutrino scattering) the neutrinos (anti-neutrinos) interact via the exchange of neutral Z^0 bosons [see Fig.1(b)] with a nucleus as

$$\begin{aligned} \nu + (A, Z) &\longrightarrow \nu' + (A, Z)^* \\ \bar{\nu} + (A, Z) &\longrightarrow \bar{\nu}' + (A, Z)^* \end{aligned} \tag{2}$$

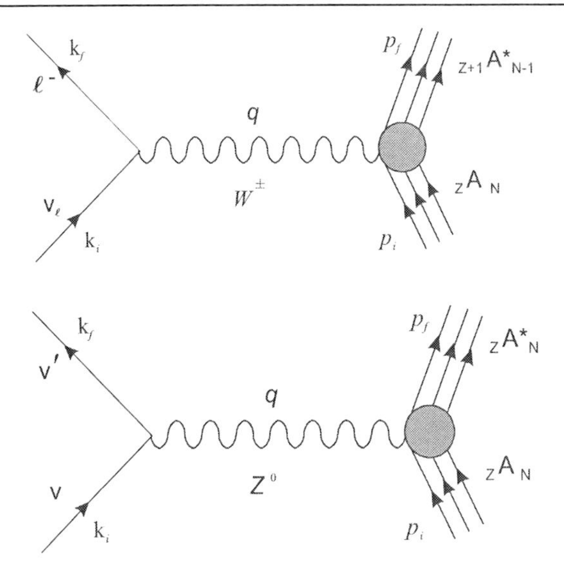

Figure 1. Feynman-diagram of lowest order for: (a) the CC neutrino-nucleus reactions $\nu_l + (A, Z) \longrightarrow l^- + (A, Z+1)^*$, and (b) the NC neutrino-nucleus processes $\nu + (A, Z) \longrightarrow \nu' + (A, Z)^*$. The diagrams which correspond to the anti-neutrino reactions are similar.

where ν ($\bar{\nu}$) denote neutrinos (anti-neutrinos) of any flavor. In neutral current reactions the final nucleus leaves mostly in an excited state lying below particle-emission thresholds (semi-inclusive processes). The transitions to energy-levels higher than the particle-bound states usually decay by emitting mostly neutrons which are likely to escape the detector without further nuclear reactions unless detectors are very large. On the other hand charge-current neutrino scattering reactions always involve a charged lepton in the exit channel accompanied by several neutron and γ signals. The most important reactions for electron neutrinos and electron antineutrinos are charged current processes on nucleons (β-processes). All other flavors undergo neutral-current interactions. Although charged-current cross sections are substantially larger than those for neutral-current scattering [27, 28, 29] the latter provide important information about the evolution of the Sun and distant stars, the neutrino-oscillations, the neutrino-interactions with matter, etc. The energy-spectra of such neutrinos could be used in stellar evolution modeling and in the interpretation of neutrinos coming from various sources [30, 31, 32, 33].

Nowadays, new experimental programms are being planned such as, SuperNEMO [34], MOON [35, 36], COBRA [37], CUORE [38] , for low-energy astrophysical neutrino searches to detect solar ($E_\nu < 20$ MeV) and supernova($E_\nu < 60 - 70$ MeV) neutrinos operating in conjunction with searches of rare event processes like the direct-detection of cold dark matter (CDM) particles and the neutrinoless double beta decay ($0\nu\beta\beta$-decay). At the same time, it became feasible to detect low-flux neutrinos (CNO-cycle neutrinos) by measuring the energy recoil of the recoiling nucleus with gaseous-detectors having very-low threshold-energy like those required for the direct-detection of CDM events [37]. These experimental programs will provide essential data on the neutrino-nucleus reactions and also help to improve the reliability of present cross-section calculations.

Various microscopic approaches have been employed in the evaluation of neutrino-nucleus reaction rates at low and intermediate energies. These include the nuclear shell model [39, 40, 41, 42], the random-phase approximation (RPA) [10, 43, 44], continuum RPA (CRPA) [45, 46, 47, 48, 49, 50], hybrid models of CRPA and shell model [43, 51], the Fermi gas model [52] and quasi-particle random phase approximation (QRPA) [22, 23, 24, 25, 26, 29].

In the present work we examine the cross sections of some nuclear isotopes involved recently in terrestrial detectors to various low-energy neutrino spectra. To this end we examine the dependence of differential and integrated neutral current neutrino-nucleus cross sections on the scattering angle and initial neutrino-energy of Fermi and Gamow-Teller type transition rates. The required many-body nuclear wave-functions are calculated in the context of quasi-particle random phase approximation (QRPA). The original code QRPA was constructed in university of Tübingen. Finally we investigate the averaged cross sections of some of these nuclear isotopes as a supernova neutrino detector assuming the Fermi-Dirac distribution for the supernova neutrino energy spectra.

2. The Formalism for Neutrino-Nucleus Cross Sections Calculations

Let us consider a neutral or charged current neutrino-nucleus interaction in which a low or intermediate energy neutrino (or antineutrino) is scattered inelastically from a nucleus (A, Z). The initial nucleus is assumed to be spherically symmetric having ground state a $|J^{\pi}\rangle = |0^{+}\rangle$ state.

The corresponding standard model effective Hamiltonian in current-current interaction form is written as

$$\mathcal{H} = \frac{G}{\sqrt{2}} j_{\mu}(\mathbf{x}) J^{\mu}(\mathbf{x}), \tag{3}$$

where $G = 1.1664 \times 10^{-5} GeV^{-2}$ is the Fermi weak coupling constant. j_{μ} and J^{μ} denote the leptonic and hadronic currents, respectively. According to V-A theory, the leptonic current takes the form

$$j_{\mu} = \bar{\psi}_{\nu_{\ell}}(x)\gamma_{\mu}(1-\gamma_5)\psi_{\nu_{\ell}}(x)\,, \tag{4}$$

where $\psi_{\nu_{\ell}}$ are the neutrino/antineutrino spinors.

From a nuclear physics point of view only the hadronic current is important. The structure for neutral current (nc) and charge current (cc) processes of both vector and axial-vector components (neglecting the pseudo-scalar contributions) is written as

$$J_{\mu}^{cc,nc} = \bar{\Psi}_N \left[F_1^{cc,nc}(q^2)\gamma_{\mu} + F_2^{cc,nc}(q^2)\frac{i\sigma_{\mu\nu}q^{\nu}}{2M} + F_A^{cc,nc}(q^2)\gamma_{\mu}\gamma_5 \right] \Psi_N \tag{5}$$

(M stands for the nucleon mass and Ψ_N denote the nucleon spinors). The form factors $F_{1,2}^{cc}(q^2)$ and $F_A^{cc}(q^2)$ are defined as

$$\begin{aligned} F_{1,2}^{cc}(q^2) &= F_{1,2}^{p}(q^2) - F_{1,2}^{n}(q^2) \\ F_A^{cc}(q^2) &= F_A(q^2) \end{aligned} \tag{6}$$

and the neutral current form factors $F^{nc}_{1,2}(q^2)$ and $F^{nc}_{A}(q^2)$ as

$$F^{nc}_{1,2}(q^2) = \left(\frac{1}{2} - sin^2\theta_W\right)\left[F^{p}_{1,2}(q^2) - F^{n}_{1,2}(q^2)\right]\tau_0 - sin^2\theta_W\left[F^{p}_{1,2}(q^2) + F^{n}_{1,2}(q^2)\right]$$
$$F^{nc}_{A}(q^2) = \frac{1}{2}F_A(q^2)\tau_0 \quad (7)$$

Here τ_0 represents the nucleon isospin operator and θ_W is the Weinberg angle ($sin^2\theta_W = 0.2325$). The detailed expressions of the nucleonic form factors $F^{p,n}_{1,2}(q^2)$ are given in Ref. [44]. In Eqs.(7) and (6) $F_A(q^2)$ stands for the axial-vector form factor for which we employ the dipole ansatz given by

$$F_A = -g_A\left(1 - q^2/M_A^2\right)^{-2}, \quad (8)$$

where $M_A = 1.05$ GeV, is the dipole mass, and $g_A = 1.258$, is the static value (at $q = 0$) of the axial form factor.

In the convention we used in the present work q^2, the square of the momentum transfer, is written as

$$q^2 = q^\mu q_\mu = \omega^2 - \mathbf{q}^2 = (\varepsilon_i - \varepsilon_f)^2 - (\mathbf{p}_i - \mathbf{p}_f)^2, \quad (9)$$

where $\omega = \varepsilon_i - \varepsilon_f$ is the excitation energy of the nucleus. ε_i denotes the energy of the incoming and ε_f that of the outgoing neutrino. $\mathbf{p}_i$, $\mathbf{p}_f$ are the corresponding 3-momenta of the incoming and outgoing neutrino/antineutrino, respectively. In Eq. (7) we have not taken into account the strange quark contributions in the form factors. In the scattering reaction considered in our work only low-momentum transfers are involved and the contributions from strangeness can be neglected [53].

The neutrino/antineutrino-nucleus differential cross section, after applying a multipole analysis of the weak hadronic current, is written

$$\left(\frac{d^2\sigma_{i\to f}}{d\theta d\omega}\right)_{\nu/\bar{\nu}} = \frac{G^2}{\pi}\frac{|\vec{p}_f|\varepsilon_f}{(2J_i+1)}F(\varepsilon_f, Z)\left(\sum_{J=0}^{\infty}\sigma^J_{CL} + \sum_{J=1}^{\infty}\sigma^J_T\right) \quad (10)$$

θ denotes the lepton scattering angle. The summations in Eq. (10) contain the contributions σ^J_{CL}, for the Coulomb $\widehat{\mathcal{M}}_J$ and longitudinal $\widehat{\mathcal{L}}_J$, and σ^J_T, for the transverse electric $\widehat{\mathcal{T}}^{el}_J$ and magnetic $\widehat{\mathcal{T}}^{mag}_J$ multipole operators defined in section 3. These operators include both polar-vector and axial-vector weak interaction components. The contributions σ^J_{CL}, and σ^J_T are written as

$$\sigma^J_{CL} = (1 + a\cos\theta)\left|\langle J_f||\widehat{\mathcal{M}}_J(q)||J_i\rangle\right|^2 + \left(1 + a\cos\theta - 2b\sin^2\theta\right)\left|\langle J_f||\widehat{\mathcal{L}}_J(q)||J_i\rangle\right|^2$$
$$+ \left[\frac{\omega}{q}(1 + a\cos\theta) + c\right] 2\,\Re e\langle J_f||\widehat{\mathcal{L}}_J(q)||J_i\rangle\langle J_f||\widehat{\mathcal{M}}_J(q)||J_i\rangle^* \quad (11)$$

$$\sigma^J_T = \left(1 - a\cos\theta + b\sin^2\theta\right)\left[\left|\langle J_f||\widehat{\mathcal{T}}^{mag}_J(q)||J_i\rangle\right|^2 + \left|\langle J_f||\widehat{\mathcal{T}}^{el}_J(q)||J_i\rangle\right|^2\right]$$
$$\mp \left[\frac{(\varepsilon_i + \varepsilon_f)}{q}(1 - a\cos\theta) - c\right] 2\,\Re e\langle J_f||\widehat{\mathcal{T}}^{mag}_J(q)||J_i\rangle\langle J_f||\widehat{\mathcal{T}}^{el}_J(q)||J_i\rangle^* \quad (12)$$

where $b = \varepsilon_i \varepsilon_f / q^2$, $a = \vec{p}_f / \varepsilon_f$ and $c = (m_f c^2)^2 / q\varepsilon_f$. For charged-current reactions the cross section Eq. (10) must be corrected for the distortion of the outgoing lepton wave function by the Coulomb field of the daughter nucleus. The cross section can either be multiplied by a Fermi function $F(\varepsilon_f, Z)$ obtained from the numerical solution of the Dirac equation for an extended nuclear charge distribution [51] or, at higher energies, the effect of the Coulomb field can be described by the effective momentum approximation (EMA) [54, 55, 56].

3. Transition Operators for the Weak Interaction

Following the Walecka and Donnelly method [57] the multipole expansion procedure results in eight independent irreducible tensor multipole operators, four of them come from the polar-vector component, $\hat{J}_\lambda = (\hat{\rho}, \hat{\mathbf{J}})$, of the hadronic current and the other four come from the axial-vector component, $\hat{J}^5_\lambda = (\hat{\rho}^5, \hat{\mathbf{J}}^5)$. These tensor multipole operators are defined as

$$\widehat{\mathcal{M}}_{JM}(q) = \hat{M}^{coul}_{JM} - \hat{M}^{coul5}_{JM} = \int d\mathbf{r} M^J_M(q\mathbf{r}) \widehat{\mathcal{J}}_0(\mathbf{r}), \tag{13}$$

$$\widehat{\mathcal{L}}_{JM}(q) = \hat{L}_{JM} - \hat{L}^5_{JM} = i \int d\mathbf{r} \left(\frac{1}{q} \nabla M^J_M(q\mathbf{r}) \right) \cdot \widehat{\mathcal{J}}(\mathbf{r}), \tag{14}$$

$$\widehat{\mathcal{T}}^{mag}_{JM}(q) = \hat{T}^{el}_{JM} - \hat{T}^{el5}_{JM} = \int d\mathbf{r} \left(\frac{1}{q} \nabla \times \mathbf{M}^{JJ}_M(q\mathbf{r}) \right) \cdot \widehat{\mathcal{J}}(\mathbf{r}), \tag{15}$$

$$\widehat{\mathcal{T}}^{el}_{JM}(q) = \hat{T}^{mag}_{JM} - \hat{T}^{mag5}_{JM} = \int d\mathbf{r} \mathbf{M}^{JJ}_M(q\mathbf{r}) \cdot \widehat{\mathcal{J}}(\mathbf{r}). \tag{16}$$

where by adopting the $V - A$ theory we have written

$$\widehat{\mathcal{J}}_\mu = \hat{J}_\mu - \hat{J}^5_\mu = (\hat{\rho}, \hat{\mathbf{J}}) - (\hat{\rho}^5, \hat{\mathbf{J}}^5) \tag{17}$$

The above eight types of irreducible tensor multipole operators [Eqs. (13)-(17)] are acting on the nuclear Hilbert space and have rank J. The components of the polar vector $(\hat{\rho}(\mathbf{r}), \hat{\mathbf{J}}(\mathbf{r}))$ and axial vector $(\hat{\rho}^5(\mathbf{r}), \hat{\mathbf{J}}^5(\mathbf{r}))$ currents are defined e.g. in Ref. [58].

In the unified description of semi-leptonic electro-weak processes in nuclei developed by Walecka and Donnelly [11, 57, 59, 60] (for a review see Ref. [1]), the calculation of the required transition strengths of $\Gamma_{i \to f} \propto |\langle f | H_{eff} | i \rangle|^2$ relies on a multipole decomposition of the hadronic current-density matrix elements leading to a set of eight independent irreducible tensor multipole operators. These operators contain spherical Bessel functions, j_l, combined with spherical harmonics, Y^L_M, or vector spherical harmonics, $\mathbf{Y}^{(L1)J}_M$ as

$$M^J_M(q\mathbf{r}) = \delta_{LJ} j_L(qr) Y^L_M(\hat{r}), \tag{18}$$

$$\mathbf{M}^{(L1)J}_M(q\mathbf{r}) \equiv \mathbf{M}^{LJ}_M(q\mathbf{r}) = j_L(qr) \mathbf{Y}^{(L1)J}_M(\hat{r}). \tag{19}$$

where

$$\mathbf{Y}^{(L1)J}_M(\hat{r}) \equiv \mathbf{Y}^{LJ}_M(\hat{r}) = \sum_{M_L, q} \langle L m_L 1 q | J M \rangle Y^L_{M_L}(\hat{r}) \hat{e}_q \tag{20}$$

In the present paper the functions $j_L(\rho)$, $Y^L_M(\hat{r})$ and $\mathbf{Y}^{(L1)J}_M(\hat{r})$ are defined by assuming the same conventions with those of Ref. [1].

In fact, the matrix elements of these seven basic operators involve momentum dependent form factors, $F^{cc,nc}_X(q^2_\mu)$, $X = 1, 2, A$ and according to Walecka-Donnelly method [1, 60, 61] seven new operators are defined. Here the latter operators are denoted as $T^{JM}_i(q\mathbf{r})$, $i = 1, 2, ..., 7$ and are defined by the expressions

$$T^{JM}_1 \equiv M^J_M(q\mathbf{r}) = \delta_{LJ}\, j_L(\rho) Y^L_M(\hat{r}), \tag{21}$$

$$T^{JM}_2 \equiv \Sigma^J_M(q\mathbf{r}) = \mathbf{M}^{JJ}_M \cdot \boldsymbol{\sigma}, \tag{22}$$

$$T^{JM}_3 \equiv \Sigma'^J_M(q\mathbf{r}) = -i\left[\frac{1}{q}\nabla \times \mathbf{M}^{JJ}_M(q\mathbf{r})\right] \cdot \boldsymbol{\sigma}, \tag{23}$$

$$T^{JM}_4 \equiv \Sigma''^J_M(q\mathbf{r}) = \left[\frac{1}{q}\nabla M^J_M(q\mathbf{r})\right] \cdot \boldsymbol{\sigma}, \tag{24}$$

$$T^{JM}_5 \equiv \Delta^J_M(q\mathbf{r}) = \mathbf{M}^{JJ}_M(q\mathbf{r}) \cdot \frac{1}{q}\nabla, \tag{25}$$

$$T^{JM}_6 \equiv \Delta'^J_M(q\mathbf{r}) = -i\left[\frac{1}{q}\nabla \times \mathbf{M}^{JJ}_M(q\mathbf{r})\right] \cdot \frac{1}{q}\nabla, \tag{26}$$

$$T^{JM}_7 \equiv \Omega^J_M(q\mathbf{r}) = M^J_M(q\mathbf{r})\boldsymbol{\sigma} \cdot \frac{1}{q}\nabla. \tag{27}$$

(for the readers convenience, in addition to our unified notation T^{JM}_i, we keep the notation of Refs. [1, 60, 61] as well). As it is well known, most physical observables describing the semi-leptonic electro-weak processes in nuclei, are (to a rather good approximation) expressed in terms of reduced nuclear matrix elements of the above basic one-body operators between two single particle energy levels $|n(l1/2)j\rangle \equiv |j\rangle$, i.e. matrix elements of the form

$$\langle n_1(l_1 1/2)j_1||T^J_i||n_2(l_2 1/2)j_2\rangle \equiv \langle j_1||T^J_i||j_2\rangle, \quad i = 1, 2, ...7. \tag{28}$$

Due to the fundamental importance of such reduced nuclear matrix elements (even though very accurate numerical integration techniques are also available) in other section below we present compact analytic expressions for their evaluation [25]. The multipole operators (13) - (16) are written in terms of the seven basic multipole operators of Eqs. (21)-(27) as:

$$\hat{M}^{coul}_{JM}(q\mathbf{r}) = F^{cc,nc}_1(q^2_\mu) M^J_M(q\mathbf{r}), \tag{29}$$

$$\hat{L}_{JM}(q\mathbf{r}) = \frac{q_0}{q}\hat{M}^{coul}_{JM}(q\mathbf{r}), \tag{30}$$

$$\hat{T}^{el}_{JM}(q\mathbf{r}) = \frac{q}{M_N}\left[F_1^{cc,nc}(q_\mu^2)\Delta'^J_M(q\mathbf{r}) + \frac{1}{2}(F_1^{cc,nc}(q_\mu^2) + 2MF_2^{cc,nc}(q_\mu^2))\Sigma^J_M(q\mathbf{r})\right], \tag{31}$$

$$i\hat{T}^{mag}_{JM}(q\mathbf{r}) = \frac{q}{M_N}\left[F_1^{cc,nc}(q_\mu^2)\Delta^J_M(q\mathbf{r}) - \frac{1}{2}(F_1^{cc,nc}(q_\mu^2) + 2MF_2^{cc,nc}(q_\mu^2))\Sigma'^J_M(q\mathbf{r})\right], \tag{32}$$

$$i\hat{M}^5_{JM}(q\mathbf{r}) = \frac{q}{M_N}\left[F_A^{cc,nc}(q_\mu^2)\Omega^J_M(q\mathbf{r}) + \frac{1}{2}F_A^{cc,nc}(q_\mu^2)\Sigma''^J_M(q\mathbf{r})\right], \tag{33}$$

$$-i\hat{L}^5_{JM}(q\mathbf{r}) = F_A^{cc,nc}(q_\mu^2)\Sigma''^J_M(q\mathbf{r}), \tag{34}$$

$$-i\hat{T}^{el5}_{JM}(q\mathbf{r}) = F_A^{cc,nc}(q_\mu^2)\Sigma'^J_M(q\mathbf{r}), \tag{35}$$

$$\hat{T}^{mag5}_{JM}(q\mathbf{r}) = F_A^{cc,nc}(q_\mu^2)\Sigma^J_M(q\mathbf{r}). \tag{36}$$

The magnitude of the three-momentum transfer $q = |\mathbf{q}|$, where $q^\mu = (q_0, \mathbf{q})$ is determined from the kinematics of the process in question.

4. The Single-Particle Reduced Matrix Elements

It is easy to see by using known identities (see Appendix 3 of Ref. [25]) that, the matrix elements of the seven basic tensor operators T_i^J can be written in terms of the following reduced matrix elements

$$\langle j_1||\mathcal{O}_i^{(LS_i)J}||j_2\rangle, \quad \mathcal{O}_i^{JM} = M_M^J,\, \mathbf{M}_M^{LJ}\cdot\boldsymbol{\sigma},\, \mathbf{M}_M^{LJ}\cdot(\nabla/q),\, M_M^J\boldsymbol{\sigma}\cdot(\nabla/q). \tag{37}$$

By applying the re-coupling relations the latter matrix elements can be written in the forms shown below.

1). For the operators $\mathcal{O}_1^{JM} \equiv M_M^J$ and $\mathcal{O}_2^{JM} \equiv \mathbf{M}_M^{LJ}\cdot\boldsymbol{\sigma}$ the reduced matrix elements $\langle j_1||\mathcal{O}_i^J||j_2\rangle$, have been previously written as [62]

$$\langle j_1||\mathcal{O}_i^{(LS_i)J}||j_2\rangle = (l_1\; L\; l_2)\,\mathcal{U}^J_{LS_i}\langle n_1 j_1|j_L(\rho)|n_2 j_2\rangle, \; i = 1, 2\,, \tag{38}$$

(the argument in the radial integrals is $\rho = qr$) with $S_1 = 0$ for Fermi type operators and $S_2 = 1$ for Gamow-Teller type operators. The symbol $(l_1\; L\; l_2)$ contains the 3-j symbol as

$$(l_1\; L\; l_2) \equiv (-)^{l_1}\frac{1}{\sqrt{4\pi}}[l_1][L][l_2]\begin{pmatrix} l_1 & L & l_2 \\ 0 & 0 & 0 \end{pmatrix}, \tag{39}$$

and $\mathcal{U}^J_{LS}$ is given by

$$\mathcal{U}^J_{LS} \equiv [j_1][j_2][J](S+1)^{1/2}(S+2)^{1/2}\begin{Bmatrix} l_1 & l_2 & L \\ 1/2 & 1/2 & S \\ j_1 & j_2 & J \end{Bmatrix}. \tag{40}$$

We note that, throughout the paper we employ the common notation $[J] = (2J+1)^{1/2}$.

2). The reduced matrix elements of $\langle j_1||\mathbf{M}^{LJ}(q\mathbf{r})\cdot(\nabla/q)||j_2\rangle$ can be cast in the form

$$\langle j_1||\mathbf{M}^{LJ}(q\mathbf{r})\cdot\frac{1}{q}\nabla||j_2\rangle = \sum_\alpha \mathcal{A}_L^\alpha(j_1 j_2; J)\langle n_1 j_1|\theta_L^\alpha(\rho)|n_2 j_2\rangle, \; \alpha = \pm\,, \tag{41}$$

with

$$\mathcal{A}_L^{\pm}(j_1 j_2; J) = \pm(-)^{l_1+L+j_2+1/2}\left[\frac{1}{2}(2l_2+1\mp 1)\right]^{1/2}[j_1][j_2][J](l_1\, L\, l_2 \mp 1) \\ \times \mathcal{W}_6(l_1, j_1, 1/2, j_2, l_2, J)\mathcal{W}_6(L, 1, J, l_2, l_1, l_2 \mp 1)\,. \quad (42)$$

The symbol $\mathcal{W}_6$ is defined as

$$\mathcal{W}_6(l_1, j_1, s, j_2, l_2, J) \equiv \begin{Bmatrix} l_1 & j_1 & s \\ j_2 & l_2 & J \end{Bmatrix}. \quad (43)$$

(the common 6-j symbol [63, 64])

3). Similarly, the reduced matrix element of $\langle j_1||M^J(q\mathbf{r})\boldsymbol{\sigma}\cdot(\nabla/q)||j_2\rangle$ reads

$$\langle j_1||M^J(q\mathbf{r})\boldsymbol{\sigma}\cdot\frac{1}{q}\nabla||j_2\rangle = \sum_{\alpha}\mathcal{B}_L^{\alpha}(j_1 j_2; J)\langle n_1 j_1|\theta_J^{\alpha}(\rho)|n_2 j_2\rangle,\ \alpha=\pm \quad (44)$$

where

$$\mathcal{B}_L^{\pm}(j_1 j_2; J) = \pm\delta_{j_2, l_2\mp\frac{1}{2}}[j_1][j_2](l_1\ J\ 2j_2-l_2)\mathcal{W}_6(l_1, j_1, \frac{1}{2}, j_2, 2j_2-l_2, J) \quad (45)$$

The three types of radial integrals entering Eqs. (38), (41) and (44) are shortly written as

$$\langle n_1 j_1|\theta_L^{\alpha}(\rho)|n_2 j_2\rangle \equiv \int dr r^2 R^*_{n_1 j_1}(r)\theta_L^{\alpha}(\rho)R_{n_2 j_2}(r),\ \alpha=0,\pm\,, \quad (46)$$

with

$$\theta_L^0(\rho) = j_L(\rho)\,, \quad (47)$$

$$\theta_L^{\pm}(\rho) = j_L(\rho)\left(\frac{d}{d\rho}\pm\frac{2l_2+1\pm 1}{2\rho}\right)\,, \quad (48)$$

and $\rho = qr$.

4.1. Compact Expressions for the Radial Integrals

For single-particle wave functions $R_{nj}(r)$ with arbitrary radial dependence it is not easy to perform analytically the integrations over r in Eq. (46), but, if a (spherical) harmonic oscillator basis is used (then the radial wave function are the same for $j = \ell \pm 1/2$ and the subscript j can be suppressed), these integrals can be simplified and take the closed expressions shown below. Thus, $\langle n_1 l_1|\theta_L^0(\rho)|n_2 l_2\rangle$ reads [62]

$$\langle n_1 l_1|j_L(\rho)|n_2 l_2\rangle = e^{-y}y^{L/2}\sum_{\mu=0}^{n_{max}}\varepsilon_{\mu}^L\, y^{\mu}\,, \quad (49)$$

where

$$y = (qb/2)^2 \quad (50)$$

(b denotes the harmonic oscillator size parameter) and

$$n_{max} = (N_1 + N_2 - L)/2, \quad (51)$$

The coefficients $\varepsilon_\mu^L(n_1 l_1 n_2 l_2)$ are defined in the Appendix A4 of Ref. [25]. The integers $N_i = 2n_i + l_i$ represent the harmonic oscillator quanta of the i_{th} major shell. (for some coefficients see Refs. [25, 64]).

Relying on Eq. (49) similar formulations for the integrals $\langle n_1 l_1 | \theta_l^\pm | n_2 l_2 \rangle$ are achieved as

$$\langle n_1 l_1 | j_L(\rho) \left(\frac{d}{d\rho} \pm \frac{2l_2 + 1 \pm 1}{2\rho} \right) | n_2 l_2 \rangle = e^{-y} y^{(L-1)/2} \sum_{\mu=0}^{n_{max}} \zeta_\mu^\pm(L)\, y^\mu, \quad (52)$$

where the coefficients $\zeta_\mu^\pm(n_1 l_1 n_2 l_2; L)$ are given in terms of $\varepsilon_\mu^L(n_1 l_1 n_2 l_2)$ as

$$\zeta_\mu^-(L) = -\frac{1}{2} \begin{cases} (n_2 + l_2 + 3/2)^{1/2} \varepsilon_\mu^L(n_1 l_1 n_2 l_2 + 1) \\ \quad + n_2^{1/2} \varepsilon_\mu^L(n_1 l_1 n_2 - 1 l_2 + 1), \; 0 \le \mu < n_{max} \\ (n_2 + l_2 + 3/2)^{1/2} \varepsilon_{n_{max}}^L(n_1 l_1 n_2 l_2 + 1), \; \mu = n_{max} \end{cases} \quad (53)$$

$$\zeta_\mu^+(L) = \frac{1}{2} \begin{cases} (n_2 + l_2 + 1/2)^{1/2} \varepsilon_\mu^L(n_1 l_1 n_2 l_2 - 1) \\ \quad + (n_2 + 1)^{1/2} \varepsilon_\mu^L(n_1 l_1 n_2 + 1 l_2 - 1), \; 0 \le \mu < n_{max} \\ (n_2 + 1)^{1/2} \varepsilon_{n_{max}}^L(n_1 l_1 n_2 + 1 l_2 - 1), \; \mu = n_{max} \end{cases} \quad (54)$$

with

$$n_{max} = (N_1 + N_2 - L + 1)/2.$$

We note that for $\mu = n_{max}$ the coefficients $\zeta_\mu^\pm$ are much simpler. (For more details see Refs. [25, 26]).

4.2. Compact Expressions for the Matrix Elements of $\mathcal{O}_i^{(L,S_i)J}$

By inserting in Eqs. (38), (41) and (44) the expressions found for the radial matrix elements (see Ref. [25]) and manipulating properly, the four types of reduced matrix elements of Eq. (37) are written in closed forms as follows:

1). For the operators $\mathcal{O}_i^{(LS_i)J}$ we have

$$\langle j_1 || \mathcal{O}_i^{(L,S_i)J} || j_2 \rangle = e^{-y} y^{L/2} \sum_{\mu=0}^{n_{max}} E_{i,\mu}^L\, y^\mu, \quad (55)$$

with

$$E_{i,\mu}^L = (l_1\, L\, l_2)\, \mathcal{U}_{LS_i}^J\, \varepsilon_\mu^L(n_1 l_1 n_2 l_2), \qquad i = 1, 2. \quad (56)$$

2). The reduced matrix element $\langle j_1 || \mathbf{M}^{LJ}(q\mathbf{r}) \cdot (\nabla/q) || j_2 \rangle$ takes the form

$$\langle j_1 || \mathbf{M}^{LJ}(q\mathbf{r}) \cdot \frac{1}{q} \nabla || j_2 \rangle = e^{-y} y^{(L-1)/2} \sum_{\mu=0}^{n_{max}} E_{3,\mu}^L\, y^\mu, \quad (57)$$

with

$$E_{3,\mu}^L = \mathcal{A}_L^-\, \zeta_\mu^-(L) + \mathcal{A}_L^+\, \zeta_\mu^+(L). \quad (58)$$

3). Similarly for $\langle j_1 || M^L(q\mathbf{r}) \boldsymbol{\sigma} \cdot \frac{1}{q} \nabla || j_2 \rangle$ we obtain

$$\langle j_1 || M^L(q\mathbf{r}) \boldsymbol{\sigma} \cdot \frac{1}{q} \nabla || j_2 \rangle = e^{-y} y^{(L-1)/2} \sum_{\mu=0}^{n_{max}} E_{4,\mu}^L\, y^\mu, \quad (59)$$

with

$$E^L_{4,\mu} = \mathcal{B}^-_L\, \zeta^-_\mu(L) + \mathcal{B}^+_L\, \zeta^+_\mu(L). \tag{60}$$

Table 1. Definition of the coefficients $\mathcal{P}^J_\mu$ entering Eq. (62), see also headline of this Table, for the seven basic operators describing any one-body semi-leptonic process in nuclei. The values for the parameter β of Eqs. (62) and (63) are also given.

$$\langle j_1||T^J||j_2\rangle = e^{-y} y^{\beta/2} \sum_{\mu=0}^{n_{max}} \mathcal{P}^J_\mu\, y^\mu, \quad y = (qb/2)^2, \quad n_{max} = (N_1 + N_2 - \beta)/2 \tag{61}$$

Operator	β	$\mathcal{P}^J_\mu, \quad 0 \le \mu \le n_{max}$
$T^J_1 = M^J = \delta_{LJ}\, j_L(\rho) Y^L_M(\hat{r})$	J	$E^J_{1,\mu} = (l_1\, J\, l_2)\, \mathcal{U}^J_{JS_1}\, \varepsilon^J_\mu(n_1 l_1 n_2 l_2)$
$T^J_2 = \Sigma^J = \mathbf{M}^{JJ}_M \cdot \boldsymbol{\sigma}$	J	$E^J_{2,\mu} = (l_1\, J\, l_2)\, \mathcal{U}^J_{JS_2}\, \varepsilon^J_\mu(n_1 l_1 n_2 l_2)$
$T^J_3 = \Sigma'^J = -i\left[\frac{1}{q}\nabla \times \mathbf{M}^{JJ}_M(q\mathbf{r})\right] \cdot \boldsymbol{\sigma}$	$J-1$	$(J+1)^{1/2} E^{J-1}_{2,\mu} - J^{1/2} E^{J+1}_{2,\mu-1}$
$T^J_4 = \Sigma''^J = \left[\frac{1}{q}\nabla M^J_M(q\mathbf{r})\right] \cdot \boldsymbol{\sigma}$	$J-1$	$J^{1/2} E^{J-1}_{2,\mu} + (J+1)^{1/2} E^{J+1}_{2,\mu-1}$
$T^J_5 = \Delta^J = \mathbf{M}^{JJ}_M(q\mathbf{r}) \cdot \frac{1}{q}\nabla$	$J-1$	$E^L_{3,\mu} = \mathcal{A}^-_L\, \zeta^-_\mu(L) + \mathcal{A}^+_L\, \zeta^+_\mu(L)$
$T^J_6 = \Delta'^J = -i\left[\frac{1}{q}\nabla \times \mathbf{M}^{JJ}_M(q\mathbf{r})\right]$	$J-2$	$(J+1)^{1/2} E^{J-1}_{3,\mu} - J^{1/2} E^{J+1}_{3,\mu-1}$
$T^J_7 = \Omega^J = M^J_M(q\mathbf{r}) \boldsymbol{\sigma} \cdot \frac{1}{q}\nabla$	J	$E^J_{4,\mu} = \mathcal{B}^-_L\, \zeta^-_\mu(L) + \mathcal{B}^+_L\, \zeta^+_\mu(L)$
$\Omega'^J = \Omega^J_M + \frac{1}{2}\Sigma''^J_M$	J -1	$E^J_{4,\mu} + \frac{1}{2}\left\{J^{1/2} E^{J-1}_{2,\mu} + (J+1)^{1/2} E^{J+1}_{2,\mu-1}\right\}$

4.3. Compact Expressions for $\langle j_1||T^J||j_2\rangle$

Using Eqs. (55), (57) and (59), one can straightforwardly deduce a general closed analytic formula for the reduced matrix elements $\langle j_1||T^J_i||j_2\rangle$ of the seven basic operators Eqs. (21)-(27) as

$$\langle j_1||T^J||j_2\rangle = e^{-y} y^{\beta/2} \sum_{\mu=0}^{n_{max}} \mathcal{P}^J_\mu\, y^\mu, \tag{62}$$

where In the latter summation the upper index n_{max} represent the maximum h.o. quanta included in the model space chosen as

$$n_{max} = (N_1 + N_2 - \beta)/2\,. \tag{63}$$

From Eq. (62) one can infer that the dependence of the matrix elements $\langle j_1||T^J||j_2\rangle$ on the momentum transfer q is simple and is uniquely determined from the momentum-independent quantities $\mathcal{P}^J_\mu$.

5. The Convolution in Astro-physical Neutrino Searches

An important research field in neutrino physics is the detection of astrophysical neutrinos (solar, supernova, atmospheric etc.) through the neutrino-nucleus reactions, i.e. by probing a nuclear target in terrestrial experiments [3, 17, 20, 36, 48]. In order to estimate the response of a nucleus to a specific source of neutrinos, the calculated cross sections (differential or total) of neutrino-nucleus induced reactions must be folded with the neutrino energy distribution of the source in question.

In the case of differential neutrino-nucleus cross sections, $d\sigma(\varepsilon_\nu, \omega)/d\omega$, the folding is defined by the expression

$$\frac{d\sigma(\omega)}{d\omega} = \int_\omega^\infty \frac{d\sigma(\varepsilon_\nu, \omega)}{d\omega} n(\varepsilon_\nu) d\varepsilon_\nu\,, \tag{64}$$

where $\omega = \epsilon_i - \epsilon_f$, represents the excitation energy of the nucleus. ϵ_f (ϵ_i) denote the final (initial) nuclear-energy and neutrino-energy (or, in general, outgoing lepton-energy), respectively. $n_(\epsilon_\nu)$ denotes the energy-spectra of supernova neutrinos. For the total neutrino-nucleus cross section, $\sigma(\varepsilon_\nu)$, the folding is rather simpler and it is defined as

$$\langle\sigma_\nu\rangle = \int_{E_{thres}}^\infty \sigma(\varepsilon_\nu) n(\varepsilon_\nu) d\varepsilon_\nu\,, \tag{65}$$

where E_{thres} denotes the energy-threshold of the nuclear detector. The result obtained from the convolution of Eq. (65), is known as the flux averaged cross section, $\langle\sigma_\nu\rangle$, of the studied nucleus with respect to the assumed neutrino source.

A very interesting neutrino source in current astro-particle physics studies is the supernova explosion which produces an enormous number of neutrinos [36]. Numerical simulations on supernova neutrino dynamics have shown [17, 20, 48] that the energy-spectra $n(\varepsilon_\nu)$ of supernova neutrinos could be well described by a two-parameter Fermi-Dirac distribution of the form

$$n_{FD[T,\eta]}(\varepsilon_\nu) = N_2(\eta)\frac{1}{T^3}\frac{\varepsilon_\nu^2}{1 + e^{(\varepsilon_\nu/T - \eta)}}\,, \tag{66}$$

where T is the neutrino temperature and $N_2(\eta)$ denotes the normalization factor depending on the degeneracy parameter η given from

$$N_k(\eta) = \left(\int_0^\infty \frac{x^k}{1 + e^{x-\eta}} dx\right)^{-1} \tag{67}$$

for $k = 2$. If the temperature of the spectrum is kept fixed, the degeneracy parameter shifts the spectrum to higher-energy values. Spectra with the same average energy tend to be reduced in their high energy tail when the degeneracy parameter becomes larger [33, 48]. The average neutrino energy $\langle \varepsilon_\nu \rangle$ can be written in terms of the functions of Eq. (67) as [65]

$$\langle \varepsilon_\nu \rangle = \frac{N_2(\eta)}{N_3(\eta)} T = (3.1515 + 0.125\eta + 0.0429\eta^2 + ...)T \tag{68}$$

Figure 2 shows the averaged neutrino energy as a function of the parameter η for various

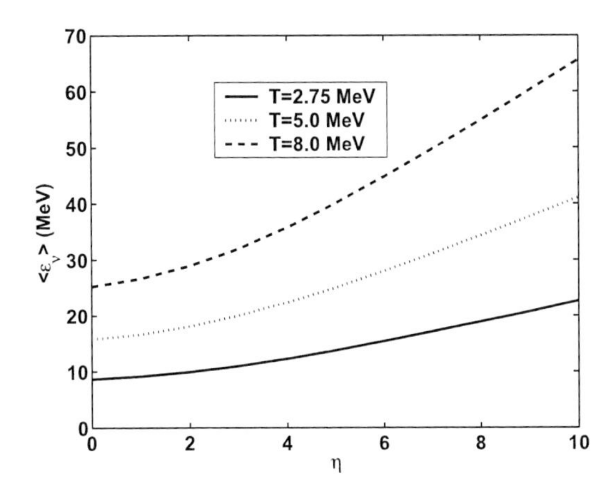

Figure 2. Averaged neutrino energy as a function of the parameter η for various temperatures T.

temperatures T. As it is seen the introduction of a chemical potential in the spectrum at fixed neutrino temperature increases the average neutrino energy.

The parameterization yield energy distributions, characterized by the mean energy $\langle \varepsilon_\nu \rangle$ or temperature T of the neutrinos and the width $w = \sqrt{\langle \varepsilon_\nu^2 \rangle - \langle \varepsilon_\nu \rangle^2}$ of the spectrum. Typical values for the mean energies obtained in numerical simulations are

$$\langle \varepsilon_\nu \rangle = \left\{ \begin{array}{lll} 10 - 12 & \mathrm{MeV} & \nu_e \\ 14 - 18 & \mathrm{MeV} & \tilde{\nu}_e \\ 18 - 24 & \mathrm{MeV} & \nu_{\mu,\tau}\tilde{\nu}_{\mu,\tau} \end{array} \right\} \tag{69i}$$

The widths of the above spectra are characterized by the parameters η or α and they are given by the equations

$$w_{FD} = w_0 \left(\frac{3\, N_3^2(\eta)}{N_2(\eta)\, N_4(\eta)} - 3 \right)^{1/2}, \tag{70}$$

where

$$w_0 = \frac{\langle \varepsilon_\nu \rangle}{\sqrt{3}} \tag{71}$$

Using the bisection method we can solve Eq. (70) for given width w and determine the parameter η. In Table 1 we give the resulting parameters for various values of w.

Table 2. Parameters for fit-functions.

	Fermi-Dirac (FD)
w	η
$w_0 = \langle \varepsilon_\nu \rangle / \sqrt{3}$	
$0.5w_0$	13.892
$0.6w_0$	6.9691
$0.7w_0$	4.4014
$0.8w_0$	2.7054
$0.9w_0$	1.1340

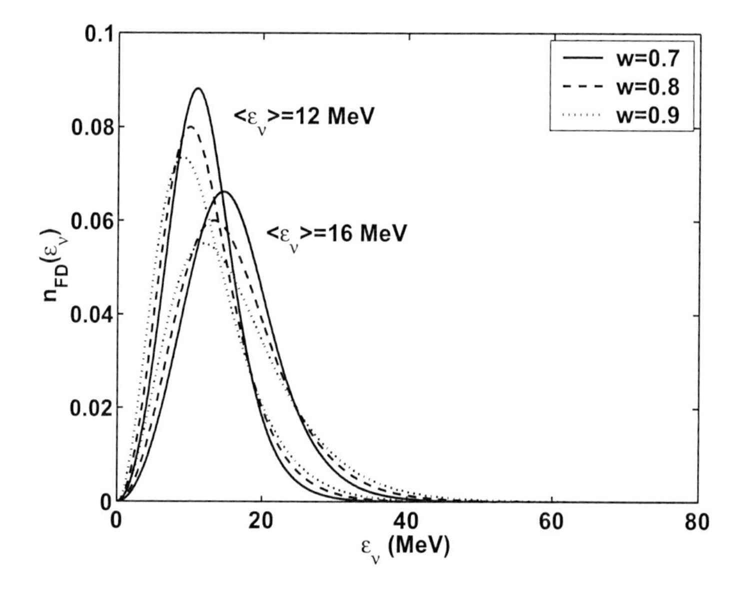

Figure 3. Neutrino energy spectra for two different averaged energies and three different widths with the Fermi-Dirac distribution

This parameterization allow the construction of spectra that are equivalent up to their second moment by adjusting the energy and width parameters. In Fig. 3 we illustrate the Fermi-Dirac spectra for two different values of average neutrino energy $\langle \varepsilon_\nu \rangle = 12, 16$ MeV and three different widths.

The precise shape of the spectrum and its tail is very important for the nuclear response to supernova neutrinos, as cross sections are rising fast with increasing neutrino energies. This is due to the fact that the supernova-neutrino energy range is probing the giant resonance region of the nuclear spectrum, where cross sections are varying fast. This makes nuclei very sensitive probes, on the one hand, but, on the other hand, also makes the concept of nuclei as supernova-neutrino detectors very sensitive to uncertainties in nuclear structure calculations.

6. Energies and Wave Functions

For neutral current neutrino-nucleus induced reactions, the ground state and the excited states of the even-even nucleus are created using the quasi-particle random phase approximation (QRPA) including two quasi-neutron and two quasi-proton excitations in the QRPA matrix [66] (hereafter denoted by pp-nn QRPA). We start by writing the A-fermion Hamiltonian H, in the occupation-number representation, as a sum of two terms. One is the sum of the single particle energies (spe) ϵ_α which runs over all values of quantum numbers $\alpha \equiv \{n_\alpha, l_\alpha, j_\alpha, m_\alpha\}$ and the second term which includes the two-body interaction V that is

$$H = \sum_\alpha \epsilon_\alpha c_\alpha^\dagger c_\alpha + \frac{1}{4} \sum_{\alpha\beta\gamma\delta} \bar{V}_{\alpha\beta\gamma\delta} c_\alpha^\dagger c_\beta^\dagger c_\delta c_\gamma \tag{72}$$

where the two-body term contains the antisymmetrised two-body interaction matrix element defined by $\bar{V}_{\alpha\beta\gamma\delta} =< \alpha\beta|V|\gamma\delta > - < \alpha\beta|V|\delta\gamma >$. The operators $c_\alpha^\dagger$ and c_α stand for the usual creation and destruction operators of nucleons in the state α.

For spherical nuclei with partially filled shells, the most important effect of the two-body force is to produce pairing correlations. The pairing interaction is taken into account by using the BCS theory [67]. The simplest way to introduce these correlations in the wave function is to perform the Bogoliubov-Valatin transformation

$$\begin{aligned} a_\alpha^\dagger &= u_\alpha c_\alpha^\dagger - v_\alpha \tilde{c}_\alpha \\ \tilde{a}_\alpha^\dagger &= u_\alpha \tilde{c}_\alpha^\dagger + v_\alpha c_\alpha \end{aligned} \tag{73}$$

where $\tilde{c}_\alpha^\dagger = c_{-\alpha}^\dagger (-1)^{j_\alpha + m_\alpha}$, $\tilde{a}_\alpha^\dagger = a_{-\alpha}^\dagger (-1)^{j_\alpha + m_\alpha}$ and $-\alpha \equiv \{n_\alpha, l_\alpha, j_\alpha, -m_\alpha\}$. The occupation amplitudes v_α and u_α are determined via variational procedure for minimizing the energy of the BCS ground state for protons and neutrons separately. In the BCS approach the ground state of an even-even nucleus is described as a superconducting medium where all the nucleons have formed pairs that effectively act as bosons. The BCS ground state is defined as

$$|BCS >= \prod_{\alpha > 0} (u_\alpha - v_\alpha c_\alpha^\dagger \tilde{c}_\alpha^\dagger) |CORE > \tag{74}$$

where $|CORE >$ represents the nuclear core (effective particle vacuum).

After the transformation (73) the Hamiltonian can be written in its quasi-particle representation as

$$H = \sum_\alpha E_\alpha a_\alpha^\dagger a_\alpha + H_{qp} \tag{75}$$

where the first term gives the single quasi-particle energies E_α and the second one includes the different components of the residual interaction.

In the present calculations we use a renormalization parameter g_{pair} which can be adjusted when doing the BCS calculations. The monopole matrix elements $< \alpha\alpha; J = 0|V|\beta\beta; J = 0 >$ of the two-body interaction are multiplied by a factor g_{pair}. The adjustment can be done by comparing the resulting lowest quasi-particle energy to the phenomenological energy gap Δ obtained from the separation energies of the neighboring doubly-even nuclei for protons and neutrons separately [22].

In the next step the excited states of the even-even reference nucleus are constructed by use of the QRPA. In the QRPA the creation operator for an excited state(QRPA phonon) has the form

$$\hat{Q}^{\dagger}(J_k^{\pi}M) = \sum_{\alpha\leq\alpha'} \left[X_{\alpha\alpha'}^{J_k^{\pi}} A^{\dagger}(\alpha\alpha'; JM) - Y_{\alpha\alpha'}^{J_k^{\pi}} \tilde{A}(\alpha\alpha'; JM) \right] \tag{76}$$

where the quasi-particle pair creation $A^{\dagger}(\alpha\alpha'; JM)$ and annihilation $\tilde{A}(\alpha\alpha'; JM)$ operators are defined as

$$A^{\dagger}(\alpha\alpha'; JM) \equiv (1 + \delta_{\alpha\alpha'})^{-\frac{1}{2}} \left[a_{\alpha}^{\dagger} a_{\alpha'}^{\dagger} \right]_{JM}, \tag{77}$$

and

$$\tilde{A}(\alpha\alpha'; JM) \equiv (-1)^{J+M} A(\alpha\alpha'; J - M), \tag{78}$$

where α and α' are either proton (p) or neutron (n) indices, M labels the magnetic sub-states and k numbers the states for particular angular momentum J and parity π.

The X and Y forward and backward going amplitudes are determined from the QRPA matrix equation

$$\begin{pmatrix} \mathcal{A} & \mathcal{B} \\ -\mathcal{B} & -\mathcal{A} \end{pmatrix} \begin{pmatrix} X^{J^{\pi}} \\ Y^{J^{\pi}} \end{pmatrix} = \omega \begin{pmatrix} X^{J^{\pi}} \\ Y^{J^{\pi}} \end{pmatrix}, \tag{79}$$

where ω denotes the excitation energies of the nuclear state $|J^{\pi}\rangle$. The QRPA-matrices $\mathcal{A}$ and $\mathcal{B}$, are deduced by the matrix elements of the double commutators of $A^{\dagger}$ and A with the nuclear hamiltonian $\hat{H}$ defined as

$$\mathcal{A}_J(\alpha\alpha'; \beta\beta') = \langle BCS|[A(\alpha\alpha'; JM), \hat{H}, A^{\dagger}(\beta\beta'; JM)]|BCS\rangle, \tag{80}$$

$$\mathcal{B}_J(\alpha\alpha'; \beta\beta') = -\langle BCS|[A(\alpha\alpha'; JM), \hat{H}, \tilde{A}(\beta\beta'; JM)]|BCS\rangle \tag{81}$$

where $2[A, B, C] = [A, [B, C]] + [[A, B], C]$. Finally the two body matrix elements of each multipolarity J^{π}, occurring in the QRPA-matrices $\mathcal{A}$ and $\mathcal{B}$, are multiplied by two phenomenological scaling constants, namely the particle-hole strength g_{ph} and the particle-particle strength g_{pp}. These parameter values are determined by comparing the resulting lowest phonon energy with the corresponding lowest collective vibrational excitation of the doubly-even nucleus and by reproducing some giant resonances which play crucial role [22].

For charge current neutrino-nucleus induced reactions the states of the odd-odd nuclei are generated using the proton-neutron QRPA(pnQRPA). The QRPA in its proton-neutron form contains phonons made out of proton-neutron pairs as follows

$$\hat{Q}^{\dagger}(J_k^{\pi}M) = \sum_{pn} \left[X_{pn}^{J_k^{\pi}} A^{\dagger}(pn; JM) - Y_{pn}^{J_k^{\pi}} \tilde{A}(pn; JM) \right] \tag{82}$$

and the matrices $\mathcal{A}$ and $\mathcal{B}$ are derived from the commutator equations

$$\mathcal{A}_J(pn; p'n') = \langle BCS|[A(pn; JM), \hat{H}, A^{\dagger}(p'n'; JM)]|BCS\rangle, \tag{83}$$

$$\mathcal{B}_J(pn; p'n') = -\langle BCS|[A(pn; JM), \hat{H}, \tilde{A}(p'n'; JM)]|BCS\rangle \tag{84}$$

7. Results and Discussion

7.1. Calculated Cross Sections

In the present work we focus on the investigation of the neutral current reaction cross sections at low and intermediate neutrino energies of various even-even isotopes which play a significant role in astrophysical neutrino studies such as Mo,Fe,O, Cd. To accomplish this aim we need to perform explicit state-by-state calculations for the nuclear transition matrix elements within the QRPA model. The initial nucleus was assumed to be spherically symmetric (having a 0^+ ground state). The single particle energies (s.p.e) were produced by a Coulomb corrected Woods-Saxon potential using the parameters of Bohr and Mottelson [68]. The two-body interaction matrix elements were obtained from the Bonn one-boson-exchange potential applying G-matrix techniques [69]. The strong pairing interaction between the nucleons can be adjusted by solving the BCS equations. The monopole matrix elements of the two-body interaction are scaled by the pairing-strength parameters g^p_{pair} and g^n_{pair} separately for protons and neutrons. The adjustment can be done by comparing the resulting lowest quasiparticle energy to reproduce the phenomenological pairing gap obtained from the linear approximation [70]

$$\Delta_n(^A_ZX) = -\frac{1}{4}[S_n(^{A+1}_ZX) - 2S_n(^A_ZX) + S_n(^{A-1}_ZX)], \tag{85}$$

$$\Delta_p(^A_ZX) = -\frac{1}{4}[S_p(^{A+1}_{Z+1}X) - 2S_p(^A_ZX) + S_p(^{A-1}_{Z-1}X)] \tag{86}$$

in which ^A_ZX stands for the doubly-even nucleus under consideration. The separation energies $S_{n/p}$ are provided, e.g., by [71].

After settling the values of the pairing parameters, two other parameters are left to fix, the overall scale of the particle-hole interaction g_{ph} and separately the particle-particle channel of the interaction g_{pp} for each multipole up to $J = 8^{\pm}$. In the present work we fixed the QRPA parameters on the bound energy spectrum, so as the low-lying excitation energies to fit the experimental spectrum. An alternative fixing of the parameters g_{ph} and g_{pp}, especially for the charged-current neutrino-nucleus reactions, could be done on the giant dipole resonance of the studied nucleus [22, 23, 24, 25, 26].

In the following we proceed with the calculation of the double ($d^2\sigma/d\theta d\epsilon_i$) and single ($d\sigma/d\Omega$) differential cross sections, for neutrino energies $\epsilon_i \leq 100$ MeV. The double differential cross section $d^2\sigma/d\theta d\epsilon_i$, is calculated by summing over partial rates of all multipole states up to $J = 8^{\pm}$. Figure 4 shows the double differential cross section versus initial neutrino energy for various scattering angle θ (step $\Delta\theta = 15^o$ from $\theta = 15^o$ to $\theta = 165^o$). In general, our results show a smooth dependence of $d^2\sigma/d\theta d\epsilon_i$ on the initial neutrino energy, ϵ_i. As can be seen, for low neutrino energies up to about $\epsilon_i \leq 8 - 12$ MeV, namely for the region of the discrete energy spectrum, the differential cross section decreases as the scattering angle increases but for higher energies the trend is reversed [29]. This effect is due to the dominance of the transverse contribution as the scattering angle increases and reminds us the similar behavior occurring in inelastic electron scattering. These results should be useful to experimentalists, as they provide basic information to nuclear response of the neutrino detector.

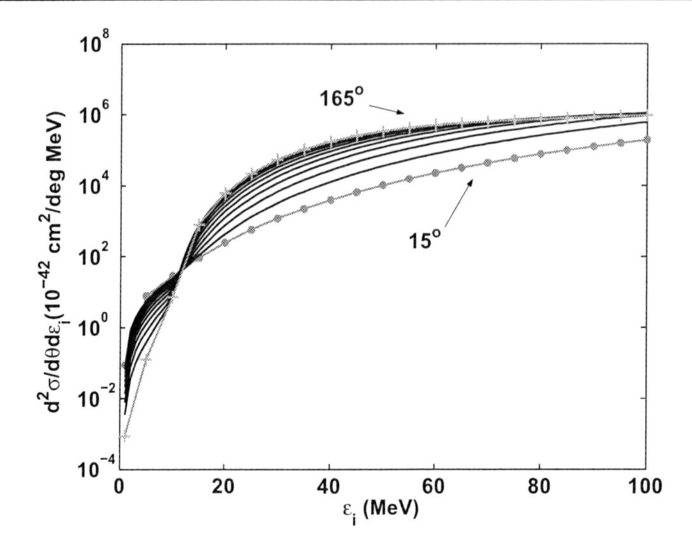

Figure 4. (Color Online) Double differential cross section $d^2\sigma/d\theta d\epsilon_i$ for the reaction $^{98}Mo(\nu,\nu')^{98}Mo^*$ as a function of the neutrino energy and the scattering direction of the lepton. Results are given for scattering angles between $\theta = 15^o$ to $\theta = 165^o$, in 15° steps [29].

The contribution of the dominant multipole states up to $J = 6^{\pm}$ to the differential cross section, $d\sigma/d\Omega$ for the reaction $^{98}Mo(\nu,\nu')^{98}Mo^*$, is illustrated in Fig. 5 and 6 for the scattering angles $\Phi = 0^o, 15^o$ as a function of the incoming neutrino energies $\varepsilon_i = 0 - 100$ MeV.

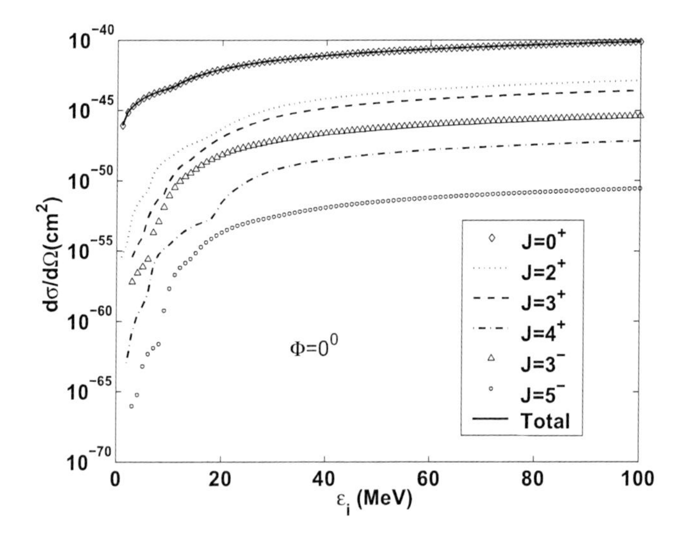

Figure 5. Contribution of the different multipolarities to the differential cross section, $d\sigma/d\Omega$, with respect to the incoming neutrino energy $\varepsilon_i = 0 - 100$ MeV for the reaction $^{98}Mo(\nu,\nu')^{98}Mo^*$ in the scattering angle $\Phi = 0^o$.

Figure 7 shows the dependence of the integrated cross section $\sigma(\varepsilon_i)$ on the incoming

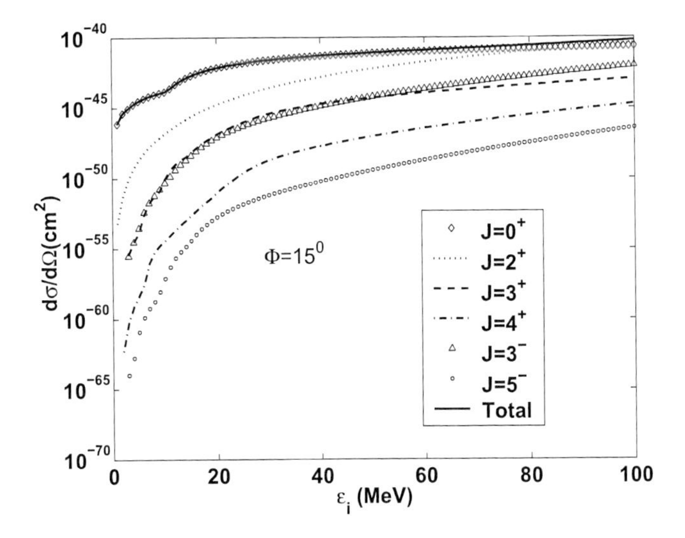

Figure 6. Contribution of the different multipolarities to the differential cross section, $d\sigma/d\Omega$, with respect to the incoming neutrino energy $\varepsilon_i = 0 - 100$ MeV for the reaction $^{98}Mo(\nu,\nu')^{98}Mo^*$ in the scattering angle $\Phi = 15^o$.

neutrino energy ε_i for the reaction $^{116}Cd(\nu,\nu')^{116}Cd^*$. Our results were obtained from the double differential cross section by summing over all possible final nuclear states and by numerical integration over angles. The general trend is that the cross section is proportional to the square of the lepton energy. The individual contributions of the polar vector and axial vector components of the current are also shown in the same Figure. It is worth mentioning that for energies greater than $\varepsilon_i \geq 15$ MeV the cross section comes from the axial vector component of the operator. The same results of the reaction $^{98}Mo(\nu,\nu')^{98}Mo^*$ are shown in the Fig. 8.

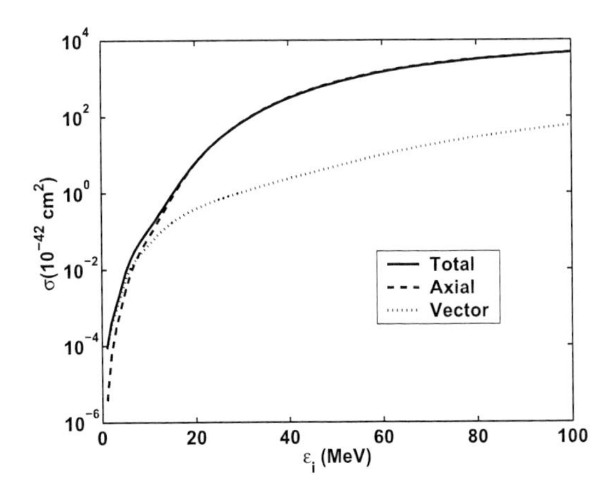

Figure 7. Individual contributions coming from vector and axial-vector components for the reaction $^{116}Cd(\nu,\nu')^{116}Cd^*$. The full line gives the integrated cross section, the dashed-dot the axial contribution and the dotted line the vector one.

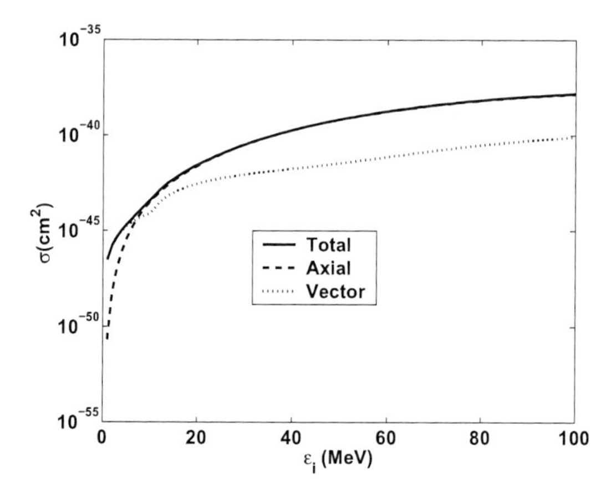

Figure 8. Individual contributions coming from vector and axial-vector components for the reaction $^{98}Mo(\nu,\nu')^{98}Mo^{*}$. The full line gives the integrated cross section, the dashed-dot the axial contribution and the dotted line the vector one [26].

In the Fig. 9 are shown the contributions coming from the integrated and the coherent cross sections for the reaction $^{98}Mo(\nu,\nu')^{98}Mo^{*}$. In the incoming neutrino energies $\varepsilon_i \leq 80$ MeV the contribution of the coherent is higher. In the Fig. 10 are shown the

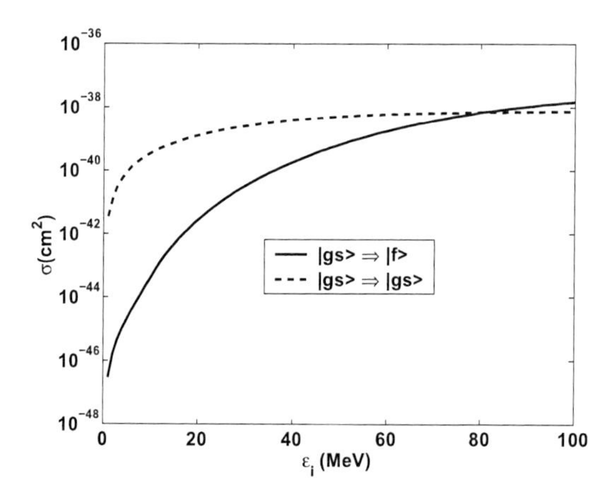

Figure 9. The contributions coming from the integrated and the coherent cross sections for the reaction $^{98}Mo(\nu,\nu')^{98}Mo^{*}$ [26]. The full line gives the $|gs\rangle \rightarrow |f\rangle$, the dashed-dot the$|gs\rangle \rightarrow |gs\rangle$.

contributions in the the integrated cross section from Coulomb and transverse components of the multipole operators for the reaction $^{98}Mo(\nu,\nu')^{98}Mo^{*}$. For the incoming neutrino energies $\varepsilon_i \geq 10$ MeV the contribution from transverse component is higher. The contributions coming from the multipole states 2^{+} in the double differential cross section for the reaction $^{98}Mo(\nu,\nu')^{98}Mo^{*}$, with respect of the scattering angle and the excitation energy ω, are shown in the Fig. 11. The contributions of these states are higher in the excitation energy $\omega \simeq 2$ and 8 MeV.

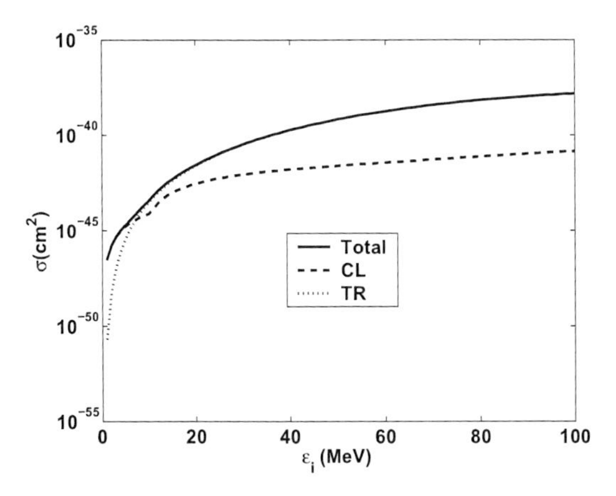

Figure 10. The contributions in the the integrated cross section from Coulomb and transverse components of the multipole operators for the reaction $^{98}Mo(\nu,\nu')^{98}Mo^*$. The full line gives the integrated cross section, the dashed the Coulomb component and the dotted the transverse component..

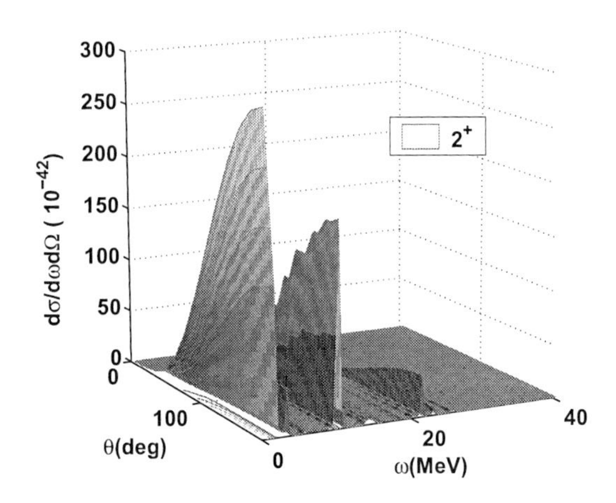

Figure 11. The contributions coming from the multipole states 2^+ in the double differential cross section for the reaction $^{98}Mo(\nu,\nu')^{98}Mo^*$, with respect of the scattering angle and the excitation energy ω.

For comparison in Table 3 we present our results for the integrated cross sections for the reaction $^{56}Fe(\nu,\nu')^{56}Fe^*$ obtained by our QRPA model (second column) and those obtained by Kolbe and Langanke [20] (third column). The range of the incoming neutrino energies is $10 \leq \varepsilon_i \leq 100$ MeV. It should be noted that results of Ref. [20] have been obtained by a hybrid approach that combines Shell Model calculations and Continuum RPA. As can be seen, our total cross sections (in 10^{-42} cm^2) are in good agreement with those of Ref. [20] (for neutrino energy less than 10 MeV there are no results of Ref. [20] to compare with). Some differences of at most 50% at low neutrino energies may be attributed to differ-

Table 3. Integrated cross sections (in $10^{-42}cm^2$) for neutral-current neutrino-nucleus reaction $^{56}Fe(\nu,\nu')^{56}Fe^*$ calculated with our QRPA method [22] (second column). The exponents are given in parentheses. We also present the integrated cross sections calculated by Kolbe and Langanke [20](third column).

Initial energy of incoming ν ε_i(MeV)	^{56}Fe[22]	^{56}Fe[20]
10	1.01(+0)	1.91(-1)
15	2.85(+0)	2.19(+0)
20	5.79(+0)	6.90(+0)
25)	1.06(+1)	1.51(+1)
30	1.87(+1)	2.85(+1)
35	3.24(+1)	4.89(+1)
40	5.51(+1)	7.86(+1)
45	9.05(+1)	1.19(+2)
50	1.43(+2)	1.72(+2)
55	2.15(+2)	2.39(+2)
60	3.09(+2)	3.20(+2)
65	4.26(+2)	4.15(+2)
70	5.63(+2)	5.25(+2)
75	7.17(+2)	6.50(+2)
80	8.82(+2)	7.89(+2)
85	1.05(+3)	9.42(+2)
90	1.22(+3)	1.11(+3)
95	1.38(+3)	1.29(+3)
100	1.52(+3)	1.49(+3)

ent theoretical approaches as well as to different parametrization in single particle energies and effective interaction. Obviously, various theoretical approaches differ in the predicted neutrino-nucleus cross sections and this will require more detailed studies of the underlying nuclear structure that contributes to the neutrino reaction rates. On the other hand, the only experimental data for $^{56}Fe(\nu,\nu')^{56}Fe^*$ reaction are from KARMEN Collaboration [72] and this result has to not yet confirmed by independent measurements.

7.2. Cross Sections for Supernova Neutrinos

In order to find the response of the nuclear detector to the neutrinos of the studied source, we must make the numerical integration involved in the convolution of Eqs. (64) and (65). In the present work, we show results of the convolution involved in Eq. (65). In Tables 4 and 5, we list the folded results for the flux-averaged cross sections of ^{98}Mo and ^{16}O,

Table 4. Averaged cross sections (in $10^{-42}cm^2$) for the isotope ^{98}Mo. We employ the Fermi-Dirac distribution of supernova neutrinos.

	$a = 0$						$a = 3$					
$\sigma \setminus T$	2.75	3.5	4.0	5.0	6.4	8.0	2.75	3.5	4.0	5.0	6.4	8.0
	^{98}Mo:			*Averaged cross sections for* ν_e								
σ_{coh}	320	502	639	943	1410	1960	469	732	928	1360	1990	2710
σ_{tot}	0.4	1.8	3.9	14.4	55.6	168.0	0.8	3.3	7.3	26.6	103.0	308.0
σ_A	0.39	1.7	3.8	14.1	54.7	165.0	0.7	3.2	7.1	26.2	101.0	302.0
σ_V	0.0	0.1	0.1	0.3	0.6	1.1	0.1	0.1	0.2	0.4	0.9	1.9
	^{98}Mo:			*Averaged cross sections for* $\bar{\nu}_e$								
σ_{coh}	320	502	639	943	1410	1960	469	732	928	1360	1990	2710
σ_{tot}	0.4	1.8	4.0	14.4	54.4	164.0	0.8	3.4	7.5	26.7	101.0	300.0
σ_A	0.4	1.7	3.8	14.1	54.4	164.0	0.7	3.2	7.1	26.2	101	300.0
σ_V	0.0	0.1	0.1	0.3	0.6	1.1	0.1	0.1	0.2	0.4	0.9	1.9

with the contributions both of ν_e and $\bar{\nu}_e$, for different temperatures T. Also, we show the contributions of the coherent, axial-vector and vector components. The values are higher for the coherent component and for the greater nucleus.

8. Conclusions

In the present work we performed detailed calculations of neutral current neutrino-nucleus cross sections on Mo, Fe, O and Cd in the low-energy range below 100 MeV neutrino energy. The above isotopes contained in the next-generation neutrino experiment searching for double beta and neutrino-less double beta decay events. The many-body nuclear matrix elements are evaluated in the context of quasi-particle random phase approximation (QRPA). It is worth mentioning that our QRPA method is based on realistic two-body forces and has been successfully used in other semi-leptonic nuclear processes. The dependence of the cross sections on the scattering angle and initial neutrino energy is investigated for both Fermi and Gamow-Teller like operators. In addition to the total neutrino-nucleus cross sections, we have also analyzed the evolution of the contributions of different multipole

Table 5. Averaged cross sections (in $10^{-42}cm^2$) for the isotope^{16}O.

	$a=0$						$a=3$					
$\sigma \setminus T$	2.75	3.5	4.0	5.0	6.4	8.0	2.75	3.5	4.0	5.0	6.4	8.0
	^{16}O: *Averaged cross sections for* ν_e											
σ_{coh}	8.8	14.2	18.3	28.1	44.6	66.4	13.1	20.9	27.0	41.3	65.0	95.9
σ_{tot}	0.0	0.1	0.2	0.7	2.5	6.9	0.0	0.2	0.4	1.4	4.6	12.4
σ_A	0.0	0.1	0.2	0.7	2.4	6.5	0.0	0.2	0.4	1.3	4.4	11.8
σ_V	0.0	0.0	0.0	0.1	0.1	0.3	0.0	0.0	0.0	0.1	0.2	0.5
	^{16}O: *Averaged cross sections for* $\bar{\nu}_e$											
σ_{coh}	8.8	14.2	18.3	28.1	44.6	66.4	13.0	20.9	27.0	41.3	65.0	95.9
σ_{tot}	0.0	0.1	0.2	0.8	2.6	6.8	0.0	0.2	0.4	1.4	4.7	12.2
σ_A	0.0	0.1	0.2	0.7	2.4	6.5	0.0	0.2	0.4	1.3	4.4	11.8
σ_V	0.0	0.0	0.0	0.1	0.1	0.3	0.0	0.0	0.0	0.0	0.2	0.5

excitations with respect to neutrino energy. It has been shown that excitations of higher multipolarities are essential for the quantitative description of neutrino-nucleus cross sections. Moreover, the double differential cross section decreases as the scattering angle increases in the neutrino energy range $\varepsilon_i \leq 8-12$ MeV but for higher energies the cross section increases with the scattering angle. Comparing the contribution of the individual vector component with the axial vector one we found that for energies $\varepsilon_i \geq 15$ MeV, the integrated cross section comes from the axial vector component. This means that for solar neutrino detection only the vector component participates in the detection signal.

Finally in order to explore the response of supernova neutrinos on Mo, and O, we convolute the integrated neutrino-nucleus cross sections with the well known Fermi-Dirac neutrino-energy distribution. The values depend from the size of nucleus.

Acknowledgments

I want to thank my colleague Dr. P. Divari for her valuable help.

References

[1] Donnelly, T.W. and Peccei, R.D. *Phys. Rep.* (1979), 50, 1-85.

[2] Davis, R. *Prog. Part. Nucl. Phys.* (1994), 32, 13-32.

[3] Kolbe, E.; Kosmas, T.S. In *Springer Trac. Mod. Phys.* (2000), 163, 199-225.

[4] Haxton, W. C. *Phys. Rev. Lett.* (1988), 60, 1999-2002.

[5] Bahcall, J.N.; Ulrich, R.K. *Rev. Mod. Phys.* (1988), 60, 297-372; Kubodera, K.; Nozawa,S. *Int. J. Mod. Phys. E* (1994), 3, 101-148.

[6] Rapaport, J. et al., *Phys. Rev. Lett.* (1981), 47 1518-1521; (1985), 54, 2325-2327; Krofcheck, D. et al., *Phys. Rev. Lett.* (1985), 55, 1051-1054; *Phys. Lett. B* (1987), 189, 299-305; Lutostansky, Yu.S.; Skulgina, N.B. *Phys. Rev. Lett.* (1991), 67, 430-432.

[7] J.W.F. Valle, E-print: hep-ph/0610247, and references therein.

[8] Bodmann, B. et al., KARMEN Collaboration, *Phys. Lett. B* (1991), 267, 321-324; B (1992) 280, 198-203; B (1994), 332, 251-257.

[9] Drexlin, G. et al., *Prog. Part. Nucl. Phys.* (1994), 32, 375-396.

[10] Volpe, C.; Auerbach, N.; Col'o, G.; Van Giai, N. *Phys. Rev. C* (2002), 65, 044603-1 - 044603-4 .

[11] Donnelly, T. W.; Walecka, J. D. *Phys. Lett. B* (1972), 41, 275-280.

[12] Donnelly, T. W. *Phys. Lett. B* (1973), 43, 93-97.

[13] Langworthy, J. B.; Lamers,B. A.; Uberall, H. *Nucl. Phys. A* (1977), 280, 351;Bugaev, E. V. *et al.*, *Nucl. Phys. A* (1979), 324, 350-364.

[14] Bell, J.S.; Llewellyn-Smith, C.H. *Nucl. Phys. B* (1971), 28, 317-340.

[15] Gaisser, T.K.; OConnell, J.S. *Phys. Rev. D* (1986), 34, 822; Kuramoto,T.; Fukugita, M.; Kohyama, Y.; Kubodera, K.; *Nucl. Phys. A* (1990), 512, 711-736.

[16] Singh, S.K.; Oset, E. *Nucl. Phys. A* (1992), 542, 587-615.

[17] Kolbe, E. *Phys. Rev. C* (1996), 54, 1741-1748.

[18] Kosmas, T.S.; Oset, E.; *Phys. Rev. C* (1996), 53, 1409–1415.

[19] Divari, P.C.; Kosmas, T.S.; Vergados, J.D.; Skouras, L.D. *Phys. Rev. C* (2000), 61, 054612-1 – 054612-12.

[20] Kolbe, E.; Langanke, K.; Vogel, P. *Phys. Rev. D* (2002), 66, 013007-1 – 013007-5.

[21] Jachowicz, N.; Heyde, K.; Rombouts, S. *Nucl. Phys. A* (2001), 688, 593-595.

[22] Chasioti, V.Ch.; Kosmas, T.S.; Divari, P.C. *Progr. Part. Nucl. Phys.* (2007), 59 481-485.

[23] Balasi, K.G; Kosmas, T.S.; Divari, P.C. *Prog. Part. Nucl.Phys.* (2010), 64, 414-416.

[24] Balasi, K.G.; Kosmas, T.S.; Divari, P.C.; Chasioti, V.Ch. *AIP Conf. Proc.* (2008), 972, 554-557.

[25] Chasioti, V.Ch.; Kosmas, T.S. *Nucl. Phys. A* (2009), 829, 234-252.

[26] V.Ch. Chasioti, PhD Thesis (p 174), Ioannina University Press, (2008).

[27] Kolbe, E.; Langanke, K. *Phys. Rev. C* (2001), 63, 025802–11.

[28] Fuller, G.; Haxton, W.C.; McLaughlim, G.C *Phys. Rev. D* (1999), 59 085005–1-15.

[29] Divari, P.C.; Chasioti, V.Ch.; Kosmas, *Phys. Scr.* (2010) 82 065201.

[30] Janka, H.T.; Muller, B. *Phys. Rep.* (1995), 256, 135-156; Janka, H.T.; Hillebrand, W. *Astron. Astrophys.* (1989), 224, 49.

[31] Zucchelli, P. *Phys. Lett. B* (2002), 532, 166-172.

[32] Volpe, C. *J.Phys. G* (2004), 30, L1.

[33] Jachowicz, N.; McLaughlin, G.C.; Volpe, C. *Phys. Rev. C* (2003), 68, 055502-1–055502-6 ; *Phys. Rev. C* (2008), 77, 055501-12.

[34] Barabash, A.S; *Phys. Atom. Nucl.* (2005), 68, 414-419 ; Mauger, F. *Journal of Physics: Conference Series* (2010), 203 012065-1 – 012065-3.

[35] Ejiri, H. *Phys. Rep.* (2000), 338 265-351.

[36] Ejiri, H. et al. *Eur. Phys. J. ST* (2008), 162, 239-250.

[37] Zuber, K. *Phys. Lett. B* (2001), 519, 1–7 ; *Prog. Part. Nucl. Phys.* (2006), 57, 235-240.

[38] Giuliani, From Cuoricino to CUORE: investigating the inverted hierarchy region of neutrino mass, *J. Phys.: Conf. Series 120* (2008) 052051-1 – 052051-7.

[39] Haxton, W. C. *Phys. Rev. D* (1987), 36, 2283-2292 .

[40] Engel, J.; Kolbe, E.; Langanke, K.; Vogel, P. *Phys. Rev. C* (1996), 54, 2740-2744.

[41] Hayes, A. C.; Towner, I. S. *Phys. Rev. C* (2000), 61, 044603-1– 044603-7.

[42] Volpe, C.; Auerbach, N.; Col'o, G.; Suzuki, T.; Van Giai, N. *Phys. Rev. C* (2000), 62, 015501-1–015501-11 .

[43] Auerbach, N.; Van Giai, N.; Vorov, O. K. *Phys. Rev. C* (1997), 56, R2368 - R2372.

[44] Singh, S. K.; Mukhopadhyay, N. C.; Oset, E. *Phys. Rev. C* (1998), 57, 2687 - 2692.

[45] Buballa, M.; Drozdz, S.; Krewald, S.; Speth, J. *Ann. Phys.* (1991), 208, 346-375.

[46] Kolbe, E.; Langanke, K.; Krewald, S.; Thielemann, F.-K. *Nucl. Phys. A* (1992), 540, 599-620.

[47] Kolbe, E.; Langanke, K.; Thielemann, F.-K.; Vogel, P. *Phys. Rev. C* (1995), 52, 3437-3441.

[48] Jachowicz, N.; Rombouts, S.; Heyde, K.; Ryckebusch, J. *Phys. Rev. C* (1999), 59, 3246-3255.

[49] Jachowicz, N.; Heyde, K.; Ryckebusch, J.; Rombouts, S. *Phys. Rev. C* (2002), 65, 025501-1–025501-7.

[50] Botrugno, A.; Co, G.; *Nucl. Phys. A* (2005), 761, 200-231.

[51] Kolbe, E.; Langanke, K.; Martinez-Pinedo, G.; Vogel, P. *J. Phys. G.* (2003), 29 2569-2596.

[52] Kuramoto, T.; Fukugita, M.; Kohyama, Y.; Kubodera, K. *Nucl. Phys. A* (1990), 512, 711-736.

[53] Meucci, A.; Giusti, C.; Pacati, F.D. *Nucl. Phys. A* (2004), 744 307-322.

[54] Engel, J. *Phys. Rev. C* (1998), 57 2004-2009.

[55] Traini, M. *Nucl. Phys. A* (2001), 694 325-336.

[56] Aste, A.; Jourdan, J. *Europhys. Lett.* (2004), 67 753-759.

[57] Donnelly, T.W.; Walecka, J.D. *Nucl. Phys. A* (1973), 201, 81-106.

[58] Zinner, N.T.; Langanke, K.; Vogel, P. *Phys. Rev. C* (2006), 74 024326-1–024326-2.

[59] Walecka, J.D. *In Muon physics*; ed. V.W. Hughes and C.S. Wu; Academic Press, N.Y.; (1975); Vol. 2, 113.

[60] Connell, J.S.; Donnelly, T.W.; Walecka, J.D. *Phys. Rev. C* (1972), 6 719-733; Donnelly, T.W.;Walecka, J.D. *Nucl. Phys. A* (1976), 274 368-412.

[61] Friar, J.L.; Haxton, W.C.; *Phys. Rev. C* (1985), 31 2027-2035.

[62] Kosmas, T.S.; Vergados, J.D. *Phys. Rev. D* (1997), 55 1752-1764.

[63] Donnelly, T.W.; Haxton, W.C. *At. Data Nucl. Data Tabl.* (1979), 23 103-176.

[64] Chasioti, V.Ch.; Kosmas, T.S. *Czech. J. Phys.* 52 (2002), 467-479; (2002). Nuclear matrix elements for exclusive neutrino-nucleus reactions. arXiv:nucl-th/0202062.

[65] Athar, M.S.; Ahmad, Shaked; Singh, S.K. *Phys. Rev. C* (2005), 71 045501-1–045501-7.

[66] Ring, P.; Schuck, P. *The Nuclear Many-Body Problem*; Springer, N.Y.; (1980).

[67] Bohr, A.; Mottelson, B.R.; Pines, D. *Phys. Rev.* (1958), 110 936.

[68] Bohr, A.; Mottelson, B.R.; *Nuclear structure*; Benjamin, N.Y.; 1969;Vol 1.

[69] Holinde, K. *Phys. Rep.* (1981), 68 121-188.

[70] Kaminski, W.A.; Faessler, A. *Nucl. Phys. A* (1991), 529 605-632.

[71] Audi, G. et al. *Nucl. Phys. A* (2003), 729 337-676.

[72] Maschuw, R. *Prog.Part.Nucl.Phys.* (1998), 40, 183-192.

In: Neutrinos: Properties, Sources and Detection
Editor: Joshua P. Greene, pp. 119-132
ISBN 978-1-61209-650-6

Chapter 4

The Double Beta Decay and the Neutrino Mass

S. Cebrián, H. Gómez and J.A. Villar[*]
Universidad de Zaragoza, Facultad de Ciencias,
C/ Pedro Cerbuna 12, 50009 Zaragoza, Spain.

Abstract

Even after the impressive confirmation of the non-zero mass of neutrinos by observing flavor oscillations in neutrinos from different sources, neutrino physics is a hot and exciting research area since many unknowns still remain. Double beta decay (DBD) is a rare nuclear process that some nuclei can undergo where a nucleus changes into an isobar with the emission of two electrons and two antineutrinos; although rare, this process has been observed for several nuclei. A non-standard version of this decay without the emission of antineutrinos has been proposed and great efforts are being devoted to its observation due to the outstanding implications of its occurrence, which can shed light on some of the pending questions in the field of neutrino physics: violation of leptonic number, confirmation of the Majorana nature of neutrinos, that is, if neutrinos and antineutrinos are the same particle, and even an estimate of the neutrino mass scale. Here, motivation and status of double beta decay searches will be reviewed. In particular, double beta decay processes will be firstly described, emphasizing the connection with the neutrino mass problem. Then, experimental searches of this nuclear process will be also revised, describing their stringent requirements, comparing the different experimental approaches and techniques followed and summarizing the relevant results and prospects.

PACS: 14.60.Pq; 23.40.-s

Keywords: Double beta decay, Neutrino mass

1. Introduction

The Standard Model of Particle Physics includes three species or "flavors" of neutrinos, linked to electrons, muons and tauons; in addition, neutrinos are considered massless and

[*]E-mail address: scebrian@unizar.es, hgomez@unizar.es, villar@unizar.es

different that antineutrinos. However, several experiments studying neutrinos from different sources (coming from the Sun, produced in the atmosphere, generated in accelerators or emitted in nuclear reactors in power plants) have confirmed the non-zero mass of neutrinos thanks to the observation of the so-called oscillations: neutrinos having a certain flavor change into another species in their propagation. This phenomenon can only be explained if neutrinos are massive particles. Although neutrino detection is very hard, due to the extremely low interaction cross sections, a great effort has been devoted in the last decades to study the intrinsic properties of neutrinos: mass, flavor mixing or if neutrinos and antineutrinos are the same (and then they are named Majorana particles) or different (being Dirac particles). On the other hand, being elusive neutrinos are excellent surveys of phenomena like supernovae or Gamma Ray Bursts, processes in the Sun or even cosmological studies. For all these reasons, Neutrino Physics is a hot research topic.

The results of oscillation experiments are outstanding, since they have provided the first evidence of Physics beyond the Standard Model. But to fix the absolute scale of neutrino masses other kind of experiments are needed and only upper limits have been found up to date. Neutrino mass could be directly estimated analyzing the shape of the end of the beta spectrum in nuclei with a low transition energy like tritium. These experiments have given an upper bound of 2.2 eV for the observable related to the electronic neutrino mass; this result could be improved one order of magnitude in the KATRIN experiment, starting operation in Karlsruhe (Germany). Other limits come from cosmological observations or analysis of supernovae emissions, but here the focus will be on DBD, which can shed light not only on the absolute mass scale but also on the neutrino nature (Dirac or Majorana) [1, 2, 3, 4].

2. The Double Beta Decay Process

Double beta decay is a second-order standard weak process, where a nucleus changes into an isobar with the spontaneous and simultaneous emission of two electrons and two antineutrinos, keeping the conservation of the leptonic number:

$$(A, Z) \rightarrow (A, Z+2) + 2e^{-} + 2\overline{\nu}_{e} \qquad (1)$$

This decay is possible for nuclei (A,Z) with an even number of both protons and neutrons for which beta transition to the isobar (A,Z+1) is energetically forbidden or at least strongly suppressed by a big change of angular momentum. At the nucleon level, DBD corresponds to the conversion of neutrons into protons, and at the quark level, quarks d change into quarks u. Figure 1 (a) shows the DBD at this quark level. But in addition to this standard channel, a process without the emission of antineutrinos has been proposed and is presented in Fig. 1 (b), admitting violation of the leptonic number:

$$(A, Z) \rightarrow (A, Z+2) + 2e^{-} \qquad (2)$$

The DBD with emission of neutrinos has been observed using different techniques for 10 nuclei: ^{48}Ca, ^{76}Ge, ^{82}Se, ^{96}Zr, ^{100}Mo, ^{116}Cd, ^{128}Te, ^{130}Te, ^{150}Nd and ^{238}U. Measured half-lives vary from approximately 10^{19} to 10^{24} years (see recommended values at [5]).

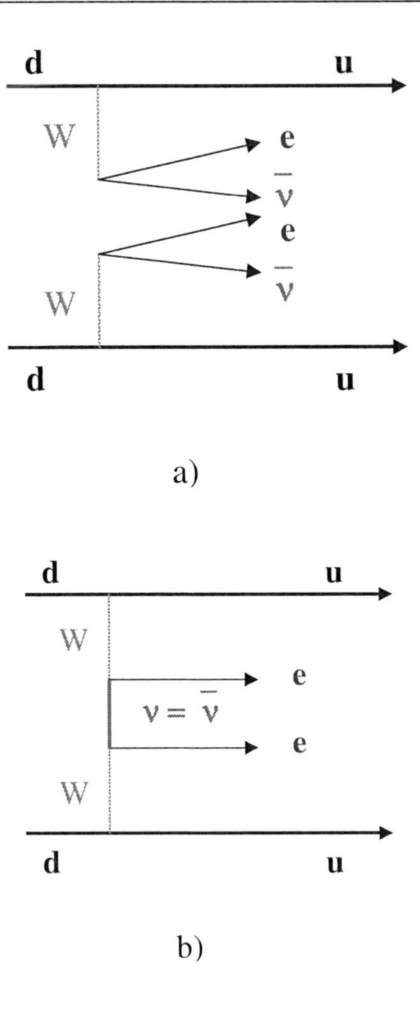

Figure 1. Double beta decay at the quark level a) with and b) without emission of antineutrinos.

An important difference between the two DBD channels is the energy spectrum of the emitted electrons. In the neutrinoless process, the two electrons share all the transition energy Q, since the recoil of the daughter nucleus is negligible. As shown in Fig. 2, the spectrum of the sum energy of the electrons in this case is a peak on the transition energy, while when the antineutrinos are emitted the energy spectrum of electrons is continuous, due to the additional particles in the final state. This difference, as commented later on, is important for the direct detection of both processes.

The relevance of observing the DBD without the emission of neutrinos, not detected up to date, comes from the fact that it would be an evidence that neutrinos are Majorana particles with a non-zero mass. Several mechanisms have been proposed for the neutrinoless DBD, some of them in the framework of supersymmetric theories, but the simplest is based on the exchanging of a light Majorana neutrino. The neutrino emitted in the transformation of a first neutron must be equal to the antineutrino and must be able to couple

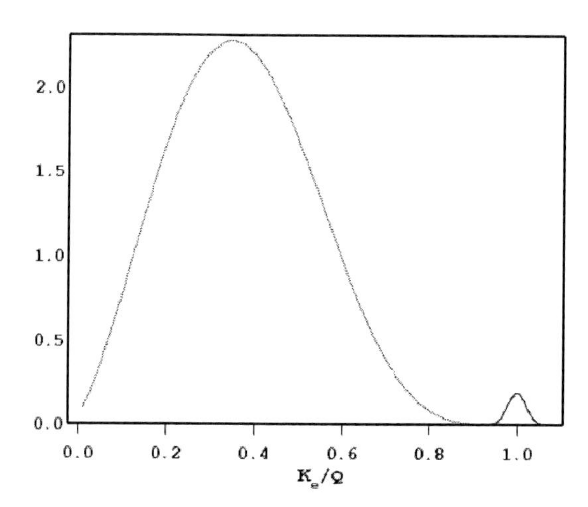

Figure 2. Spectrum of the kinetic energy K_e of the two electrons (divided by the transition energy Q) for the two channels of DBD. The one with the emission of neutrinos is continuous up to Q while the neutrinoless process should appear as a peak at Q.

to the helicity of the neutrino absorbed in the transformation of the second neutron (see Fig. 1 (b)); this implies the existence of a mass term and/or the existence of non-standard V+A interactions in the weak hamiltonian. In any case, it has been shown that whatever the mechanism is, the existence of neutrinoless DBD implies the existence of a Majorana mass term for the neutrino [6]. These are the reasons why the searching for the neutrinoless DBD allows to explore the Physics beyond the Standard Model. The particular mechanism occuring in neutrinoless DBD could be identified by measuring the angular distribution of emitted electrons or the spectrum of each of these electrons.

Historically, the DBD with emission of neutrinos was already studied in 1935 by M. Goeppert-Mayer [7], just some years after the proposal of the existence of neutrinos and the Fermi theory of beta decay. In 1939 W. H. Furry [8] took into consideration the neutrinoless channel proposed by Racah. In the eighties a third channel was proposed, considering the emission together with the electrons of a Goldstone boson coupled to the exchanged virtual neutrino, called Majoron, M, which appears in the spontaneous breaking of the symmetry generating the neutrino mass:

$$(A, Z) \rightarrow (A, Z + 2) + 2e^{-} + M \tag{3}$$

In this case, having four bodies in the final state, the energy spectrum of electrons is also continuous as in the standard channel, but with the maximum shifted to the right.

Processes of DBD with the generation of electrons are the most commonly studied; but as in ordinary beta decay with the emission of electrons, positrons or electron capture, DBD has other possibilities where the nuclear charge is diminished by two units. Typically, DBD occurs between ground states; as in even-even nuclei these states have spin and parity 0^{+}, the transition $0^{+} \rightarrow 0^{+}$ is always possible, but transitions to excited states can also occur if are kinematically allowed. The disadvantage to detect transitions to excited states is that rates are even lower, but on the other hand they can be better identified thanks to the gamma emissions in coincidence. This kind of processed have been observed for ^{100}Mo and ^{150}Nd.

3. Decay Rates and Relationship with Neutrino Mass

The theoretical calculation of decay rates of DBD emitters is in some way analogous to that of ordinary beta emitters, based on the Fermi Golden Rule [9, 10, 11]. The transition probability, inversely proportional to the half-life $T_{1/2}$, is typically written for the DBD with emission on neutrinos as:

$$(T_{1/2}^{2\nu})^{-1} = G_{2\nu}|M_{GT}^{2\nu}|^2 \tag{4}$$

with $G_{2\nu}$ a kinematical factor and $M_{GT}^{2\nu}$ the Gamow-Teller matrix element. For the neutrinoless channel, which has a larger phase space, the corresponding expression (considering only the mass term) is:

$$(T_{1/2}^{0\nu})^{-1} = G_{0\nu}|M^{0\nu}|^2 \frac{\langle m_\nu \rangle^2}{m_e^2} \tag{5}$$

with $G_{0\nu}|M^{0\nu}|^2 = F_N$ the so-called nuclear factor-of-merit, being $G_{0\nu}$ the integral of the phase space for this process and $M^{0\nu} = M_{GT}^{0\nu} - (g_V/g_A)^2 M_F^{0\nu}$, with $M_{GT}^{0\nu}$ and $M_F^{0\nu}$ the Gamow-Teller and Fermi nuclear matrix elements respectively, an g_V y g_A the vectorial and axial-vectorial coupling constants. The factor m_e is the electron mass and

$$\langle m_\nu \rangle = \sum_{j=1}^{3} \lambda_j m_j |U_{ej}|^2 \tag{6}$$

is the neutrino effective mass, where m_j is the mass eigenvalue of state j, U_{ej} are the elements of the matrix describing the mixing of neutrinos j with the electronic neutrino and λ_j CP phase factor. The effective mass is sometimes written also as $\langle m_{\beta\beta} \rangle$. Calculations of the kinematical factor $G_{0\nu}$ are not problematic, but those of the nuclear matrix elements quantifying the transition probability are complex and depend strongly on the nuclear model considered. It must be noted that in other kind of experiments trying also to measure the absolute value of the neutrino mass (based on single beta decay or on cosmology) the observable is a different combination of neutrino mass eigenvalues.

Expressions below show how from the decay rate of neutrinoless DBD is possible to derive a value for the neutrino effective mass:

$$\langle m_\nu \rangle = m_e (F_N T_{1/2}^{0\nu})^{-1/2} \tag{7}$$

There might be cancelations in $\langle m_\nu \rangle$, but the measured value should be lower than the heaviest mass eigenvalue, and this means that at least one neutrino mass eigenstate has a mass equal or greater than the observed effective mass. It must be noted that because of the uncertainties in the nuclear factor of merit, a restrictive limit to the half-life of a DBD emitter does not guarantee for all the nuclei a restrictive bound to the effective mass. Due to these uncertainties, it would be advisable to extend the searches for neutrinoless DBD decay to different emitters.

DBD experiments can inform not only on the absolute neutrino mass but also on the mass hierarchy (normal, inverted or degenerated, according to the relationship between the mass eigenvalues m_j, j =1-3) and even on the CP violation [12, 13, 14, 15, 16, 17, 18]. Taking into account the dependence of the effective neutrino mass on the lightest mass eigenvalue for the different hierarchies and the expected sensitivities of DBD projects ongoing (see next section) if the masses of the eigenstates are degenerated or following the inverted hierarchy, they can be explored in the proposed experiments; however, masses for normal hierarchy seem to be out of reach for the time being.

Calculations of the nuclear matrix elements quantifying transition probabilities are based mainly on two different approaches, the shell model and QRPA ("Quasiparticle Random Phase Approximation"). Last decade has seen a great progress on shell model calculations [19], which can now be used for almost all double beta decay emitters. Within QRPA, it has been discussed the tuning of some parameters for the neutrinoless channel by means of the agreement of model predictions with experimental results for the channel with neutrino emission [20, 21]. Other approaches have been also used, like the interacting boson model [22]. A encouraging convergence between different calculations is taking place in the last years; different effects, as for instance deformation of nuclei, are being taken into account [23, 24, 25, 26, 27, 28, 29] and also specific measurements are made to help nuclear calculations [30, 31]. Figure 3, taken using values from Table II in [32], compares for different nuclei predictions for matrix elements of neutrinoless DBD according to different models: ISM ("Interacting Shell Model") [32], QRPA from [21] and QRPA from [23].

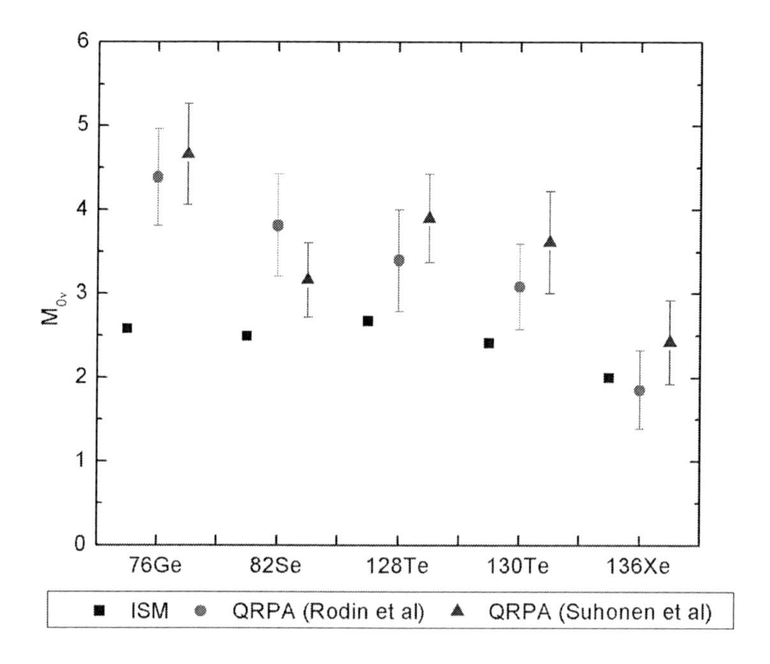

Figure 3. Values of nuclear matrix elements for neutrinoless DBD of several nuclei from different models.

It must be noted that, as the parameter $\langle m_\nu \rangle$ coming from measurements of half-lives and nuclear matrix elements informs on neutrino mass, it is possible to take information on V+A couplings in weak interactions by means of parameters $\langle\lambda\rangle = \lambda\sum_i U_{ei}V_{ei}$ and $\langle\eta\rangle = \eta\sum_i U_{ei}V_{ei}$, with λ and η describing respectively the coupling between right-handed leptonic currents and right-handed or left-handed quark currents and V the mixing matrix analogous to U but for right-handed neutrinos.

4. Detection of Double Beta Decay

Three kind of experiments have been carried out to investigate DBD:

1. *Geochemical experiments*: by means of isotopical analysis, an abnormal excess of daughter nuclei (A,Z+2) accumulated in very large times is searched for in rocks containing nuclei (A,Z). This method has been used to see DBD in ^{82}Se, ^{128}Te, ^{130}Te, ^{96}Zr and more recently ^{100}Mo. In fact, this kind of experiments gave the first evidence of DBD in the fifties.

2. *Radiochemical experiments*: in these experiments the daughter nuclei of a DBD emitter are accumulated, taken out and counted. These daughter nuclei must therefore be radioactive and produce a distinctive signal. This method has been used for ^{238}U.

3. *Direct counting experiments*: in this case, the energy spectrum of electrons emitted by DBD is studied. This method has the advantage of discriminating the DBD channel. Proposed in the sixties [33], the first evidence of DBD with the emission of neutrinos came in the eighties, for ^{82}Se and using a Time Projection Chamber. Most of the experimental efforts are devoted now to this direct search. Due to the extremely low expected rates of DBD, very sensitive detectors are mandatory, accumulating big masses of DBD emitters and taking data for very long periods of time. A good energy resolution is also necessary to identify the neutrinoless DBD signal, to avoid the interference of the final part of the signal of DBD with neutrinos.

The approach detector=source, being the DBD emitters inside the detector, is specially interesting for increasing significantly the efficiency to the signal. Concerning the emitters, the most relevant features are the transition energy Q and the natural isotopic abundance. The phase space in neutrinoless DBD grows as Q^5 and given the natural radioactive background, the higher Q, the lower background will entangle the region of interest where the signal is expected to appear. If the natural isotopic abundance of the DBD emitter is low, enrichment techniques must be applied, but they are usually complicated and costly.

Different kinds of detectors have been used in counting experiments: semiconductors, scintillators, chambers and cryogenic detectors [1, 2]. Cryogenic detectors, where energy deposits produce tiny temperature increases which can be measured operating at extremely low temperatures (not higher than a few tens of mK), offer a wide range of targets, which makes easy to follow the detector=source approach. Drifting chambers allow to measure not only the energy deposit but also the electron tracks to discriminate the signal from the background.

But all DBD experiments have a same challenge: to reduce the detector background to achieve enough sensitivity to the signal. Underground operation, under hundreds of rock shielding cosmic rays, is mandatory. But even underground, primordial or cosmogenic activities in materials, radon from the air, environmental neutrons or residual muons can induce events of the same energy than the signal; active and passive shieldings against these radiations and extreme control of the radiopurity of the materials in the set-up are necessary [34, 35]. In addition, sophisticated methods to discriminate DBD signal and backgrounds are typically used.

A detector factor-of-merit F_D, giving an estimate of the limit (at 1σ confidence level) to $T_{1/2}$ achievable for the neutrinoless DBD, has been defined. In the detector=source approach, F_D expressed in years can be written as:

$$F_D = 4.17 \times 10^{26} \times \frac{a}{A}\sqrt{\frac{Mt}{b\Gamma}} \times \epsilon \tag{8}$$

being A the atomic mass, a the isotopic abundance, M the detector mass in kg, t the time of measurement in years, b the background as c/(keV kg y), Γ the energy resolution (Full Width at Half Maximum, FWHM, in keV) and ϵ the detection efficiency. A thorough analysis of the sensitivity evaluation of DBD projects, based on different parameters, is made in [36].

In the rest of this section, some of the most outstanding experiments made or underway to directly detect DBD are described. Detection techniques are presented, showing pros and cons, and the relevant results and prospects are commented.

- Experiments using **germanium detectors** are operating for more than four decades, investigating the DBD of ^{76}Ge (Q=2039 keV, 7.4 % of natural isotopic abundance) [37]. Exploiting the detector=source method, these experiments have very high detection efficiency for signal and as semiconductor detectors, they enjoy excellent energy resolution and are based on well established technologies. In addition, germanium is a very radiopure material. Background discrimination techniques have been developed analyzing electric pulse shapes from detector; on an event-by-event basis, it can be deduced if energy have been deposited only at one point, as expected for DBD, or at different places, as it is the case for many kinds of background [38, 39, 40, 41, 42]. Transition energy Q [43] is quite high and the nuclear matrix element is favorable. The main disadvantage of germanium DBD experiments is the need of enrichment of ^{76}Ge, due to its low abundance.

  The study of the half-life of this isotope for neutrinoless DBD using HPGe detectors has given the best bound to the effective neutrino mass for years. There have been two remarkable experiments (both finished):

  - The "International Germanium EXperiment" (IGEX) [44], with the collaboration of American, Sovietic and Spanish groups, used three crystals of $\sim$ 2 kg of germanium enriched in ^{76}Ge to 86% at the Canfranc Underground Laboratory, in Spain, and other smaller in Baksan, Russia. No positive signal for the neutrinoless DBD was found, giving a limit of $T^{0\nu}_{1/2} > 1.57 \times 10^{25}$ y (90 % C.L.). The corresponding bound to the effective neutrino mass, depending on the nuclear matrix elements used, was $\langle m_\nu \rangle < (0.33 - 1.35)$ eV [45].
  - The Heidelberg-Moscow (HM) experiment [46, 47] used five detector with a total mass of 10.96 kg of germanium, equally enriched in ^{76}Ge, at the Gran

Sasso Underground Laboratory, in Italy. From the analysis of the neutrinoless channel of DBD, the following limits were presented: $T^{0\nu}_{1/2} \geq 1.9 \times 10^{25}$ y and $\langle m_\nu \rangle \leq 0.35$ eV, at 90 % C.L. [48]. Soon after, a part of the collaboration after a reanalysis of data informed of an evidence [49] discussed by the community [50, 51, 52]. Once finished the data taking, analyzing more than 13 y of data and after a new analysis [53], final results point to hints of a positive signal [54, 55], pending to be confirmed or refuted by other experiments. Authors consider a 4.2σ confidence level for the signal, corresponding to a half-life $T^{0\nu}_{1/2} = 1.2 \times 10^{25}$ y and an effective neutrino mass $\langle m_\nu \rangle = 0.2 - 0.6$ eV.

These experiments seem to have achieved the limit in material radiopurity, so the new projects are developing new strategies to get rid of the radioactive background. The "GERmanium Detector Array" (GERDA) [56] is based on the operation of naked germanium crystals in cryogenic liquid with the aim to eliminate material around the detectors. The first phase underway is operating at the Gran Sasso Laboratory the detectors used by IGEX and HM ($\sim$ 18 kg). In a second phase, using segmented or the so-called broad-energy germanium detectors, GERDA will accumulate up to 40 kg to reach a sensitivity to the effective neutrino mass even below 100 meV. In a longer term, masses at the scale ton would be necessary to explore masses below 20 meV. The American collaboration Majorana [57] plans also to use segmented germanium detectors to discriminate signal and background, using up to half a ton at SNOlab, in Canada. Many efforts have been devoted in the last years to apply segmented detectors to DBD experiments, applying more sophisticated techniques of background rejection than in conventional germanium detectors [58, 59, 60, 61, 62, 63, 64, 65, 66, 67, 68]. Development of broad-energy detectors could open new possibilities for a further improvement of this kind of experiments [69, 70].

- Other DBD emitter investigated is ^{130}Te. Its main advantage is that due to a high natural isotopic abundance (33.8%), detectors of tellurium oxide are not expensive; they are also free of radioactive impurities but crystals are fragile and as bolometers must operate at very low temperatures, which makes difficult stable operation for long time (cryogenic detectors were proposed for DBD searches in 1984 [71]). Transition energy is quite high (Q=2529 keV), which gives a reduced background in the region of interest. Experiments can be based on the detector=source approach optimizing efficiency and bolometers of tellurite have reached very good energy resolutions.

  A group from Milan is leading for years the direct detection of DBD in tellurium at the Gran Sasso Underground Laboratory, increasing steadily the mass of TeO_2 used [72, 73]. The CUORICINO experiment operated 40.7 kg at a temperature of $\sim$10 mK, using cubic crystals with a side of 5 cm. No positive signal for the neutrinoless DBD of ^{130}Te, deriving the following limits: $T^{0\nu}_{1/2} \geq 3.0 \times 10^{24}$ y and $\langle m_\nu \rangle \leq 0.19 - 0.68$ eV at 90% C. L. [74]. CUORICINO was the first phase of the "Cryogenic Underground Observatory for Rare Events" (CUORE) project, using 750 kg of TeO_2 [75, 76]. The challenge of cooling such a big mass is important but data taking is expected to start soon, to achieve a sensitivity to the effective neutrino mass of some tens of meV.

- Other experiments, following different strategies, have been made or are ongoing to detect DBD in different isotopes: ^{100}Mo, ^{82}Se, ^{116}Cd, ^{136}Xe, ...

The "Neutrino Ettore Majorana Observatory" (NEMO) experiment operates at the Modane Underground Laboratory (France), finishing the third phase in 2010. In this case, the source of DBD emitters is outside the detector, in very thin sheets. The detection system is made of gaseous detectors to register tracks, surrounded of scintillators acting as calorimeters, using a magnetic field of 25 Gauss. The main advantage of this approach is the possibility of identifying the kind of particle interacting and registering the tracks of electrons from DBD; this allows to reject background and even having angular information of events. In addition, different emitters can be studied at the same set-up. However, detection efficiency is quite low and energy resolution very limited. NEMO3 has taken data using 10 kg of emitters ($\sim$7 kg of ^{100}Mo and $\sim$1 kg of ^{82}Se, and smaller amounts of ^{150}Nd, ^{130}Te, ^{116}Cd, ^{96}Zr and ^{48}Ca) [77]. The DBD process with emission of neutrinos has been studied with unprecedented statistics for several isotopes (^{100}Mo, ^{116}Cd, ^{82}Se and ^{96}Zr) [78]. For the neutrinoless channel, no positive signal have been found, and the following limits have been obtained at 90% C. L. [79, 80]: $T^{0\nu}_{1/2} > 4.6 \times 10^{23}$ y and $\langle m_\nu \rangle < 0.7 - 2.8$ eV for ^{100}**Mo** with Q=3034 keV, $T^{0\nu}_{1/2} > 1.0 \times 10^{23}$ y and $\langle m_\nu \rangle < 1.7 - 4.9$ eV for ^{82}**Se** with Q=2295 keV, and $T^{0\nu}_{1/2} > 9.2 \times 10^{21}$ y and $\langle m_\nu \rangle < 7.2 - 19.5$ eV for ^{96}**Zr** with Q=3350 keV. Results have been derived also for different DBD modes [81, 82]. The project SuperNEMO [83] is based on the same approach, increasing the mass of the sources (^{82}Se, ^{150}Nd) to the scale of hundreds of kg, which requires a change in the detector geometry. Improving some experimental parameters, the sensitivity to the effective neutrino mass could reach $\sim$50 meV. Control of the required radiopurity in the DBD source is a real challenge, and a specific detector based on the detector of BiPo events from the natural radioactive chains is under development [84]. One module of the SuperNEMO project could be in operation in some years.

The NEXT project is developing a novel detection concept for investigating the neutrinoless DBD of ^{136}Xe [85]. This concept is based on a Time Projection Chamber (TPC) filled with high-pressure gaseous xenon, and with separated-function capabilities for calorimetry and tracking. Thanks to its excellent energy resolution, together with its powerful background rejection provided by the distinct DBD topological signature, it could be competitive with other next-generation neutrinoless DBD experiments. They plan to construct a detector with 100 kg fiducial mass in isotope ^{136}Xe, to be installed in the Canfranc Underground Laboratory, in Spain. Also the EXO experiment in USA is focused on the study of ^{136}Xe.

There are many other DBD experiments at different stages of development, like MOON (^{100}Mo) and CANDLES (^{48}Ca) in Japan [86], COBRA (^{116}Cd and ^{130}Te) in Italy [87], CARVEL (^{48}Ca) in Ucraine [88] or SNO++ (^{150}Nd) in Canada [89].

5. Summary

The DBD with the emission of neutrinos is a rare nuclear process observed for several nuclei, but the neutrinoless channel, implying the violation of leptonic number conservation, has not been evidenced. Its identification would be outstanding in Neutrino Physics, since this would confirm the Majorana nature of neutrinos, inform on the neutrino mass scale and shed light on CP violation.

DBD experiments with various emitters use different techniques for semiconductors, scintillators, chambers and cryogenic detectors. The extremely low expected rates impose underground operation and the development of specific background rejection techniques. Each approach has pros and cons, what make different experiments necessary (also due to the nuclear uncertainties). Big projects like GERDA, Majorana, CUORE, SuperNEMO, EXO or NEXT could in next years give evidence of neutrinoless DBD and then unveil some of the unknown neutrino properties.

References

[1] A. Morales, *Nucl. Phys. B (Proc. Suppl.)* **77** (1999) 335.

[2] S. R. Elliot y P. Vogel, *Ann. Rev. Nucl. Part. Sci.* **52** (2002) 115-151. S.R. Elliott y J. Engel, *J.Phys. G* **30** (2004) R183.

[3] F. T. Avignone III et al, *Rev. Mod. Phys.* **80** (2008) 481.

[4] H. V. Klapdor-Kleingrothaus, "Sixty Years of Double Beta Decay: From Nuclear Physics to Beyond Standard Model Particle Physics", World Scientific, 2001.

[5] A.S. Barabash, *Phys. Rev. C* **81** (2010) 035501.

[6] J. Schechter and J. W. F. Valle, *Phys. Rev. D* **25** (1982) 2951.

[7] M. Goeppert-Mayer, *Phys. Rev.* **48** (1935) 512.

[8] W. H. Furry, *Phys. Rev.* **56** (1939) 1184.

[9] Ch. W. Kim y A. Pevsner, "Neutrinos in physics and astrophysics", Harwood Academic Press, Switzerland, 1993.

[10] F. Boehm y P. Vogel, "Phyisics of massive neutrinos", Cambridge University Press, 1992.

[11] J. Suhonen y O. Civitarese, *Phys. Rep.* **300** (1998) 123-214.

[12] J. N. Bahcall et al, *Phys. Rev. D* **70** (2004) 033012.

[13] S.M. Bilenky et al, *Phys. Rev. D* **70** (2004) 033003.

[14] S.M. Bilenky et al, *Phys. Rev. D* **72** (2005) 053015.

[15] A. Strumia y F. Vissani, *Nucl. Phys. B* **726** (2005) 294-316.

[16] S. Pascoli et al, *Nucl. Phys. B* **734** (2006) 24-49.

[17] S. Pascoli and S. T. Petcov, *Phys. Rev. D* **77** (2008) 113003.

[18] G. L. Fogli et al, *Phys. Rev. D* **78** (2008) 033010.

[19] E. Caurier et al, *Reviews of Modern Physics* **77** (2005) 427-488.

[20] J. Suhonen, *Phys. Lett. B* **607** (2005) 87-95.

[21] V. A. Rodin et al, Nucl. Phys. A 766 (2006) 107-131. Erratum, *Nucl. Phys. A* **793** (2007) 213-215.

[22] J. Barea and F. Iachello, *Phys. Rev. C* **79** (2009) 044301.

[23] M. Kortelainen et al, *Phys. Lett. B* **647** (2007) 128-132. M. Kortelainen and J. Suhonen, *Phys. Rev. C* **76** (2007) 024315.

[24] R. Alvarez-Rodriguez et al, *Phys. Rev. C* **70** (2004) 064309.

[25] E. Caurier et al, *Phys. Rev. Lett.* **100** (2008) 052503.

[26] A. Faessler et al, *J. Phys. G: Nucl. Part. Phys.* **35** (2008) 075104.

[27] F. Simkovic et al, *Phys. Rev. C* **77** (2008) 045503.

[28] F. Simkovic et al, *Phys. Rev. C* **79** (2009) 055501.

[29] J. Suhonen and O. Civitarese, *Nucl. Phys. A* **847** (2010) 207-232.

[30] K. Amos et al, *Phys. Rev. C* **76** (2007) 014604.

[31] J. P. Schiffer et al, *Phys. Rev. Lett.* **100** (2008) 12501.

[32] E. Caurier et al, *Eur. Phys. J. A* **36** (2008) 195-200.

[33] G. F. dell'Antonio y E. Fiorini, Suppl. Nuovo Cimento 17 (1960) 132. E. Fiorini et al, *Phys. Lett.* **25B** (1967) 602.

[34] G. Heusser, *Annu. Rev. Nucl. Part. Sci.* **45** (1995) 543.

[35] J. A. Formaggio and C. J. Martoff, *Annu. Rev. Nucl. Part. Sci.* **54** (2004) 361.

[36] F. T. Avignone III et al, *New Journal of Physics* **7** (2005) 6.

[37] A. Morales y J. Morales, *Nucl. Phys. B (Proc. Suppl.)* **114** (2003) 141-157.

[38] D. González et al, *Nucl. Instrum. and Meth. A* **515** (2003) 634.

[39] B. Majorovits, H.V. Klapdor-Kleingrothaus, *Eur. Phys. J. A* **6** (1999) 463.

[40] J. Hellmig and H.V. Klapdor-Kleingrothaus, *Nucl. Instrum. Meth. A* **455** (2000) 638.

[41] H. V. Klapdor-Kleingrothaus et al, *Phys. Lett. B* **632** (2006) 623-631.

[42] H. V. Klapdor-Kleingrothaus et al, *Phys. Rev. D* **73** (2006) 013010.

[43] G. Douysset et al, *Phys. Rev. Lett.* **86** (2001) 4259.

[44] C. E. Aalseth et al, *Phys. Rev. C* **59** (1999) 2108.

[45] C. E. Aalseth et al, *Phys. Rev. D* **65** (2002) 092007.

[46] M. Gunther et al, *Phys. Rev. D* **55** (1997) 54.

[47] L. Baudis et al, *Phys. Rev. Lett.* **83** (1999) 41.

[48] H. V. Klapdor-Kleingrothaus et al, *Eur. Phys. J. A* **12** (2001) 147-154.

[49] H. V. Klapdor-Kleingrothaus et al, *Mod. Phys. Lett. A* **16** (2001) 2409-2420.

[50] C. E. Aalseth et al, *Mod. Phys. A* **17** (2002) 1475.

[51] Yu. G. Zdesenko et al, *Phys. Lett. B* **546** (2002) 206.

[52] A.M. Bakalyarov et al, *Phys. Part. Nucl. Lett.* **2** (2005) 77-81.

[53] H. V. Klapdor-Kleingrothaus et al, *Nucl. Instrum. Meth. A* **522** (2004) 371-406.

[54] H. V. Klapdor-Kleingrothaus et al, *Phys. Lett. B* **586** (2004) 198.

[55] H. V. Klapdor-Kleingrothaus et al, *Phys. Lett. B* **578** (2004) 54-62.

[56] I. Abt et al, GERDA Letter of Intent, A new 76Ge Double Beta Decay Experiment at LNGS, [arXiv:hep-ex/0404039]. J. Jochum, *Prog. Part. Nucl. Phys.* **64** (2010) 261-263.

[57] The Majorana Collaboration, White Paper on the Majorana Zero- Neutrino Double-Beta Decay Experiment, [arXiv:nucl-ex/0311013]. F.T. Avignone III, *Prog. Part. Nucl. Phys.* **64** (2010) 258-260.

[58] S.R. Elliot et al, *Nucl. Instrum. Meth. A* **558** (2006) 504.

[59] I. Abt et al, *Nucl. Instrum. Meth. A* **577** (2007) 574.

[60] I. Abt et al, *Nucl. Instrum. Meth. A* **583** (2007) 332.

[61] I. Abt et al, *Eur. Phys. J. A* **26** (2008) 139.

[62] B. Majorovits, *Prog. Part. Nucl. Phys.* **64** (2010) 264.

[63] H. Gómez et al, *Astropart. Phys.* **28** (2007) 435.

[64] I. Abt et al, *Nucl. Instrum. Meth. A* **570** (2007) 479.

[65] I. Abt et al, *Eur. Phys. J. C* **52** (2007) 19.

[66] I. Abt et al, *Eur. Phys. J. C* **54** (2008) 425.

[67] D. B. Campbell et al, *Nucl. Instrum. Meth. A* **587** (2008) 60.

[68] I. Abt et al, *Eur. Phys. J. C* **68** (2010) 609.

[69] P. S. Barbeau, J.I. Collar and O. Tench, *JCAP* **09** (2007) 009.

[70] D. Budjas et al, *Journal of Instrumentation* **4** (2009) P10007.

[71] E. Fiorini, T. Niinikoski, *Nucl. Instrum. Meth.* **224** (1984) 83.

[72] A. Alessandrello et al, *Phys. Lett. B* **486** (2000) 13.

[73] C. Arnaboldi et al, *Phys. Rev. Lett.* **95** (2005) 142501.

[74] C. Arnaboldi et al, *Phys. Rev. C* **778** (2008) 035502.

[75] C. Arnaboldi et al, *Nucl. Instrum. and Meth. A* **518** (2004) 775-798.

[76] C. Arnaboldi et al, *Astropart. Phys.* **20** (2003) 91.

[77] R. Arnold et al, *Nucl. Instrum. and Meth. A* **536** (2005) 79-122.

[78] R. Arnold et al, *JETP Lett.* **80** (2004) 429.

[79] R. Arnold et al, *Phys. Rev. Lett.* **95** (2005) 182302.

[80] J. Argyriades et al, *Nucl. Phys. A* **847** (2010) 168-179.

[81] R. Arnold et al, *Nucl. Phys. A* **765** (2006) 483-494.

[82] R. Arnold et al, *Nucl. Phys. A* **781** (2007) 209-226.

[83] R. Arnold et al, Eur. Phys. J. C in press, [arXiv:1005.1241]. L. Simard et al, *Prog. Part. Nucl. Phys.* **64** (2010) 270-272.

[84] J. Argyriades et al, *Nucl. Instrum. and Meth. A* **622** (2010) 120.

[85] NEXT Collaboration, Letter of Intent to the Canfranc Underground Laboratory scientific Committee, [arXiv.org/0907.4054].

[86] T. Kishimoto, *Int. J. Mod. Phys. E* **18** (2009) 2129-2133.

[87] J. V. Dawson et al, *Phys. Rev. C* **80** (2009) 025502.

[88] Yu.G. Zdesenko et al, *Astropart. Phys.* **23** (2005) 249-263.

[89] Ch. Kraus, and S. J.M. Peeters, *Prog. Part. Nucl. Phys.* **64** (2010) 273-277.

In: Neutrinos: Properties, Sources and Detection
Editor: Joshua P. Greene, pp. 133-144
ISBN 978-1-61209-650-6

Chapter 5

STERILE NEUTRINOS IN THE UNIVERSE

M.H. Chan [*] ***and M.-C. Chu*** [†]
The Chinese University of Hong Kong

Abstract

In this chapter, we investigate observable consequences of sterile neutrinos, in galactic, cluster, and cosmological scales. We first review some basic physics of the formation of sterile neutrinos from the oscillation of active neutrinos in the early universe. Then we discuss the possibility of having sterile neutrino halos in the Milky Way and galaxy clusters. The radiative decay of the sterile neutrinos gives energy to heat up the interstellar medium which can simultaneously explain the cooling flow problem in galaxy clusters and the hot gas problem in the Milky Way. Also, this model enables us to obtain the possible ranges of rest mass, decay rate and mixing angle of sterile neutrinos. The mixing angle obtained is not as small as we expected before so that the oscillation of active-sterile neutrino may be visible in near future experiments. However, the existence of the sterile neutrinos cannot be treated as the major component of the dark matter in the universe.

PACS 14.60.St, 98.35.Gi, 98.65.Cw.

Keywords: Sterile Neutrinos.

Key Words: Milky Way, Galaxy Clusters

1. Introduction

The observations of neutrino oscillations indicate that active neutrinos have non-zero rest mass. This implies that they have rest mass and right-handed neutrinos should exist, which may indeed be massive sterile neutrinos. Sterile neutrino is a kind of neutrinos which does not interact with ordinary matter except through gravity. They have been proposed to be a dark matter candidate even though they have not been observed directly.

[*] E-mail address: mhchan@phy.cuhk.edu.hk
[†] E-mail address: mcchu@phy.cuhk.edu.hk

In this chapter, we first review some basic physics of sterile neutrinos, including their possible formation mechanisms. Then, we discuss the possibility of the existence of sterile neutrinos as well as their observational consequences in the Milky Way and galaxy clusters. Also, we will obtain the ranges of rest mass m_s, radiative decay rate Γ and the mixing angle $\sin^2 2\theta$ of the sterile neutrinos from the observational bounds.

1.1. Production Mechanism of Sterile Neutrinos

Two production mechanisms of sterile neutrinos have been proposed: (1) non-resonant mechanism and (2) resonant mechanism. The non-resonant production mechanism is first proposed by Dodelson and Widrow [9]. They assume that the sterile neutrinos were produced by active-sterile neutrino oscillation $\nu_a \longleftrightarrow \nu_s$, which was driven by finite-temperature matter effects at the Big Bang Nucleosynthesis (BBN) epoch, with temperature of order 100 MeV. The ratio of the mean momentum of these sterile neutrinos to temperature $< p/T >$, is approximately equal to 1. By using the Boltzmann equation for sterile neutrinos, we have [9]

$$\left(\frac{\partial}{\partial t} - HE\frac{\partial}{\partial E}\right) f_s(E,t) = \left(\frac{1}{2}\sin^2[2\theta(E,t)]\Gamma_r(E,t)\right) f_a(E,t), \tag{1}$$

where E is the energy of active neutrinos, Γ_r is the collision rate between sterile and active neutrinos, θ is the mixing angle, H is the Hubble parameter, and f_s and f_a are the distribution functions of the sterile and active neutrinos respectively. The mixing angle and the collision rate are given by [9]

$$\sin^2 2\theta \approx \frac{\mu_a^2}{\mu_a^2 + [(26\Gamma_r E/m_s) + (m_s/2)]^2} \tag{2}$$

and

$$\Gamma_r \approx \frac{7\pi}{24} G_F^2 T^4 E, \tag{3}$$

where μ_a is a parameter, T is the temperature and G_F is the Fermi constant. Integrating Eq. (1) over momenta and using $\partial f_s/\partial t = -HT\partial f_s/\partial T$, we have

$$f_s \approx 2\left(\frac{\mu_a}{1\text{ eV}}\right)^2\left(\frac{1\text{ keV}}{m_s}\right) f_a. \tag{4}$$

Writing the above equation in terms of the cosmological density parameter of sterile neutrinos Ω_s, we have

$$\Omega_s h^2 \approx 0.2\left(\frac{\mu_a}{0.1\text{ eV}}\right)^2, \tag{5}$$

where h is the dimensionless Hubble parameter.

only on page 2!

In this mechanism, Ω_s does not depend on the sterile neutrino mass. Since $\Omega_s h^2 \leq 0.137$, μ_a should not exceed 0.1 eV. This implies that the mixing angle is very small, $\sin^2 2\theta \sim (\mu_a/m_s)^2 \sim 10^{-8}$, for m_s of order of 1 keV.

In the second production mechanism, sterile neutrinos were produced resonantly in the presence of a large lepton asymmetry. The density matrix of neutrinos (mixed between active and sterile) can be expanded in Pauli matrices with coefficients $(P_0, \mathbf{P})$ [24, 23]:

$$\rho_\nu = \begin{pmatrix} \rho_{aa} & \rho_{as} \\ \rho_{sa} & \rho_{ss} \end{pmatrix} = \frac{P_0 I + \mathbf{P} \cdot \boldsymbol{\sigma}}{2}, \tag{6}$$

where $\boldsymbol{\sigma}$ are the Pauli matrices, and the number densities of sterile and active neutrinos are given by $n_s = (P_0 - P_z)/2$ and $n_a = (P_0 + P_z)/2$ respectively. The active-sterile oscillation with energy E in the BBN epoch is described by [24]

$$\dot{\mathbf{P}} = \mathbf{V} \times \mathbf{P} + \dot{P}_0 \hat{z} - D(P_x \hat{x} + P_y \hat{y}), \tag{7}$$

where D is the quantum damping, and the vector $\mathbf{V} \approx m_s^2 \sin 2\theta / 2E \hat{x} - (m_s^2 \cos 2\theta / 2E \pm 0.35 G_F T^3 L - 100 G_F^2 E T^4) \hat{z}$ is the effective potential of the oscillation, with L the matter-antimatter asymmetry, which includes contributions from baryons, electron-positrons, and active neutrinos [24]. The $\nu_a \longleftrightarrow \nu_s$ oscillation encounters a resonance when $V_z = 0$ while the antineutrino oscillation $\bar{\nu}_a \longleftrightarrow \bar{\nu}_s$ is suppressed, or vice versa depending on the sign of the asymmetry term in the effective potential. The resonant temperature for the neutrino system is [24]

$$T_{\text{res}} \approx 9 \left(\frac{m_s}{100 \text{ eV}} \right)^{1/2} \left(\frac{L}{0.1} \right)^{-1/4} \epsilon^{-1/4} \text{ MeV}, \tag{8}$$

where $\epsilon = E/T$. When the active neutrinos are transformed into sterile neutrinos, their number density n_a is decreasing while that of the active antineutrinos $\bar{n}_a$ is unchanged. The value of L decreases until $L = 0$ $(n_a = \bar{n}_a)$. Let the total change of $\nu_a - \bar{\nu}_a$ be ΔL_a. We then have

$$\Delta L_a = \frac{n_a - \bar{n}_a}{n_\gamma} = \frac{n_s}{n_\gamma} = \frac{n_a}{n_\gamma} F, \tag{9}$$

where n_s and n_γ are the number density of sterile neutrinos and photons respectively, and $F = n_s / n_a$. Since $n_a / n_\gamma = 3/4$, the cosmological density parameter of sterile neutrino is given by

$$\Omega_s h^2 = F \left(\frac{m_s}{92 \text{ eV}} \right) = 0.73 \left(\frac{m_s}{1 \text{ keV}} \right) \left(\frac{L}{0.1} \right). \tag{10}$$

In this mechanism, Ω_s depends on m_s. For $\Omega_s h^2 \leq 0.137$, $m_s L \leq 0.019$ keV. Also, the mixing angle in this mechanism is $\sin^2 2\theta \geq 10^{-9}$ [24].

1.2. Low Reheating Temperature Scenario

In the above discussion, we follow the standard cosmology, which assumes that the inflation was immediately followed by a reheating process, which is the radiation dominated era with a reheating temperature T_R higher than 100 MeV. Therefore, most of the neutrinos (and sterile neutrinos) were produced during the radiation dominated era. However, several recent articles have proposed that the reheating process may not be started immediately after inflation, and it could be dominated by a component other than radiation [13, 12, 11]. This component may be some unstable massive particle species. During this epoch, it reaches a maximum temperature $T_{\max}$ and the Hubble expansion rate scales like $H \sim T^4 / T_R^2$ instead of the radiation dominated one $H \sim T^2$. Therefore, the expansion is faster for smaller reheating temperature. When the temperature decreases from $T_{\max}$ to T_R, the universe enters the radiation phase.

Both resonant and non-resonant production mechanisms require the temperature of the thermal bath to be $10 - 100$ MeV. If the reheating temperature is low enough, say, 5 MeV, the universe expands in a faster rate than in the radiation phase, which decreases the thermal scattering rate. Therefore, there will be less the active and sterile neutrinos produced than in the standard scenario. Using [26]

$$H = \frac{3T^4}{T_R^2} m_{\text{Planck}} \left(\frac{8\pi^3 g_*}{90} \right)^{1/2} \tag{11}$$

in Eq. (1), where g_* is the effective number of degrees of relativistic freedom, one obtains [12]

$$f_s = 3.2 d_a \left(\frac{T_R}{5 \text{ MeV}} \right)^3 \left(\frac{E}{T} \right) \sin^2 2\theta f_a, \tag{12}$$

where $d_a = 1.13$ for electron neutrinos and $d_a = 0.79$ for muon and tau neutrinos. Integrating Eq. (12) over momenta, we get

$$n_s = 10 d_a \sin^2 2\theta \left(\frac{T_R}{5 \text{ MeV}} \right)^3 n_a. \tag{13}$$

Therefore, the cosmological density parameter in the non-resonant production mechanism in the low reheating temperature scenario is given by [12]

$$\Omega_s h^2 = 0.1 d_a \left(\frac{\sin^2 2\theta}{10^{-3}} \right) \left(\frac{m_s}{1 \text{ keV}} \right) \left(\frac{T_R}{5 \text{ MeV}} \right)^3. \tag{14}$$

For the resonant production mechanism, we do not have a simple analytic expression relating n_s and n_a. Numerical calculations indicate that the number density of active electron neutrinos produced in the low reheating temperature scenario is suppressed by a factor of 0.1 (0.027 for muon and tau neutrinos) if $T_R = 1$ MeV [13]. Therefore, the allowed range for m_s or $\sin^2 2\theta$ can be larger by around 10 times.

From Eq. (14), we can notice that $\sin^2 2\theta$ needs not be very small in this scenario. For $\Omega_s h^2 \leq 0.137$, $\sin^2 2\theta$ can be as large as 10^{-3} for $m_s = 1$ keV. The large mixing angle lies within the experimental bound in β decay experiments if the mass-squared difference $\Delta m^2 \sim \text{keV}^2$ [11]. If $\sin^2 2\theta \sim 0.001 - 0.1$, the sterile neutrinos are potentially visible in near future experiments [12].

2. Radiative Decay of Sterile Neutrinos

Sterile neutrinos may decay. There are many possible decay channels, such as radiative channel $\nu_s \to \nu_a + \gamma$ and active neutrino channel $\nu_s \to 3\nu_a$. If $m_s \geq 1$ MeV, sterile neutrinos may also decay into electron-positron pairs. In the following, we only consider

$m_s \leq 1$ MeV. The decay rates of sterile neutrinos by radiative and active neutrino channels are respectively given by [1]

$$\Gamma_\gamma = \frac{9\alpha G_F^2}{1024\pi^4} \sin^2 2\theta m_s^5 = 1.38 \times 10^{-22} \sin^2 2\theta \left(\frac{m_s}{1 \text{ keV}}\right)^5 \tag{15}$$

and

$$\Gamma_{3\nu} = \frac{8\pi}{27\alpha} \Gamma_\gamma = 1.77 \times 10^{-20} \sin^2 2\theta \left(\frac{m_s}{1 \text{ keV}}\right)^5, \tag{16}$$

where α is the electromagnetic fine-structure constant. The total decay rate of sterile neutrinos is $\Gamma = \Gamma_\gamma + \Gamma_{3\nu} \approx \Gamma_{3\nu}$. In the standard cosmology, the mixing angles of sterile neutrinos in the two production mechanisms are very small, $\sin^2 2\theta \sim 10^{-8}$, so that even for $m_s \sim 10$ keV, $\Gamma \sim 10^{-23}$ s^{-1}. This decay lifetime is very long compared to the age of the universe t_0 ($\Gamma^{-1} \gg t_0$), and most of the sterile neutrinos are stable. Nevertheless, in the low reheating temperature scenario, the mixing angle can be as large as 10^{-3}. For $m_s \sim 10$ keV, Γ can be of order 10^{-18} s^{-1}. Therefore, a significant amount of sterile neutrinos may have decayed into active neutrinos and emitted photons to heat up the matter in the universe. This effect may be observable in galaxy clusters or even in Milky Way Galaxy. In the following sections, we discuss how this large decay rate can solve the high temperature problem in Milky Way Galaxy and the low cooling flow problem in galaxy clusters. Also, we will obtain the allowed ranges on rest mass, decay rate and mixing angle of sterile neutrinos by using observational data.

2.1. Decaying Sterile Neutrinos in Milky Way Galaxy

Recently, a large amount of diffuse X-ray data have been obtained by Chandra and XMM-Newton [17, 20, 15], giving a complex picture near the Milky Way center. The data indicate that there exists an extremely high temperature ($\sim$ keV) hot gas near the Milky Way center. Sakano et al. analysed the XMM-Newton data to get 1 and 4 keV hot-gas components in Sgr A East, a supernova remnant located close to the Milky Way center [20]. Later Muno et al. used the data from Chandra to model the temperatures of the two components within 20 pc as 0.8 and 8 keV [15]. The origin of such high temperature hot gas is still a mystery. About 3×10^{36} erg s^{-1} is needed to maintain the temperature of the soft component of hot gas ($T_1 \sim 0.8$ keV), which is equivalent to 1% of the kinetic energy by one supernova explosion in every 10^5 yr near the Milky Way center [15]. Since the Milky Way supernova rate is thought to be about 1 per 100 yr, the expected rate within 20 pc of the Milky Way is adequate for powering soft component of hot gas. However, the power needed to sustain the high temperature of the hard component ($T_2 \sim 8$ keV) is about 10^{40} erg s^{-1}, which is too high to be explained by supernova explosions alone. Muno et al. suggested that the magnetic reconnection that is driven by the turbulence that supernovae generate in the interstellar medium may provide such a high power. The temperature that can be heated by this mechanism is given by [15]

$$T \sim \frac{B^2}{8\pi n}, \tag{17}$$

where B and $n \sim 0.1$ cm^{-3} are the magnetic field strength and number density of the hot gas in the Milky Way center respectively. For $T \sim 8$ keV, we need $B \sim 0.2$ mG. However, such a high magnetic field strength in the Milky Way center has not been confirmed yet. Moreover, the emission of the hard component is distributed much more uniformly than the soft component, and the intensity of the two components are correlated, which suggest that they are produced by related physical processes [15].

Recently, the Suzaku X-ray mission has observed emission lines above 6 keV near the Milky Way center [14, 16]. Prokhorov and Silk find an excess in Lyγ (8.7 keV) photons which cannot be explained by ionization and recombination processes [18]. The excess Lyγ intensity is $(1.1 \pm 0.6) \times 10^{-5}$ ph cm^{-2} s^{-1} [18].

In the following, we proposed that there exists a decaying sterile neutrino halo in the Milky Way center. The sterile neutrinos in the halo decay into active neutrinos and emit photons to heat up the surrounding gas cloud and constitute the excess Lyγ photons. Since the energy of the decayed photons $E_s \approx m_s/2$, the rest mass of the sterile neutrinos should be around 17.4 keV. The existence of a small size keV sterile neutrino halo is first suggested by Bilic et al. [3]. The radius of a self-gravitating degenerate neutrino halo is given by [2]

$$R_s = 0.012 \left(\frac{M_s}{10^6 M_\odot} \right)^{-1/3} \left(\frac{m_s}{16 \text{ keV}} \right)^{-8/3} \text{pc}, \tag{18}$$

where M_s is the total mass of the halo. The halo is assumed to be located near the center of a 20-pc radius gas cloud. The major heating mechanism is the bound-free collisions between the decayed photons and the ions in the cloud. Since the intensity of the excess Lyγ photons is not so strong, the optical depth τ for the decayed photons must be greater than 1. Near the center of the gas cloud, which is an optically thick region $\tau \geq 1$, the hot gas particles are in photoionization equilibrium. Then, the energy absorbed from the decayed photons is subsequently transferred to the surrounding gas [6]. The power loss of the hot gas is mainly due to adiabatic expansion [1]:

$$\dot{W} = PV^{2/3} c_s = PV^{2/3} \left(\frac{\gamma T}{m_g} \right)^{1/2}, \tag{19}$$

where P, V, c_s, γ and m_g are the pressure, volume, sound speed, adiabatic index of the hot gas and mean mass of the hot gas particles, respectively. If $T = 8$ keV within 20 pc, then $\dot{W} = 10^{40}$ erg s^{-1}, which is the energy needed to maintain the hard component. Therefore, the total energy emitted by the decayed photons must be greater than or equal to the cooling rate of the hard component:

$$\sum_i N_s \Gamma_\gamma (E_s - E_i) P_i \geq 10^{40} \text{ erg s}^{-1}, \tag{20}$$

where N_s is the total number of sterile neutrinos in the halo, E_i and P_i are the ionization potential and probability of photon absorption by ith type ions in the hot gas:

$$P_i = \frac{a_i \sigma_i}{\sum_i a_i \sigma_i}, \tag{21}$$

[1]The power loss by bremsstrahlung radiation is only 0.3 % of the adiabatic cooling [15].

with a_i the ratio of the number of ith type ions to the total number of ions at 0.8 keV temperature [2]. The absorption cross-section is largest for H-like and He-like ions, which is given by [8]

$$\sigma_i = 10^{-18}\sigma_{\mathrm{th},i}\left[\alpha_i\left(\frac{E_{\mathrm{th},i}}{E_s}\right)^{s_i} + (1-\alpha_i)\left(\frac{E_{\mathrm{th},i}}{E_s}\right)^{s_i+1}\right]\ \mathrm{cm}^2, \qquad (22)$$

where σ_i, α_i, $E_{\mathrm{th},i}$ and s_i are fitted parameters of a particular ith type ion (see Table 1). The total effective cross-section $\sum_i a_i\sigma_i$ depends on the metallicity Z_{metal} near the Milky Way center. The metal abundances we have used in the calculation are shown in Table 2 [20, 6].

Table 1. The parameters we have used in Eq. (22) [8].

Z	$\sigma_{\mathrm{th},i}$ $(10^{-18}\ \mathrm{cm}^{-2})$	$E_{\mathrm{th},i}$ (eV)	α_i	s_i
6(H-like)	0.194	490	1.287	2.95
6(He-like)	0.526	392	1.325	2.76
7(H-like)	0.142	666	1.287	2.95
7(He-like)	0.371	552	1.314	2.79
8(H-like)	0.109	870	1.287	2.95
8(He-like)	0.275	739	1.308	2.81
10(H-like)	0.075	1360	1.25	2.90
10(He-like)	0.18	1195	1.28	2.95
12(H-like)	0.055	1960	1.22	2.90
12(He-like)	0.13	1761	1.25	2.90
14(H-like)	0.044	2670	1.20	2.90
14(He-like)	0.10	2430	1.20	3.0
16(H-like)	0.035	3480	1.19	2.90
16(He-like)	0.08	3210	1.20	3.0
18(H-like)	0.056	4426	1.20	2.99
18(He-like)	0.060	4121	1.20	2.98
20(H-like)	0.045	5470	1.20	3.04
20(He-like)	0.048	5129	1.20	3.02
26(H-like)	0.027	9278	1.20	3.15
26(He-like)	0.028	8828	1.20	3.14

The upper limit of the excess Lyγ photons within the field of view of Suzaku ($R' \approx 20$ pc) near the Milky Way center is 7×10^{37} erg s^{-1} [18]. Therefore, we have

$$N_s\Gamma_\gamma E_s e^{-\tau} \approx 7 \times 10^{37}\ \mathrm{erg\ s}^{-1}, \qquad (23)$$

[2]The number density of the soft component is about 10 times that of the hard component if they are in equilibrium [15].

Table 2. Metal abundances we have used in the model [6].

Element	Atomic number Z	Metallicity Z_{metal}	$a_i(10^{-4})$(H-like)	$a_i(10^{-4})$(He-like)
C	6	3	6.9	2.7
N	7	3	1.7	0.84
O	8	3	16	9.6
Ne	10	3	2.5	1.9
Mg	12	3	1.0	0.89
Si	14	8.9	3.0	2.8
S	16	2.7	0.43	0.42
Ar	18	1.8	0.066	0.066
Ca	20	2.5	0.056	0.056
Fe	26	3.8	1.8	1.8

where the optical depth τ is given by

$$\tau = \int_{R_s}^{R'} n(r) \sum_i a_i \sigma_i dr. \tag{24}$$

The mass density profile of Milky Way center is nearly isothermal ($\rho \sim r^{-1.8}$) and with a core radius of $r_c = 0.34$ pc [22]; therefore the number density of hot gas near the Milky Way center is given by

$$n(r) = \frac{n_0}{(1 + r/r_c)^{1.8}}. \tag{25}$$

Including the cold molecular gas, warm atomic gas and hot gas in the Milky Way center, the total optical depth is $\tau \approx 1.25 \sum_i a_i \sigma_i r_c \approx 4.3 \pm 0.6$ [7]. Therefore, the predicted power of excess 8.7 keV photons should be $10^{40} e^{-\tau} = (7 - 20) \times 10^{37}$ erg s^{-1}, which is consistent with the observed power (7×10^{37} erg s^{-1}).

On the other hand, orbital data of stars S1 and S2 constrain $M_s \leq 2 \times 10^5 M_\odot$ [22]. From Eq. (20), $N_s \Gamma_\gamma (E_s - E_i) \approx N_s \Gamma_\gamma m_s/2 = M_s \Gamma_\gamma /2$. Therefore, we get $\Gamma_\gamma \geq 5 \times 10^{-20}$ s^{-1}. Since $\Gamma \approx \Gamma_{3\nu} \approx 128 \Gamma_\gamma$, we have $\Gamma \geq 6 \times 10^{-18}$ s^{-1} and $\sin^2 2\theta \geq 2 \times 10^{-4}$, which does not match the requirements of both standard resonant and non-resonant production mechanisms. Nevertheless, it is allowed in the low reheating temperature scenario.

3. Decaying Sterile Neutrinos in Galaxy Clusters

Recent observations indicate that hot gas exists in most galaxy clusters [21]. The temperature of the hot gas T_g lies between 1-10 keV ($T_g = 10^7 - 10^8$ K). The X-ray luminosities of the galaxy clusters can reach $10^{43} - 10^{44}$ erg s^{-1}. The cooling time scale of the galaxy clusters can be defined as

$$t_{\text{cooling}} = \left(\frac{d \ln T_g}{dt} \right)^{-1}. \tag{26}$$

If the hot gas cools isobarically, the cooling time is [21]

$$t_{\text{cooling}} = 8.5 \times 10^{10} \left(\frac{n_g}{10^{-3}\ \text{cm}^{-3}} \right)^{-1} \left(\frac{T_g}{10^8\ \text{K}} \right)^{1/2}, \tag{27}$$

where n_g is the number density of the hot gas. This cooling time scale is of the same order of magnitude as the ages of many galaxy clusters. However, at the centers of some galaxy clusters, the cooling time is shorter than their ages (Hubble time), and therefore these galaxy clusters should have cooling flows, with flow rates of more than $100 M_\odot$ yr^{-1} in order to match the observational data [25]. However, recent X-ray observations failed to reveal such high rate of cooling flows. Therefore, some heating sources are needed to maintain the high temperature of the hot gas and suppress the cooling flows. In the following, we assume that the decayed photons from sterile neutrinos provide energy to heat up the surrounding gas and obtain ranges of parameters that match the observational data. In this model, the sterile neutrinos can be hidden deeply inside galaxies (just like the sterile neutrino halo near the Milky Way center in the previous section). The energy gained by the hot gas is then transferred to the surrounding gas particles by conduction [5]. In galaxy clusters, the mean free path for conduction is [21]

$$\bar{l} = 23 \left(\frac{T_g}{10^8\ \text{K}} \right)^2 \left(\frac{n_g}{10^{-3}\ \text{cm}^{-3}} \right)^{-1} \text{kpc}, \tag{28}$$

which is smaller than the length scale of a typical galaxy cluster (1 Mpc).

Suppose the energy emitted by the decaying sterile neutrinos is equal to the energy loss of a galaxy cluster (X-ray luminosity). We have

$$L_X = \frac{1}{2} M_{sc} \Gamma_\gamma, \tag{29}$$

where M_{sc} is the total mass of sterile neutrinos in the galaxy cluster at redshift z. The initial mass of sterile neutrinos in the galaxy cluster is $M_{s0} = M_{sc} e^{\Gamma t(z)}$. Assume that the initial ratio of the cosmological density parameters of sterile neutrinos to total mass is a constant for all galaxy clusters ($\Omega_s/\Omega_m = M_{s0}/M_t =$ constant), where M_t is the total mass of a galaxy cluster. Eq. (29) then becomes

$$L_X = \frac{1}{2} \frac{\Omega_s M_t e^{-\Gamma t(z)} \Gamma_\gamma}{\Omega_m}, \tag{30}$$

or

$$\ln \left(\frac{L_X}{M_t} \right) = \text{constant} - \Gamma t(z). \tag{31}$$

In the ΛCDM model, $t(z)$ is given by

$$t(z) = \frac{2}{3H_0\sqrt{1-\Omega_m}} \ln \left[\left(\frac{1+z_{\text{m}\Lambda}}{1+z} \right)^{3/2} + \sqrt{1 + \left(\frac{1+z_{\text{m}\Lambda}}{1+z} \right)^3} \right], \tag{32}$$

where $z_{\text{m}\Lambda} = (\Omega_\Lambda/\Omega_m)^{1/3} - 1$ and Ω_Λ is the cosmological density parameter of dark energy. By plotting $\ln(L_X/M_t)$ against $t(z)$ for those low cooling flow galaxy clusters, a

Table 3. X-ray Luminosities(L_X), hot gas masses(M_g), total masses(M_t) and redshifts(z) of 12 clusters [19].

Cluster	L_X(0.01-40 keV)(10^{44} ergs^{-1})	M_g (10^{14} $M_\odot$)	M_t (10^{14} $M_\odot$)	z
A 119	3.814	0.42	8.74	0.0440
A 133	2.749	0.17	4.79	0.0569
A 399	9.083	1.2	11.6	0.0715
A 401	17.38	0.64	11.58	0.0748
A 754	6.106	0.63	15.67	0.0528
A 539	0.987	0.08	4.31	0.0288
A 1367	1.092	1.30	5.77	0.0216
A 1775	2.926	1.20	5.86	0.0757
A 2065	6.261	0.53	14.44	0.0721
A 2255	6.999	3.45	12.53	0.0800
A 2256	11.588	3.65	13.81	0.0601
A 2634	0.930	2.20	5.55	0.0312

straight line should be obtained, and Γ can be found by the slope. Fig. 1 shows the relation between $\ln(L_X/M_t)$ and $t(z)$ for 12 low cooling flow galaxy clusters. The decay rate $\Gamma = (6 \pm 3) \times 10^{-17}$ s^{-1} (95 % confidence level). Due to the large uncertainties in M_t, L_X and z of the galaxy clusters, we may only conclude that $\Gamma \sim 10^{-17}$ s^{-1}. In this model, E_s must be greater than T_g, or else the energy of the decayed photons cannot be transferred to hot gas particles. The temperatures of the 12 low cooling flow galaxy clusters range from $3 - 9$ keV. Therefore, it requires $m_s \geq 18$ keV. From Eq. (15), the mixing angle $\sin^2 2\theta$ can be as large as $10^{-4} - 10^{-3}$. This result agrees with the low reheating temperature scenario again.

4. Conclusion

We have discussed possible production mechanisms of sterile neutrinos. By using two simple models in solving the high temperature problem in the Milky Way center and low cooling flow problem in galaxy clusters, we get possible ranges of the rest mass, decay rate and mixing angle of the sterile neutrinos. Combining the results of the above sections, we have $m_s \approx 17 - 18$ keV, $\Gamma \sim 10^{-17}$ s^{-1} and $\sin^2 2\theta \sim 10^{-4} - 10^{-3}$. These parameters are consistent with current upper bounds from various experiments and agree with the prediction in the low reheating temperature scenario [11]. From Eq. (14), the reheating temperature should be around 2 MeV. On the other hand, by using the recent analysis of the X-ray background from HEAO-1 and XMM-Newton, one can obtain an upper bound on the present sterile neutrino cosmological density parameter, which is given by [26]

$$\Omega_{s0} \leq 1 \times 10^{-7} \left(\frac{m_s}{18\ \text{keV}}\right)^{-5} \left(\frac{\sin^2 2\theta}{10^{-4}}\right)^{-1}. \quad (33)$$

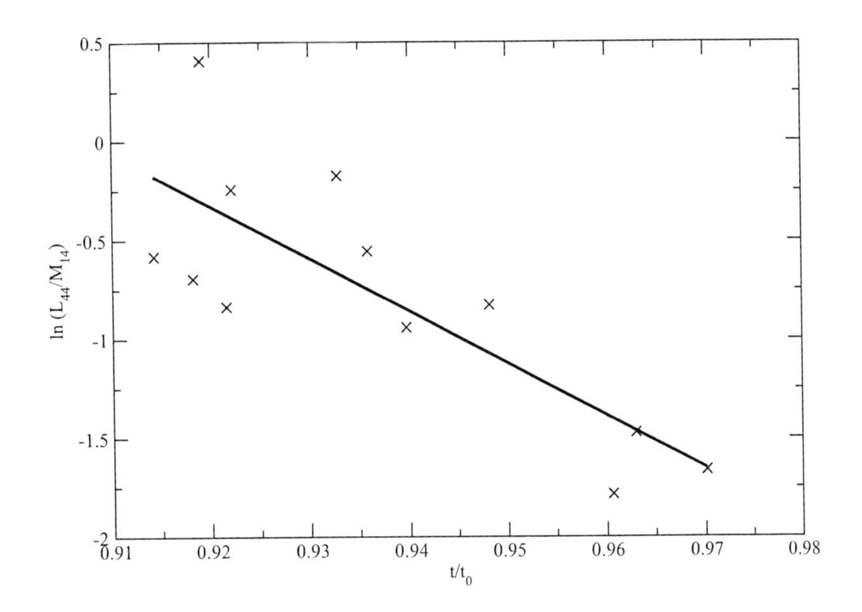

Figure 1. $\ln(L_{44}/M_{14})$ vs. t/t_0 for 12 different galaxy clusters, where $L_{44} = L_X/10^{44}$ ergs^{-1}, $M_{14} = M_t/10^{14}M_\odot$ and t_0 is the age of the universe since the Big Bang. The solid line is the best-fit line, with $\Gamma_{3\nu} = (6 \pm 3) \times 10^{-17}$ s^{-1} (95 % confidence level).

Therefore, $\Omega_{s0} \ll \Omega_m$, and sterile neutrinos is only a minor component of the dark matter. This work is partially supported by grants from the Research Grant Council of the Hong Kong Special Administrative Region, China (Project Nos. 400805 and 400910).

References

[1] Barger, V., Phillips, R. J. N. and Sarkar, S. 1995, *Phys. Lett. B,* **352**, 365.

[2] Bilic, N. *et al.* 2001, *Phys. Lett. B,* **515**, 105.

[3] Bilic, N., Munyaneza, F. and Viollier, R. D. 1999, *Phys. Rev. D,* **59**, 024003.

[4] Carroll, B. W. and Ostlie, D. A. 2007, *An Introduction to Modern Astrophysics,* (US: Pearson Addison Wesley).

[5] Chan, M. H. and Chu, M. -C. 2007, *ApJ,* **658**, 859.

[6] Chan, M. H. and Chu, M. -C. 2008, *MNRAS*, **389**, 297.

[7] Chan, M. H. and Chu, M. -C. 2011, *ApJ*, **727**, L47.

[8] Daltabuit, E. and Cox, D. P. 1972, *ApJ*, **177**, 855.

[9] Dodelson, S and Widrow, L. M. 1994, *Phys. Rev. Lett.*, **72**, 17.

[10] Ferriere, K. M. 2001, *Rev. Mod. Phys.,* **73**, 1031.

[11] Gelmini, G. 2005, *Int. J. Mod. Phys. A*, **20**, 4670.

[12] Gelmini. G., Palomares-Ruiz, S. and Pascoli, S. 2004, *Phys. Rev. Lett.* **93**, 081302.

[13] Giudice, G. F. *et al.* 2001, *Phys. Rev. D*, **64**, 043512.

[14] Koyama, K. *et al.* 2007, PASJ, **59**, 245.

[15] Muno, M. P. *et al.* 2004, ApJ, **613**, 326.

[16] Nobukawa, M. *et al.* 2010, *PASJ*, **62**, 423.

[17] Park, S. *et al.* 2003, *ApJ*, **603**, 548.

[18] Prokhorov, D. A. and Silk, J. 2010, *ApJ*, **725**, L121.

[19] Reiprich, T. H. and Böhringer, H. 2002, *ApJ*, **567**, 716.

[20] Sakano, M., Warwick, R. S., Decourchelle, A. and Predehl, P. 2004, *MNRAS*, **350**, 129.

[21] Sarazin, C. L. 1988, *X-Ray Emission from Cluster of Galaxies* (New York: Cambridge Univ. Press).

[22] Schödel, R. *et al.* 2002, *Nature*, **419**, 694.

[23] Shi, X. 1996, *Phys. Rev. D*, **54**, 2753.

[24] Shi, X. and Fuller, G. M. 1999, *Phys. Rev. Lett.*, **82**, 2832.

[25] Totani, T. 2004, *Phys. Rev. Lett.* **92**, 191301.

[26] Yaguna, C. E. 2007, *JHEP*, **0706**, 002 (hep-ph/0706.0178).

In: Neutrinos: Properties, Sources and Detection
Editor: Joshua P. Greene, pp. 145-150
ISBN 978-1-61209-650-6

Chapter 6

TOWARD A BETTER EVALUATION OF NEUTRINO-NUCLEUS REACTION CROSS SECTIONS

***Toshio Suzuki*[1,2,*] *and Michio Honma*[3]**
[1] Department of Physics and Graduate School of Integrated Basic Sciences, College of Humanities and Sciences, Nihon University, Sakurajosui 3-25-40, Setagaya-ku, Tokyo 156-8550, Japan,
[2] Center for Nuclear Study, University of Tokyo, Hirosawa, Wako-shi, Saitama, 351-0198, Japan
[3] Center for Mathematical Sciences, University of Aizu Aizu-Wakamatsu, Fukushima 965-8580, Japan

Abstract

An improvement of the evaluation of GT transitions in neutrino-nucleus reactions is done. The transition strengths at each finite-momentum transfers are obtained by shell model diagonalizations, and used to evaluate the neutrino-induced reaction cross sections. The method is applied to charge-exchange reaction on ^{56}Fe, and reliable reaction cross sections are obtained up to $E_\nu \cong 100$ MeV.

PACS 21.60.Cs, 25.30.Pt, 25.40.Kv.

Keywords: shell model, neutrino-nucleus reactions, Gamow-Teller strength.

Since the first historic detection of supernova neutrinos from Magellanic Cloud, SN1987A, at Kamioka[1] and IMB[2], further observations of supernova neutrinos from our Galaxy as well as outer galaxies became one of possible powerful tools to reveal more of the properties of neutrinos, in particular, the oscillation properties. The SNO heavy water detector succeeded to observe neutrino oscillations by measuring both the charged and neutral-current reactions on deuterium[3]. It is important to use heavier targets such as lead and iron, since the neutrino-nucleus reaction cross sections increase as the target mass increases. Accurate evaluations of the neutrino-nucleus cross sections on these nuclei is

*E-mail address: suzuki@chs.nihon-u.ac.jp

indispensable to design appropriate experimental set-ups and also to draw reliable information from the observables. It is also important to investigate nucleosynthsis by neutrino processes in supernova explosions[4].

Direct experimental information on neutrino-nucleus reactions is quite few. Reactions on deuteron are measured at SNO using solar neutrinos. Laboratory experiments using accelerators have been done only for a very few targets; on ^{12}C at LSND[5] and ^{56}Fe at KARMEN[6]. Reactions induced by decay-at-rest (DAR) and decay-in-flight (DIF) neutrinos were measured for ^{12}C. Total charge-exchange reaction cross section on ^{56}Fe induced by DAR ν_e neutrinos were obtained.

Many theoretical works on neutrino-nucleus reactions have been done for ^{12}C[7, 8], ^{56}Fe[4, 9, 10, 11, 12] and ^{208}Pb[10, 12, 13] targets. Shell model studies of neutrino induced reactions on ^{12}C and ^{4}He have been done with improved shell model Hamiltonians which describe well the spin properties of the nuclei, and applied to nucleosynthesis of light elements in supernova explosions[7]. In heavier nuclei, calculations by random-phase-approximation (RPA) and quasi-particle RPA (QRPA), both non-relativistic and relativistic versions, have been done. A hybrid model, in which the Gamow-Teller (GT) transitions are treated by shell-model while RPA-type calculations are employed for other multipoles, is another method of calculations often used. The reason why other multipoles except for 1^+ are not treated by the shell-model is just because it is still hard to carry out calculations in larger configuration space due to the limitation of computational abilities. It is obvious that shell-model is absolutely better to obtain more accurate reaction cross sections as important many-body correlations are taken into account. It is rather difficult to take full acount of correlations beyond one-particle one-hole correlations in case of RPA-type calculations. Due to recent advances in shell model studies of fp-shell nuclei with new shell model Hamiltonians[14, 15], it is now possible to make more reliable evaluations of GT transition strengths in the nuclei than before. In ref.[11], new cross sections are applied to re-evaluate abundances of heavy elements such as Mn in population III stars.

Here, we restrict ourselves to shell model calculations of GT transitions. Even in this case, it is usually necessary to make an approximation for heavy nuclei. A method usually used when we have GT transition strength but do not have wave functions in heavy nuclei is to just put the transition strength in cross section formulae. The GT strength obtained by Lanczos method in shell model calculations[16] is defined by

$$B(GT) = \frac{1}{2J_i + 1} | < f||\sigma t_-||i > |^2. \tag{1}$$

This is a quantity at zero momentum transfer. In reactions, however, the transitions occur at finite-momentum transfers, $q \neq 0$, and we needd to make a correction to take into this fact. One way is to reduce the cross section by multiplying a factor which is determined by comparing the corresponding RPA calculations at finite and zero-momentum transfers[9]. The latter is obtained by taking q to be zero in $j_0(qr)$ and $j_2(qr)$ in the formula for calculations of transition matrix elements. In this method, we need to evaluate cross sections in other models, that is, in RPA. Another way is to replace $j_0(qr)$ by $j_0(qR)$ where R is the radius of the nucleus: $< f||j_0(qr)\sigma t_-||i > \rightarrow < f||j_0(qR)\sigma t_-||i >$ in the evaluation of the transition matrix elements (see eq. (2))[11]. Both the methods give similar approximations for low energy reactions such as DAR neutrino cases. In the latter method, the accuracy of the approximation becomes gradually worse as the incident neutrino energy increases and the momentum transfer becomes as large as $q \sim \pi/R$; $q \sim 1$ fm^{1} for $R \sim 3$ fm.

In this chapter, we will improve the method to be applicable for high energy and large q cases. The following transition matrix elements are evaluated to get neutrino-nucleus reaction cross sections;

$$M_a = \sqrt{\frac{2}{3}} < f||j_0(qr)[Y^0 \times \sigma]^1 t_-||i> - \sqrt{\frac{1}{3}} < f||j_2(qr)[Y^2 \times \sigma]^1 t_-||i>$$
$$M_b = \sqrt{\frac{1}{3}} < f||j_0(qr)[Y^0 \times \sigma]^1 t_-||i> + \sqrt{\frac{2}{3}} < f||j_2(qr)[Y^2 \times \sigma]^1 t_-||i> \quad (2)$$

Here, both terms involving $j_0(rq)$ and $j_2(qr)$ are evaluated for each q up to q =1.5 fm^{-1}, and the two transition strengths in eq. (2) are calculated by Lanczos method in the shell model calculations. The size parameter and oscillator frequency for radial wave functions are determined from the experimental charge radius, R_c =3.75 fm, of the nucleus[17]; b =1.960 fm and $\hbar\omega$ = 10.80 MeV.

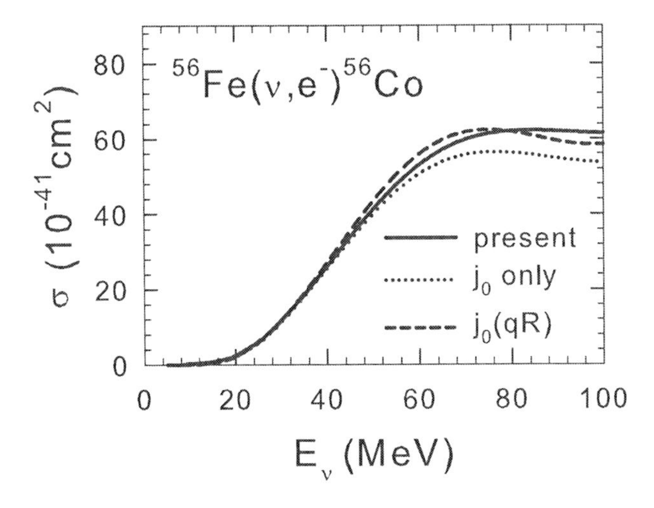

Figure 1. Calculated reaction cross sections for ^{56}Fe (ν, e^-) ^{56}Co induced by neutrinos with energy E_ν. Dotted points denote values obtained only with $j_0(qr)$ terms. Dashed line shows the result of the approximation using $j_0(qR)$.

Calculated cross sections obtained in the present method are shown in Fig. 1 for ^{56}Fe (ν, e^-) ^{56}Co. The axial-E1 (E_1^5), magnetic dipole (M1), axial-Coulomb (C_1^5) and axial-longitudinal (L_1^5) components as well as the interference term between the E_1^5 and M1 parts are included in the calculation[18]. The E_1^5 compoent gives the dominant contributions. The E_1^5 and M1 involve the matrix element M_a while the C_1^5 and L_1^5 are concerned with the matrix element M_b in eq. (2). A shell model Hamiltonian in fp-shell, GXPF1J[14] is used for the shell model calculations. A unversal quenching of g_A^{eff}/g_A =0.74 is adopted[19].

As we see from Fig. 1, the contributions from $j_2(qr)$ terms are not negligible at higher energies. The present results can be considered to be good enough at $E_\nu \leq 100$ MeV. The method using $j_0(qR)$ looks to be rather good up to ~80 MeV. The nuclear radius is taken to be R = $f_R \times R_c$ with f_R =1.0~ 1.05. A case for f_R =1.03 is shown in Fig. 1.

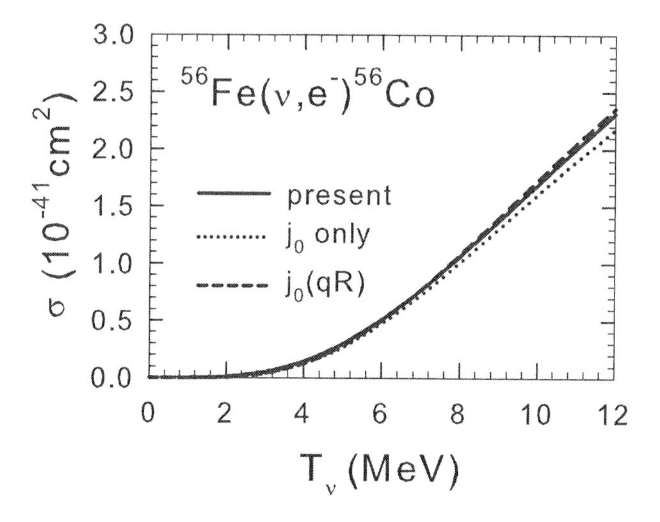

Figure 2. Calculated reaction cross sections for ^{56}Fe (ν, e^-) ^{56}Co induced by supernova neutrinos with temperature T_ν.

Table 1. Comparison of calculated total cross sections for ^{56}Fe (ν, e^-) ^{56}Co by the present method with the approximation using $j_0(R)$: cases of R $=R_c$, 1.03$\times R_c$ and 1.05 $\times R_c$ are shown. Cross sections induced by DAR neutrinos and supernova neutrinos with the temperature T_ν = 6, 8 and 10 MeV as well as the case of T_ν =6.26 MeV ans α=3 are given in units of 10^{-41} cm^2.

Neutrino spectrum	present method	$j_0(R)$ R=R$_c$	$j_0(R)$ R=1.03$\cdot R_c$	$j_0(R)$ R=1.05$\cdot R_c$
DAR	16.17	16.90	16.57	16.35
(T_ν, α) = (6 MeV, 0)	5.10	5.09	5.00	4.95
(T_ν, α) = (8 MeV, 0)	10.64	11.04	10.78	10.59
(T_ν, α) = (10 MeV, 0)	16.93	17.73	17.30	16.92
(T_ν, α) = (6.26 MeV, 3)	9.98	10.25	10.04	9.91

Cross sections for supernova neutrinos with temperature T_ν are given in Fig. 2. They are obtained by folding the cross sections in Fig. 1 with the Fermi-Dirac distributions with the temperature T_ν and a parameter α for the chemical potential;

$$f(E_\nu) = \frac{c(\alpha)}{T_\nu^3} \frac{E_\nu^2}{exp[(E_\nu/T_\nu) - \alpha] + 1} \tag{3}$$

with $C(\alpha)$ the normalization factor.

As we see from Fig. 2, errors in the folded cross sections are rather small compared to the difference in the cross sections in Fig. 1. Once folded, errors remain only within 2% at T_ν = 10 MeV. In Table 1, calculated cross sections for DAR (decay-at-rest) neutrinos as well as those for supernova neutrinos at certain temperatures are given. The approximation using $j_0(qR)$ works rather well within an error of 4% when folded over the neutrino spectra.

This approximation can be applied to much heavier nuclei such as lead[20]. We show in

Table 2 calculated cross sections for ^{208}Pb (ν, e^-) ^{208}Bi obtained by using the GT transition strength which includes the $2p-2h$ configuration mixing and the ground state correlations beyond the RPA[21]. Shell model calculations in large-space configurations are not yet accessible for ^{208}Pb at present. The charge radius of ^{208}Pb is taken to be 5.50 fm[17].

Table 2. The same as in Table I for ^{208}Pb (ν, e^-) ^{208}Bi in units of 10^{-40} cm^2. Only the total cross sections obtained by the approximate method are shown.

Neutrino spectrum	$j_0(R)$ R=R$_c$	$j_0(R)$ R=1.03$\cdot R_c$	$j_0(R)$ R=1.05$\cdot R_c$
DAR	20.61	19.51	18.81
(T_ν, α) = (6 MeV, 0)	11.79	11.33	11.04
(T_ν, α) = (8 MeV, 0)	15.68	14.98	14.52
(T_ν, α) = (10 MeV, 0)	18.31	17.41	16.85
(T_ν, α) = (6.26 MeV, 3)	17.44	16.68	16.19

In summary, we improved the evaluation of GT transitions in neutrino-nucleus reactions. The transition strengths at each finite-momentum transfers are obtained by shell model diagonalizations, and used to evaluate the reaction cross sections. The charge-exchange reaction cross sections on ^{56}Fe are obtained in a reliable way up to $E_\nu \cong 100$ MeV. When folded over neutrino spectra, the approximate method using $j_0(qR)$ is found to be rather good up to $T_\nu = 12$ MeV. It would be interesting to carry out more accurate mesurement of neutrino-induced reaction on ^{56}Fe. The applications of the present method to other heavy nuclei as well as treatment of other multipoles including the spin-dipoe transitions are also interesting issues for future investigations.

References

[1] K. S. Hirata, T. Kajita, M. Koshiba et al., *Phys. Rev. Lett.* **58**, 1490 (1987).

[2] R. M. Bionta et al., *Phys. Rev. Lett.* **58**, 1494 (1987).

[3] SNO Collaborations, Q. R. Ahmad et al., *Phys. Rev. Lett.* **89**, 011301 (2002); **89**, 011302 (2002); **87**, 071301 (2001).

[4] S. E. Woosley, D. H. Hartmann, R. D. Hoffman, and W. C. Haxton, *Astrophys. J.* **356**, 272 (1990).

[5] C. Athanassopoulos and the LSND Collaborations, *Phys. Rev. C* **55**, 2078 (1999).

[6] R. Maschuw, *Prog. Part. Nucl. Phys.* **40**, 183 (1998).

[7] T. Suzuki, S. Chiba, T. Yoshida, T. Kajino and T. Otsuka, *Phys. Rev. C* **74**, 034307 (2006);
T. Yoshida, T. Suzuki, S. Chiba, T. Kajino, H.Yokomakura, K. Kimura, A. Takamura and D. H. Hartmann, *Astrophys. J.* **686**, 448 (2008).

[8] A. C. Heyes and I. S. Towner, *Phys. Rev. C* **61**, 044603 (2000);
C. Volpe, N. Auerbach, G. Colo, T. Suzuki and N. Van Giai, *Phys. Rev. C* **62**, 015501 (2000).

[9] E. Kolbe, K. Langanke and G. Martinez-Pinedo, *Phys. Rev. C* **60**, 052801 (1999).

[10] E. Kolbe and K. Langanke, *Phys. Rev. C* **63**, 025802 (2001).

[11] T. Suzuki, M. Honma, K. Higashiyama, T. Yoshida, T. Kajino, T. Otsuka, H. Umeda and K.-I. Nomoto, *Phys. Rev. C* **79**, 061603(R) (2009).

[12] N. Paar, D. Vretenar, T. Marketin and P. Ring, *Phys. Rev. C* **77**, 024608 (2008).

[13] G. M. Fuller, W. C. Haxton and G. C. McLaughlin, *Phys. Rev. D* **59**, 085005 (1999):
C. Volpe, N. Auerbach, G. Colo abd N. Van Giai, *Phys. Rev. C* **65**, 044603 (2002).

[14] M. Honma, T. Otsuka, B. A. Brown, and T. Mizusaki, *Phys. Rev. C* **65**, 061301(R) (2002); **69** 034335 (2005);
M. Honma et al., *J. Phys. Conf. Ser.* **20** 7, (2005).

[15] E. Caurier, G. Martínez-Pinedo, F. Nowacki, A. Poves and P. Zuker, *Rev. Mod. Phys.* **77** 427 (2005).

[16] R. R. Whitehead, in *Moment Methods in Many Fermion Systems*, ed. by B. J. Dalton et al. (Plenum, New York, 1980), p.235.

[17] H. de Vries, C. W. de Jager and C. de Vries, *Atomic and Nuclear Data Tables* **36**, 495 (1987).

[18] J. D. Walecka, in *Muon Physics*, edited by V. H. Highes and C. S, Wu (Academic, New York, 1975), Vol. II;
J. S. O'Connell, T. W. Donnelly, and J. D. Walecka, *Phys. Rev. C* **6**, 719 (1972);
T. W. Donnelly and J. D. Walecka, *Nucl. Phys.* **A274**, 368 (1976);
T. W. Donnelly and W. C. Haxton, *Atomic Data Nucl. Data Tables* **23**, 103 (1979).

[19] G. Martínez-Pinedo, A. Poves, E. Caurier and A. P. Zuker, *Phys. Rev. C* **53**, R2602 (1996).

[20] T. Suzuki and H. Sagawa, *Nucl. Phys.* **A718**, 446c (2003).

[21] N. Dinh Dang, A. Arima, T. Suzuki and S. Yamaji, *Phys. Rev. Lett.* **79**, 1638 (1997).

In: Neutrinos: Properties, Sources and Detection
Editor: Joshua P. Greene, pp. 151-175
ISBN 978-1-61209-650-6

Chapter 7

NUCLEAR RESPONSES TO SUPERNOVA NEUTRINOS FOR THE STABLE MOLYBDENUM ISOTOPES

E. Ydrefors[a,*]***, K.G. Balasi***[b,†]***, J. Suhonen***[a,‡] ***and T.S. Kosmas***[b,§]
[a]Department of Physics,
P.O. Box 35 (YFL), FI-40014 University of Jyväskylä, Finland
[b]Department of Physics,
University of Physics, The University of Ioannina, 45110 Ioannina, Greece

Abstract

Detection of supernova neutrinos and their properties is very important for astrophysical applications. In this chapter the theoretical framework for neutral-current neutrino-nucleus scattering calculations is reviewed. We then use the formalism to calculate cross sections for the stable ($A = 92, 94, 95, 96, 97, 98, 100$) molybdenum isotopes. Both the coherent and incoherent contributions to the cross sections are considered. We use the quasiparticle random-phase approximation (QRPA) to construct the final and initial states of the even-even isotopes and for the odd isotopes the microscopic quasiparticle-phonon model (MQPM) is employed. The computed cross sections are folded with a two-parameter Fermi-Dirac distribution to obtain realistic estimates of the responses to supernova neutrinos.

PACS 23.40.Bw, 25.30.Pt, 21.60.Jz., 26.30.Jk

Keywords: Neutrino-nucleus scattering, Supernova neutrinos, Quasiparticle random-phase approximation, Microscopic quasiparticle-phonon model.

1. Introduction

Neutrinos are key particles for modern astrophysical studies because of their capability to reach terrestrial detectors from e.g. distant stars. Supernova neutrinos constitute very

*E-mail address: emanuel.ydrefors@jyu.fi
†E-mail adress: dbalasi@hotmail.com
‡E-mail address: jouni.suhonen@phys.jyu.fi
§E-mail adress: hkosmas@uoi.gr

sensitive probes of supernova explosions and neutrino properties. This was shown by the observation of neutrinos from the supernova SN1987 [7]. From the particle-physics point of view neutrino-nucleus reactions constitute important tools for investigations of fundamental neutrino properties such as e.g. neutrino masses and neutrino oscillations. Neutrinos can be detected with Earth-bound detectors via charged-current and neutral-current neutrino-nucleus scattering experiments. One possibility for such measurements is the MOON (Majorana/Mo Observatory Of Neutrinos) project [13].

Theoretical estimates of nuclear responses to neutrinos are important. In this chapter we therefore review the theoretical framework for calculations of neutral-current neutrino-nucleus scattering cross sections. In [11] the Donnelly-Walecka method for treatment of semi-leptonic processes in nuclei was written in terms of the isospin formalism. In the present chapter we review the neutrino-scattering theory in terms of the proton-neutron formalism which is commonly used in e.g. QRPA (quasiparticle random-phase approximation) calculations. Nucleon form factors play an important role in realistic calculations of cross sections and we therefore devote a special effort to discuss them. We subsequently use the formalism to perform calculations of cross sections for neutral-current neutrino-nucleus scattering off the stable ($A = 92, 94, 95, 96, 97, 98, 100$) molybdenum isotopes. In [25] we used the microscopic quasiparticle-phonon model (MQPM) [23] to construct the nuclear states of ^{95}Mo and ^{97}Mo. These wave functions are employed for the odd isotopes and for the even-even nuclei the quasiparticle random-phase approximation (QRPA) is adopted. We compute nuclear responses to supernova neutrinos by folding the cross sections with a two-parameter Fermi-Dirac distribution [16].

There is currently a large interest in double-beta decay as a tool for determining the mass and the properties of the neutrino [21]. ^{100}Mo is presently used in double-beta decay experiments such as the MOON [13] and the NEMO [5]. The large nuclear response to solar neutrinos [12] and the low threshold energy implies that ^{100}Mo is one of the most interesting nuclei for solar neutrino studies through charged-current neutrino-nucleus scattering. Studies of charged-current and neutral-current neutrino-nucleus reactions on this nucleus and the other stable molybdenum isotopes are therefore very important.

This chapter is organized as follows. In Sec. 2. the theoretical framework used for neutral-current neutrino-nucleus scattering calculations is outlined. The QRPA and the MQPM are also introduced. In Sec. 3. we discuss our results and finally in Sec. 4. we draw the conclusions.

2. Theory

2.1. Neutral-Current Neutrino-Nucleus Scattering

We consider in the present chapter neutral-current neutrino-nucleus scattering off a target nucleus (A, Z) with A nucleons and Z protons described by the reaction

$$\nu + (A, Z) \longrightarrow \nu' + (A, Z)^*. \tag{1}$$

This process which is mediated by an exchange of a neutral Z^0 boson is shown in Fig. 1. In the figure k_μ and k'_μ are the four-momenta of the incoming and outgoing neutrinos and p_μ and p'_μ are the four momenta of the initial and final nuclear states.

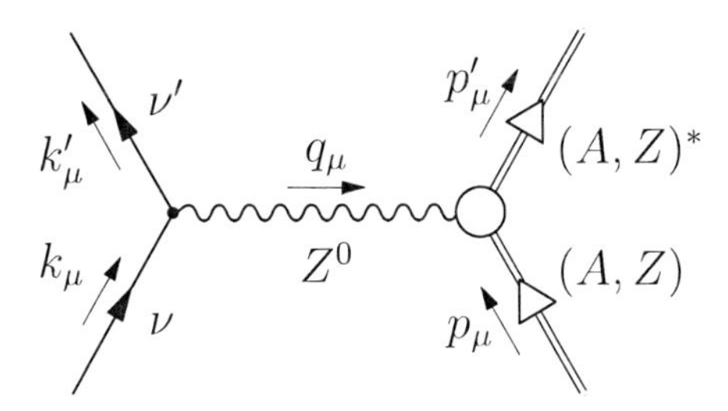

Figure 1. Neutral-current neutrino-nucleus scattering off a nucleus (A, Z). The transferred four momentum is $q_\mu = k'_\mu - k_\mu = p_\mu - p'_\mu$. More details are given in the text.

The final nuclear state in (1) can either be the ground state of the nucleus or an excited state (denoted by an asterisk). The former case is in the following referred to as coherent scattering and the latter as incoherent scattering. The total cross section can therefore be written

$$\sigma(E_k) = \sigma_{\text{coh}}(E_k) + \sigma_{\text{incoh}}(E_k), \tag{2}$$

where $\sigma_{\text{coh}}(E_k)$ and $\sigma_{\text{incoh}}(E_k)$ are the coherent and incoherent contributions respectively and E_k is the energy of the incoming neutrino.

The semi-leptonic process in Fig. 1 is described by a current-current interaction and the effective Hamiltonian therefore takes the form

$$\mathcal{H}_{\text{eff}} = \frac{G}{\sqrt{2}} \int \mathrm{d}^3\mathbf{x} j_\mu^{\text{lept}}(\mathbf{x}) \mathcal{J}^\mu(\mathbf{x}), \tag{3}$$

where $j_\mu^{\text{lept}}(\mathbf{x})$ is the lepton current and $\mathcal{J}^\mu(\mathbf{x})$ is the hadron current density and the coupling constant for neutral-current processes is given by $G = G_F = 1.1664 \times 10^{-5}$ GeV. A four-vector a^μ is here written as $a^\mu \equiv (a_0, \mathbf{a})$ with the metric $g^{\mu\nu} = g_{\mu\nu} \equiv \text{diag}(1, -1, -1, -1)$. We therefore have e.g. $k_\mu = (k_0, -\mathbf{k}) = (E_k, -\mathbf{k})$ and $k'_\mu = (k'_0, -\mathbf{k}') = (E_{k'}, -\mathbf{k}')$ where $\mathbf{k}$ and $\mathbf{k}'$ are the three-momenta and E_k and $E_{k'}$ are the energies of the incoming and outgoing neutrinos. Throughout the chapter we will also use the conventions

$$\begin{aligned}
\hbar &= c = 1, \\
\gamma_5 &= \gamma^5 \equiv \mathrm{i}\gamma^0\gamma^1\gamma^2\gamma^3, \\
\sigma^{\mu\nu} &\equiv \tfrac{\mathrm{i}}{2}[\gamma^\mu, \gamma^\nu], \\
\bar{\psi} &\equiv \psi^\dagger\gamma^0,
\end{aligned} \tag{4}$$

where the gamma matrices are given by

$$\gamma^0 \equiv \begin{pmatrix} 1 & 0 \\ 0 & -1 \end{pmatrix}, \qquad \gamma^i \equiv \begin{pmatrix} 0 & \sigma^i \\ -\sigma^i & 0 \end{pmatrix}, i = 1, 2, 3. \tag{5}$$

We consider in this chapter neutrino-nucleus scattering off nuclei with transferred four momentum $-q_\mu q^\mu \ll M_Z^2$ (M_Z denotes the mass of the Z^0 boson) so that the coupling becomes point-like and the lepton part of (3) can thus be treated to lowest order in the coupling constant G. The nuclear matrix element of the effective Hamiltonian consequently takes the form

$$
\begin{aligned}
\langle f|\mathcal{H}_{\text{eff}}|i\rangle =&\frac{G}{\sqrt{2}}l_\mu \int \mathrm{d}^3\mathbf{x}\mathrm{e}^{-\mathrm{i}\mathbf{q}\cdot\mathbf{x}}\langle f|\mathcal{J}^\mu(\mathbf{x})|i\rangle \\
=&\frac{G}{\sqrt{2}}\int \mathrm{d}^3\mathbf{x}\mathrm{e}^{-\mathrm{i}\mathbf{q}\cdot\mathbf{x}}[l_0\mathcal{J}_0(\mathbf{x})_{\text{fi}} - \mathbf{l}\cdot\boldsymbol{\mathcal{J}}(\mathbf{x})_{\text{fi}}],
\end{aligned} \tag{6}
$$

where we have written the matrix element of the lepton current as

$$
\langle f|j_\mu^{\text{lept}}(\mathbf{x})|i\rangle = l_\mu \mathrm{e}^{-\mathrm{i}\mathbf{q}\cdot\mathbf{x}}. \tag{7}
$$

and

$$
\begin{aligned}
\mathcal{J}_0(\mathbf{x})_{\text{fi}} \equiv&\langle f|\mathcal{J}_0(\mathbf{x})|i\rangle, \\
\boldsymbol{\mathcal{J}}(\mathbf{x})_{\text{fi}} \equiv&\langle f|\boldsymbol{\mathcal{J}}(\mathbf{x})|i\rangle.
\end{aligned} \tag{8}
$$

We use next the formalism outlined in [24] to derive a formula for the neutral-current neutrino-nucleus scattering cross section.

2.1.1. Cross-Section Formula for Neutral-Current Neutrino-Nucleus Scattering

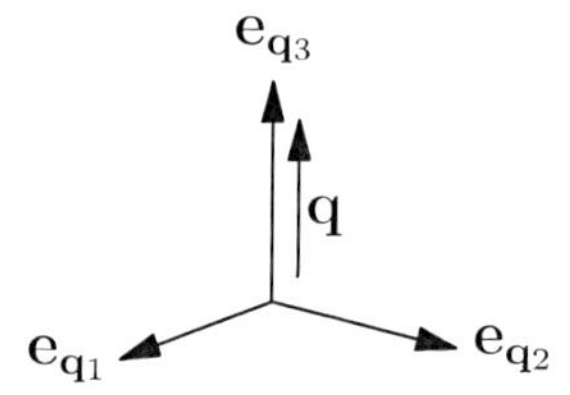

Figure 2. The Cartesian unit basis vectors $\mathbf{e}_{\mathbf{q}_1}$, $\mathbf{e}_{\mathbf{q}_2}$ and $\mathbf{e}_{\mathbf{q}_3}$ adopted in the text. The direction of the three-momentum transfer $\mathbf{q}$ is also indicated.

It is assumed in this chapter that the initial (i) and final (f) nuclear states in (1) have well-defined angular momenta J_i and J_f respectively and parities π_i and π_f. To take advantage of that we perform as the first step a multipole expansion of the matrix element (6).

We choose the 3-axis in the same direction as the three-momentum transfer $\mathbf{q}$ and the Cartesian unit basis vectors are thus oriented as shown in Fig. 2. The spherical components of these basis vectors are defined as

$$
\begin{aligned}
\mathbf{e}_{\pm 1} \equiv& \mp \frac{1}{\sqrt{2}}(\mathbf{e}_{\mathbf{q}_1} \pm \mathrm{i}\mathbf{e}_{\mathbf{q}_2}), \\
\mathbf{e}_0 \equiv&\mathbf{q}/|\mathbf{q}| \equiv \mathbf{e}_{\mathbf{q}_3}.
\end{aligned} \tag{9}
$$

The basis vectors $\mathbf{e}_\lambda$ ($\lambda = 0, \pm 1$) then obey the relations

$$
\begin{aligned}
\mathbf{e}_\lambda^\dagger \cdot \mathbf{e}_{\lambda'} =&\delta_{\lambda,\lambda'}, \\
\mathbf{e}_\lambda^\dagger =&(-1)^\lambda \mathbf{e}_{-\lambda},
\end{aligned} \tag{10}
$$

and an arbitrary vector $\mathbf{l}$ can thus be expanded as

$$\mathbf{l} = \sum_{\lambda=0,\pm1} l_\lambda \mathbf{e}_\lambda^\dagger. \tag{11}$$

But to avoid confusion with the time component we will denote $l_{\lambda=0} = l_3$.

The plane wave $e^{i\mathbf{q}\cdot\mathbf{x}}$ can be expanded in terms of spherical harmonics Y_{lm} and spherical Bessel functions j_l as (see e.g. [24])

$$e^{i\mathbf{q}\cdot\mathbf{x}} = \sum_l i^l \sqrt{4\pi}\widehat{l} j_l(\rho) Y_{l0}(\Omega), \tag{12}$$

where $\rho = q|\mathbf{x}|$, $q = |\mathbf{q}|$ and $\widehat{l} = \sqrt{2l+1}$. We can therefore write

$$\mathbf{e}_\lambda e^{i\mathbf{q}\cdot\mathbf{x}} = \sum_{lJ} i^l \sqrt{4\pi}\widehat{l} j_l(\rho)(l\ 0\ 1\ \lambda \mid J\ \lambda)\boldsymbol{\mathcal{Y}}^\lambda_{Jl1}(\Omega), \tag{13}$$

where the vector spherical harmonics $\boldsymbol{\mathcal{Y}}^M_{Jl1}$ are defined as

$$\boldsymbol{\mathcal{Y}}^M_{Jl1}(\Omega) = \sum_{m\lambda}(l\ m\ 1\ \lambda \mid J\ M) Y_{lm}(\Omega)\mathbf{e}_\lambda, \quad \Omega = (\theta,\phi). \tag{14}$$

With relations for the vector spherical harmonics given in [24] it is then straightforward to show that

$$\mathbf{e}_\lambda e^{i\mathbf{q}\cdot\mathbf{x}} = -\sum_{J\geqslant1} i^J \sqrt{2\pi}\widehat{J}\big[\lambda\mathbf{M}^\lambda_{JJ}(q) + \frac{1}{q}\boldsymbol{\nabla}\times\mathbf{M}^\lambda_{JJ}(q)\big] \quad ; \text{ for } \lambda = \pm1 \tag{15}$$

and

$$\mathbf{e}_0 e^{i\mathbf{q}\cdot\mathbf{x}} = -\frac{i}{q}\sum_{J\geqslant0} i^J \sqrt{4\pi}\widehat{J}\boldsymbol{\nabla} M_{J0}(q), \tag{16}$$

where $M_{JM}(q)$ and $\mathbf{M}^M_{JL}(q)$ are defined by

$$\begin{aligned} M_{JM}(q) &\equiv j_J(\rho) Y_{JM}(\Omega), \\ \mathbf{M}^M_{JL}(q) &\equiv j_L(\rho)\boldsymbol{\mathcal{Y}}^M_{JL1}(\Omega). \end{aligned} \tag{17}$$

The adjoints of (15) and (16) are given by

$$\begin{aligned} \mathbf{e}_\lambda^\dagger e^{-i\mathbf{q}\cdot\mathbf{x}} &= -\sum_{J\geqslant1}(-i)^J \sqrt{2\pi}\widehat{J}\big[\lambda\mathbf{M}^{-\lambda}_{JJ}(q) + \frac{1}{q}\boldsymbol{\nabla}\times\mathbf{M}^{-\lambda}_{JJ}(q)\big] \quad ; \text{ for } \lambda = \pm1, \\ \mathbf{e}_0^\dagger e^{i\mathbf{q}\cdot\mathbf{x}} &= \frac{i}{q}\sum_{J\geqslant0}(-i)^J \sqrt{4\pi}\widehat{J}\boldsymbol{\nabla} M_{J0}(q). \end{aligned} \tag{18}$$

The matrix element (6) consequently takes the form

$$\begin{aligned} \langle f|\mathcal{H}_{\text{eff}}|i\rangle =& \frac{G}{\sqrt{2}}\langle f|\Big\{\sum_{J\geqslant0}\sqrt{4\pi}\widehat{J}(-i)^J\big[l_0\mathcal{M}_{J0}(q) - l_3\mathcal{L}_{J0}(q)\big] + \\ & \sum_{J\geqslant1}\sqrt{2\pi}\widehat{J}(-i)^J \sum_{\lambda=\pm1} l_\lambda\big[\lambda\mathcal{T}^{\text{mag}}_{J,-\lambda}(q) + \mathcal{T}^{\text{el}}_{J,-\lambda}(q)\big]\Big\}|i\rangle, \end{aligned} \tag{19}$$

where the multipole operators $\mathcal{M}_{JM}$, $\mathcal{L}_{JM}$, $\mathcal{T}^{\rm el}_{JM}$ and $\mathcal{T}^{\rm mag}_{JM}$ are given by

$$\begin{aligned}
\mathcal{M}_{JM}(q) \equiv & M^{\rm V}_{JM}(q) - M^{\rm A}_{JM}(q) = \int {\rm d}^3\mathbf{x} M_{JM}(q)\mathcal{J}_0(\mathbf{x}), \\
\mathcal{L}_{JM}(q) \equiv & L^{\rm V}_{JM}(q) - L^{\rm A}_{JM}(q) = \frac{\rm i}{q}\int {\rm d}^3\mathbf{x}\big[\boldsymbol{\nabla} M_{JM}(q)\big]\cdot\boldsymbol{\mathcal{J}}(\mathbf{x}), \\
\mathcal{T}^{\rm el}_{JM}(q) \equiv & T^{\rm el,V}_{JM}(q) - T^{\rm el,A}_{JM}(q) = \frac{1}{q}\int {\rm d}^3\mathbf{x}\big[\boldsymbol{\nabla}\times\mathbf{M}^M_{JJ}(q)\big]\cdot\boldsymbol{\mathcal{J}}(\mathbf{x}), \\
\mathcal{T}^{\rm mag}_{JM}(q) \equiv & T^{\rm mag,V}_{JM}(q) - T^{\rm mag,A}_{JM}(q) = \int {\rm d}^3\mathbf{x}\big[\mathbf{M}^M_{JJ}(q)\big]\cdot\boldsymbol{\mathcal{J}}(\mathbf{x}).
\end{aligned} \tag{20}$$

The index V (A) denotes here an vector (axial-vector) operator and the hadron current density operator $\mathcal{J}^\mu(\mathbf{x}) = (\mathcal{J}_0(\mathbf{x}), \boldsymbol{\mathcal{J}}(\mathbf{x}))$ can be written in terms of its vector $(J^{{\rm V},\mu}(\mathbf{x}))$ and axial-vector $(J^{{\rm A},\mu}(\mathbf{x}))$ pieces as

$$\mathcal{J}^\mu(\mathbf{x}) = J^{{\rm V},\mu}(\mathbf{x}) - J^{{\rm A},\mu}(\mathbf{x}). \tag{21}$$

From the conservation of the vector current one has that the operators $L^{\rm V}_{JM}(q)$ and $M^{\rm V}_{JM}(q)$ are related through

$$L^{\rm V}_{JM}(q) = -\frac{E_{\rm exc}}{q}M^{\rm V}_{JM}(q), \tag{22}$$

where $E_{\rm exc} = E_k - E_{k'}$ is the excitation energy of the nucleus.

The nuclear targets are assumed to be unoriented and unobserved and we therefore sum over the final states and average over the initial states and we can then derive the important result that

$$\begin{aligned}
\frac{1}{2J_i+1}\sum_{M_f M_i}\big|\langle f|\mathcal{H}_{\rm eff}|i\rangle\big|^2 = & \frac{4\pi}{2J_i+1}\frac{G^2}{2} \\
\times & \Big\{\sum_{J\geqslant 0}\Big[l_0 l_0^*\big|(J_f\|\mathcal{M}_J(q)\|J_i)\big|^2 + l_3 l_3^*\big|(J_f\|\mathcal{L}_J(q)\|J_i)\big|^2 \\
- & 2{\rm Re}\big(l_3 l_0^*(J_f\|\mathcal{L}_J(q)\|J_i)(J_f\|\mathcal{M}_J(q)\|J_i)^*\big)\Big] \\
+ & \sum_{J\geqslant 1}\Big[\frac{1}{2}(\mathbf{l}\cdot\mathbf{l}^* - l_3 l_3^*)\big(\big|(J_f\|\mathcal{T}^{\rm mag}_J(q)\|J_i)\big|^2 + \big|(J_f\|\mathcal{T}^{\rm el}_J(q)\|J_i)\big|^2\big) \\
- & \frac{\rm i}{2}(\mathbf{l}\times\mathbf{l}^*)_3 2{\rm Re}\big((J_f\|\mathcal{T}^{\rm mag}_J(q)\|J_i)(J_f\|\mathcal{T}^{\rm el}_J(q)\|J_i)^*\big)\Big]\Big\},
\end{aligned} \tag{23}$$

where the reduced matrix element of an operator $\mathcal{T}_{JM}$ is defined through the Wigner-Eckart theorem as

$$\langle J_f M_f|\mathcal{T}_{JM}|J_i M_i\rangle = (-1)^{J_f - M_f}\begin{pmatrix} J_f & J & J_i \\ -M_f & M & M_i\end{pmatrix}(J_f\|\mathcal{T}_J\|J_i). \tag{24}$$

In (23) we have for convenience omitted the other quantum numbers required to fully specify the final and initial nuclear states. The differential cross section is then given by the Fermi's golden rule as [24]

$$\frac{{\rm d}\sigma}{{\rm d}\Omega} = \frac{2|\mathbf{k}'|E_{k'}}{(2\pi)^2}\Big(\frac{V^2}{2}\sum_{\text{lepton spins}}\frac{1}{2J_i+1}\sum_{M_f M_i}\big|\langle f|\mathcal{H}_{\rm eff}|i\rangle\big|^2\Big), \tag{25}$$

where the particles are quantized in a box with volume V. The lepton matrix elements defined by (7) are for neutral-current (anti)neutrino scattering of the form [24]

$$Vl_\mu = \begin{cases} \bar{u}(\mathbf{k}')\gamma_\mu(1-\gamma_5)u(\mathbf{k}) & ;\text{ for neutrinos,} \\ -\bar{v}(-\mathbf{k})\gamma_\mu(1-\gamma_5)v(-\mathbf{k}') & ;\text{ for antineutrinos,} \end{cases} \tag{26}$$

where $u(\mathbf{k})$ and $v(-\mathbf{k})$ are the Dirac spinors for a neutrino and antineutrino respectively. The required lepton traces therefore become

$$\begin{aligned}
\frac{V^2}{2}\sum_{\text{lepton spins}} l_0 l_0^* &= 1 + \frac{1}{E_k E_{k'}}\mathbf{k}\cdot\mathbf{k}', \\
\frac{V^2}{2}\sum_{\text{lepton spins}} l_3 l_3^* &= 1 - \frac{1}{E_k E_{k'}}\mathbf{k}\cdot\mathbf{k}' + 2\frac{1}{E_k E_{k'} q^2}(\mathbf{k}\cdot\mathbf{q})(\mathbf{k}'\cdot\mathbf{q}), \\
\frac{V^2}{2}\sum_{\text{lepton spins}} l_3 l_0^* &= \frac{1}{q}\mathbf{q}\cdot\left(\frac{1}{E_k}\mathbf{k} + \frac{1}{E_{k'}}\mathbf{k}'\right), \\
\frac{V^2}{2}\sum_{\text{lepton spins}} \frac{1}{2}(\mathbf{l}\cdot\mathbf{l}^* - l_3 l_3^*) &= 1 - \frac{1}{E_k E_{k'} q^2}(\mathbf{k}\cdot\mathbf{q})(\mathbf{k}'\cdot\mathbf{q}), \\
-\frac{\mathrm{i}}{2}\frac{V^2}{2}\sum_{\text{lepton spins}} (\mathbf{l}\times\mathbf{l}^*)_3 &= \pm\frac{1}{q}\mathbf{q}\cdot\left(\frac{1}{E_k}\mathbf{k} - \frac{1}{E_{k'}}\mathbf{k}'\right),
\end{aligned} \tag{27}$$

where the kinematical variables $\mathbf{k}$, $\mathbf{k}'$, E_k and $E_{k'}$ were defined in Sec. 2.1. and the three-momentum transfer is given by $\mathbf{q} = \mathbf{k}' - \mathbf{k}$. In (27) the plus sign refers to neutrinos and the minus sign to antineutrinos. With (23), (25) and (27) we thus obtain that the double-differential cross section for neutral-current neutrino-nucleus scattering from an initial state (i) with angular momentum J_i to a final state (f) with angular momentum J_f and excitation energy $E_{\text{exc}} = E_k - E_{k'}$ is given by

$$\left(\frac{\mathrm{d}^2\sigma_{i\to f}}{\mathrm{d}\Omega \mathrm{d}E_{\text{exc}}}\right)_{\nu/\bar{\nu}} = \frac{G_{\mathrm{F}}^2|\mathbf{k}'|\epsilon'}{\pi(2J_i+1)}\left(\sum_{J\geqslant 0}\sigma_{\mathrm{CL}}^J + \sum_{J\geqslant 1}\sigma_{\mathrm{T}}^J\right), \tag{28}$$

where the Coulomb-longitudinal (σ_{CL}^J) and transverse (σ_{T}^J) contributions are defined as

$$\begin{aligned}
\sigma_{\mathrm{CL}}^J \equiv &(1+\cos\theta)\left|\langle J_f\|\mathcal{M}_J(q)\|J_i\rangle\right|^2 + (1+\cos\theta - 2b\sin^2\theta) \\
&\times \left|\langle J_f\|\mathcal{L}_J(q)\|J_i\rangle\right|^2 + qE_{\text{exc}}(1+\cos\theta) \\
&\times 2\mathrm{Re}\left\{\langle J_f\|\mathcal{M}_J(q)\|J_i\rangle^*\langle J_f\|\mathcal{L}_J(q)\|J_i\rangle\right\},
\end{aligned} \tag{29}$$

and

$$\begin{aligned}
\sigma_{\mathrm{T}}^J \equiv &(1-\cos\theta + b\sin^2\theta)\left[\left|\langle J_f\|\mathcal{T}_J^{\text{mag}}(q)\|J_i\rangle\right|^2 + \left|\langle J_f\|\mathcal{T}_J^{\text{el}}(q)\|J_i\rangle\right|^2\right] \\
&\mp \frac{E_k + E_{k'}}{q}(1-\cos\theta)2\mathrm{Re}\left\{\langle J_f\|\mathcal{T}_J^{\text{mag}}(q)\|J_i\rangle\langle J_f\|\mathcal{T}_J^{\text{el}}(q)\|J_i\rangle^*\right\}.
\end{aligned} \tag{30}$$

In (30) the minus sign is used for the neutrino cross section and the plus sign for the antineutrino cross section. In the expressions above the scattering angle θ is the angle between the incoming and outgoing neutrinos, and the magnitude of the three-momentum transfer is given by

$$q \equiv |\mathbf{q}| = \sqrt{E_{\rm exc}^2 + 2E_k E_{k'}(1-\cos\theta)}. \quad (31)$$

In (29) and (30) we have also introduced the parameter $b = E_k E_{k'}/q^2$.

2.1.2. Construction of the Multipole Operators in Second Quantization

We assume that the impulse approximation is valid and the multipole operators (20) are then one-body operators and any of them can thus be expanded as (see e.g. [20])

$$\begin{aligned} T_{JM}(q) &= \sum_{ab} \langle a|T_{JM}(q)|b\rangle c_\alpha^\dagger \tilde{c}_\beta = \widehat{J}^{-1} \sum_{ab} (a\|T_J(q)\|b)[c_a^\dagger \tilde{c}_b]_{JM} = \\ & \widehat{J}^{-1}\Big(\sum_{\substack{ab \\ \text{protons}}} (a\|T_J^{\rm p}(q)\|b)[c_a^\dagger \tilde{c}_b]_{JM} + \sum_{\substack{ab \\ \text{neutrons}}} (a\|T_J^{\rm n}(q)\|b)[c_a^\dagger \tilde{c}_b]_{JM}\Big), \end{aligned} \quad (32)$$

where the index p (n) denotes an operator acting on a proton (neutron). The index a contains here the single-particle quantum numbers n_a, l_a and j_a and the index α includes additionally the magnetic quantum number m_α. The indices b and β are defined correspondingly. Here $c_\alpha^\dagger$ is the particle creation operator and the particle annihilation operator $\tilde{c}_\alpha$ is defined as $\tilde{c}_\alpha = (-1)^{j_a+m_\alpha} c_{-\alpha}$ where $-\alpha = (a, -m_\alpha)$. The expansion (32) is exact as long as one uses a complete set of single-particle states but in practice one has to truncate to a limited number of proton and neutron orbitals. Any of the reduced matrix elements in (23) can then be written on the form

$$\begin{aligned} (J_f\|T_J(q)\|J_i) = \widehat{J}^{-1}\Big(& \sum_{\substack{ab \\ \text{protons}}} (a\|T_J^{\rm p}(q)\|b)(J_f\|[c_a^\dagger \tilde{c}_b]_J\|J_i) \\ & + \sum_{\substack{ab \\ \text{neutrons}}} (a\|T_J^{\rm n}(q)\|b)(J_f\|[c_a^\dagger \tilde{c}_b]_J\|J_i)\Big), \end{aligned} \quad (33)$$

where the reduced one-body transition densities $(J_f\|[c_a^\dagger \tilde{c}_b]_J\|J_i)$ are calculated from the nuclear model employed to construct the many-body states. In the derivation of the formula (28) for the cross section we have only presumed the existence of a nuclear current density of the form $\mathcal{J}^\mu(\mathbf{x}) = J^{{\rm V},\mu}(\mathbf{x}) - J^{{\rm A},\mu}(\mathbf{x})$. With the requirements of Lorentz covariance, conservation of parity, time-reversal invariance and isospin invariance one can derive the most general forms of the one-nucleon matrix elements at the origin ($\mathbf{x} = \mathbf{0}$)

$$\begin{aligned} {}_{\rm p(n)}\langle \mathbf{p}'\sigma'|J^{{\rm V},\mu}(0)|\mathbf{p}\sigma\rangle_{\rm p(n)} &= \frac{\bar{u}(\mathbf{p}',\sigma')}{V}\Big[F_1^{\rm NC;p(n)}(Q^2)\gamma^\mu - \\ & \frac{\rm i}{m_{\rm N}}\sigma^{\mu\nu}q_\nu F_2^{\rm NC;p(n)}(Q^2)\Big]u(\mathbf{p},\sigma), \\ {}_{\rm p(n)}\langle \mathbf{p}'\sigma'|J^{{\rm A},\mu}(0)|\mathbf{p}\sigma\rangle_{\rm p(n)} &= \frac{\bar{u}(\mathbf{p}',\sigma')}{V} F_{\rm A}^{\rm NC;p(n)}(Q^2)\gamma_5\gamma^\mu u(\mathbf{p},\sigma). \end{aligned} \quad (34)$$

Here $Q^2 = -q_\mu q^\mu$ and $m_{\rm N}$ denotes the nucleon mass. In (34) $|\mathbf{p}\sigma\rangle_{\rm p(n)}$ denotes a state vector of a free proton (neutron) with three momentum $\mathbf{p}$ and spin σ. In the expressions above the so-called induced scalar and tensor currents have been neglected. The nucleon form factors $F_{1,2}^{\mathrm{NC;p(n)}}(Q^2)$ and $F_A^{\mathrm{NC;p(n)}}(Q^2)$ will be discussed in Sec. 2.1.4.. As the next step we therefore assume that the nuclear current density operators $J^{\mathrm{V/A},\mu}(\mathbf{x})$ at the origin are given by

$$\begin{aligned} J^{\mathrm{V/A},\mu}(0) =& J_{\mathrm{p}}^{\mathrm{V/A},\mu}(0) + J_{\mathrm{n}}^{\mathrm{V/A},\mu}(0) = \sum_{\substack{\mathbf{p}\sigma \\ \text{protons}}} \sum_{\substack{\mathbf{p}'\sigma' \\ \text{protons}}} {}_{\mathrm{p}}\langle \mathbf{p}'\sigma' | J^{\mathrm{V/A},\mu}(0) | \mathbf{p}\sigma \rangle_{\mathrm{p}} c^{\dagger}_{\mathbf{p}'\sigma'} c_{\mathbf{p}\sigma} + \\ & \sum_{\substack{\mathbf{p}\sigma \\ \text{neutrons}}} \sum_{\substack{\mathbf{p}'\sigma' \\ \text{neutrons}}} {}_{\mathrm{n}}\langle \mathbf{p}'\sigma' | J^{\mathrm{V/A},\mu}(0) | \mathbf{p}\sigma \rangle_{\mathrm{n}} c^{\dagger}_{\mathbf{p}'\sigma'} c_{\mathbf{p}\sigma}, \end{aligned} \tag{35}$$

where the matrix elements are the ones of (34).

The energy E_k of the incoming neutrino in Fig. (1) is for supernova neutrinos typically of the order $E_k \lesssim 100$ MeV $\ll m_{\mathrm{N}}c^2$ and it is therefore fair to assume that

$$|\mathbf{p}| \ll m_{\mathrm{N}}, \quad |\mathbf{p}'| \ll m_{\mathrm{N}}. \tag{36}$$

The Dirac spinor for a free nucleon is given by

$$u(\mathbf{p},\sigma) = \sqrt{\frac{E_{\mathrm{p}} + m_{\mathrm{N}}}{2E_{\mathrm{p}}}} \begin{pmatrix} \chi_\sigma \\ \frac{\boldsymbol{\sigma}\cdot\mathbf{p}}{E_{\mathrm{p}}+m_{\mathrm{N}}}\chi_\sigma \end{pmatrix}, \tag{37}$$

where χ_σ is the Pauli spinor for a particle with spin σ and E_{p} denotes the energy of a free relativistic particle. By using this explicit form one can expand the single-nucleon matrix elements (34) through order $1/m_{\mathrm{N}}$ with the result

$${}_{\mathrm{p(n)}}\langle \mathbf{p}'\sigma' | J^{\mathrm{V/A},\mu}(0) | \mathbf{p}\sigma \rangle_{\mathrm{p(n)}} = \frac{1}{V}\chi^{\dagger}_{\sigma} M_{\mathrm{p(n)}}^{\mathrm{V/A},\mu}(Q^2)\chi_{\sigma'} + O\Big(\frac{1}{m_N^2}\Big), \tag{38}$$

where $M_{\mathrm{p(n)}}^{\mathrm{V/A},\mu}(Q^2) = \big(M_{\mathrm{p(n)}}^{\mathrm{V/A},0}(Q^2), \mathbf{M}_{\mathrm{p(n)}}^{\mathrm{V/A}}(Q^2)\big)$ and

$$\begin{aligned} M_{\mathrm{p(n)}}^{\mathrm{V},0}(Q^2) =& F_1^{\mathrm{NC;p(n)}}(Q^2), \\ \mathbf{M}_{\mathrm{p(n)}}^{\mathrm{V}}(Q^2) =& \frac{F_1^{\mathrm{NC;p(n)}}(Q^2)}{2m_{\mathrm{N}}}(\mathbf{p}+\mathbf{p}') + \frac{F_1^{\mathrm{NC;p(n)}}(Q^2) + F_2^{\mathrm{NC;p(n)}}(Q^2)}{2m_{\mathrm{N}}}\mathrm{i}\mathbf{q}\times\boldsymbol{\sigma}, \\ M_{\mathrm{p(n)}}^{\mathrm{A},0}(Q^2) =& -\frac{F_{\mathrm{A}}^{\mathrm{NC;p(n)}}(Q^2)}{2m_{\mathrm{N}}}\boldsymbol{\sigma}\cdot(\mathbf{p}+\mathbf{p}'), \\ \mathbf{M}_{\mathrm{p(n)}}^{\mathrm{A}}(Q^2) =& -F_{\mathrm{A}}^{\mathrm{NC;p(n)}}(Q^2)\boldsymbol{\sigma}. \end{aligned} \tag{39}$$

From (6) we have that

$$\begin{aligned} -\mathrm{i}\mathbf{q}\langle f|\mathcal{H}_{\mathrm{eff}}|i\rangle =& \frac{G}{\sqrt{2}}l_\mu \int \mathrm{d}^3\mathbf{x}(-\mathrm{i}\mathbf{q})\mathrm{e}^{-\mathrm{i}\mathbf{q}\cdot\mathbf{x}}\langle f|\mathcal{J}^\mu(\mathbf{x})|i\rangle \\ =& \frac{G}{\sqrt{2}}l_\mu \int \mathrm{d}^3\mathbf{x}\boldsymbol{\nabla}(\mathrm{e}^{-\mathrm{i}\mathbf{q}\cdot\mathbf{x}})\langle f|\mathcal{J}^\mu(\mathbf{x})|i\rangle \\ =& \frac{G}{\sqrt{2}}l_\mu\Big(\int \mathrm{d}^3\mathbf{x}\boldsymbol{\nabla}(\mathrm{e}^{-\mathrm{i}\mathbf{q}\cdot\mathbf{x}}\langle f|\mathcal{J}^\mu(\mathbf{x})|i\rangle) - \int \mathrm{d}^3\mathbf{x}\mathrm{e}^{-\mathrm{i}\mathbf{q}\cdot\mathbf{x}}\boldsymbol{\nabla}\langle f|\mathcal{J}^\mu(\mathbf{x})|i\rangle\Big), \end{aligned} \tag{40}$$

and consequently

$$\int d^3\mathbf{x}(-i\mathbf{q})e^{-i\mathbf{q}\cdot\mathbf{x}}\langle f|\mathcal{J}^\mu(\mathbf{x})|i\rangle = \int d^3\mathbf{x}\nabla(e^{-i\mathbf{q}\cdot\mathbf{x}}\langle f|\mathcal{J}^\mu(\mathbf{x})|i\rangle) - \int d^3\mathbf{x}e^{-i\mathbf{q}\cdot\mathbf{x}}\nabla\langle f|\mathcal{J}^\mu(\mathbf{x})|i\rangle. \quad (41)$$

The first integral on the right side of (41) can be converted into a surface integral far away from the nucleus by exploiting the identity

$$\int_V d^3\mathbf{x}\nabla f(\mathbf{x}) = \int_S d^2\mathbf{x} f(\mathbf{x})\mathbf{n}, \quad (42)$$

where $\mathbf{n}$ is the normal vector to the closed surface S. The nuclear target is assumed to be localized and one has therefore that the nuclear matrix element $\langle f|\mathcal{J}^\mu(\mathbf{x})|i\rangle$ in (41) vanishes for large values of $|\mathbf{x}|$. From (41) and (42) we thus obtain that

$$\int d^3\mathbf{x}(i\mathbf{q})e^{-i\mathbf{q}\cdot\mathbf{x}}\langle f|\mathcal{J}^\mu(\mathbf{x})|i\rangle = \int d^3\mathbf{x}e^{-i\mathbf{q}\cdot\mathbf{x}}\nabla\langle f|\mathcal{J}^\mu(\mathbf{x})|i\rangle. \quad (43)$$

and one can consequently make the replacement $\mathbf{q} = -i\nabla$ in (39). The nuclear current density operators take in first quantization the forms

$$J^{V/A,\mu}(\mathbf{x}) = J_p^{V/A,\mu}(\mathbf{x}) + J_n^{V/A,\mu}(\mathbf{x}), \quad (44)$$

where

$$J_{p(n)}^{V/A,\mu}(\mathbf{x}) = \sum_{k=1}^{\tau_{p(n)}} J_{p(n)}^{V/A,\mu}(\mathbf{x})\delta^{(3)}(\mathbf{x}-\mathbf{x}_k), \quad (45)$$

and τ_p (τ_n) denotes the number of protons (neutrons), i.e. $\tau_p = Z$ and $\tau_n = N$. It is then well-known from basic quantum mechanics that the corresponding second-quantized operators are given by

$$J_{p(n)}^{V/A,\mu}(\mathbf{x}) = \sum_{\mathbf{p}\sigma}\sum_{\mathbf{p}'\sigma'} c^\dagger_{\mathbf{p}'\sigma'}c_{\mathbf{p}\sigma}\langle\mathbf{p}'\sigma'|J_{p(n)}^{V/A,\mu}(\mathbf{x})|\mathbf{p}\sigma\rangle, \quad (46)$$

with the single-particle matrix elements

$$\langle\mathbf{p}'\sigma'|J_{p(n)}^{V/A,\mu}(\mathbf{x})|\mathbf{p}\sigma\rangle = \int d^3y\phi^\dagger_{\mathbf{p}'\sigma'}(\mathbf{y})J_{p(n)}^{V/A,\mu}(\mathbf{x})\delta^{(3)}(\mathbf{x}-\mathbf{y})\phi_{\mathbf{p}\sigma}(\mathbf{y}), \quad (47)$$

where the single-particle wave function is that of a free non-relativistic particle and thus

$$\phi_{\mathbf{p}\sigma}(\mathbf{x}) = \frac{1}{\sqrt{V}}e^{i\mathbf{p}\cdot\mathbf{x}}\chi_\sigma. \quad (48)$$

By comparing the matrix elements (47) at $\mathbf{x} = 0$ with the ones in (38) we subsequently obtain for the nuclear current operators $J_{p(n)}^{V/A,\mu}(\mathbf{x}) = (J_{p(n)}^{V/A,0}(\mathbf{x}), \mathbf{J}_{p(n)}^{V/A}(\mathbf{x}))$ in first quantization the expressions

$$\begin{aligned} J_{p(n)}^{V,0}(\mathbf{x}) &= \rho_{p(n)}^V(\mathbf{x}), \quad \mathbf{J}_{p(n)}^V(\mathbf{x}) = \mathbf{J}_{C,p(n)}(\mathbf{x}) + \nabla\times\boldsymbol{\mu}_{p(n)}(\mathbf{x}) \\ J_{p(n)}^{A,0}(\mathbf{x}) &= -\rho_{p(n)}^A(\mathbf{x}), \quad \mathbf{J}_{p(n)}^V(\mathbf{x}) = -\mathbf{A}_{p(n)}(\mathbf{x}), \end{aligned} \quad (49)$$

where we have used the abbreviations

$$\begin{aligned}
\rho^{\mathrm{V}}_{\mathrm{p(n)}}(\mathbf{x}) =& F_1^{\mathrm{NC;p(n)}}(Q^2)\sum_{k=1}^{\tau_{\mathrm{p(n)}}}\delta^{(3)}(\mathbf{x}-\mathbf{x}_k),\\
\mathbf{J}_{\mathrm{C,p(n)}}(\mathbf{x}) =& F_1^{\mathrm{NC;p(n)}}(Q^2)\sum_{k=1}^{\tau_{\mathrm{p(n)}}}\Big\{\frac{\mathbf{p}(k)}{2m_{\mathrm{N}}},\delta^{(3)}(\mathbf{x}-\mathbf{x}_k)\Big\},\\
\boldsymbol{\mu}_{\mathrm{p(n)}}(\mathbf{x}) =& \frac{\mu^{\mathrm{NC:p(n)}}(Q^2)}{2m_{\mathrm{N}}}\sum_{k=1}^{\tau_{\mathrm{p(n)}}}\boldsymbol{\sigma}(k)\delta^{(3)}(\mathbf{x}-\mathbf{x}_k),\\
\rho^{\mathrm{A}}_{\mathrm{p(n)}}(\mathbf{x}) =& F_{\mathrm{A}}^{\mathrm{NC;p(n)}}(Q^2)\sum_{k=1}^{\tau_{\mathrm{p(n)}}}\boldsymbol{\sigma}(k)\cdot\Big\{\frac{\mathbf{p}(k)}{2m_{\mathrm{N}}},\delta^{(3)}(\mathbf{x}-\mathbf{x}_k)\Big\},\\
\mathbf{A}_{\mathrm{p(n)}}(\mathbf{x}) =& F_{\mathrm{A}}^{\mathrm{NC;p(n)}}(Q^2)\sum_{k=1}^{\tau_{\mathrm{p(n)}}}\boldsymbol{\sigma}(k)\delta^{(3)}(\mathbf{x}-\mathbf{x}_k).
\end{aligned}\tag{50}$$

In (50) we have defined the anti-commutator $\{A,B\}$ as $\{A,B\}=AB+BA$ and the form factors $\mu^{\mathrm{NC:p(n)}}(Q^2)$ are given by

$$\mu^{\mathrm{NC:p(n)}}(Q^2)\equiv F_1^{\mathrm{NC:p(n)}}(Q^2)+F_2^{\mathrm{NC:p(n)}}(Q^2).\tag{51}$$

It is then straightforward to derive expressions for the first-quantized versions of the multipole operators by using the definitions in (20). The corresponding second-quantized operators are subsequently obtained from (32) and they can consequently be written in terms of the one-body operators

$$\begin{aligned}
M_{JM}^{\mathrm{V:p(n)}}(q) =& F_1^{\mathrm{NC:p(n)}}(Q^2)M_{JM}(q),\\
T_{JM}^{\mathrm{el,V:p(n)}}(q) =& (q/m_{\mathrm{N}})\big[F_1^{\mathrm{NC:p(n)}}(Q^2)\Delta'_{JM}(q)+\tfrac{1}{2}\mu^{\mathrm{NC:p(n)}}(Q^2)\Sigma_{JM}(q)\big],\\
T_{JM}^{\mathrm{mag,V:p(n)}}(q) =& -\mathrm{i}(q/m_{\mathrm{N}})\big[F_1^{\mathrm{NC:p(n)}}(Q^2)\Delta_{JM}(q)-\tfrac{1}{2}\mu^{\mathrm{NC:p(n)}}(Q^2)\Sigma'_{JM}(q)\big],\\
M_{JM}^{\mathrm{A:p(n)}}(q) =& \mathrm{i}(q/m_{\mathrm{N}})F_{\mathrm{A}}^{\mathrm{NC:p(n)}}(Q^2)\big[\Omega_{JM}(q)+\tfrac{1}{2}\Sigma''_{JM}(q)\big],\\
L_{JM}^{\mathrm{A:p(n)}}(q) =& -\mathrm{i}F_{\mathrm{A}}^{\mathrm{NC:p(n)}}(Q^2)\Sigma''_{JM}(q),\\
T_{JM}^{\mathrm{el,A:p(n)}}(q) =& -\mathrm{i}F_{\mathrm{A}}^{\mathrm{NC:p(n)}}(Q^2)\Sigma'_{JM}(q),\\
T_{JM}^{\mathrm{mag,A:p(n)}}(q) =& -F_{\mathrm{A}}^{\mathrm{NC:p(n)}}(Q^2)\Sigma_{JM}(q),
\end{aligned}\tag{52}$$

where we have defined the following basic operators

$$\begin{aligned}
\Delta_{JM}(q) \equiv& \mathbf{M}_{JJ}^{M}(q) \cdot \frac{1}{q}\boldsymbol{\nabla}, \\
\Delta'_{JM}(q) \equiv& -\mathrm{i}\left[\frac{1}{q}\boldsymbol{\nabla} \times \mathbf{M}_{JJ}^{M}(q)\right] \cdot \frac{1}{q}\boldsymbol{\nabla} = \\
& \widehat{J}^{-1}\left[-\sqrt{J}\mathbf{M}_{JJ+1}^{M}(q) + \sqrt{J+1}\mathbf{M}_{JJ-1}^{M}(q)\right] \cdot \frac{1}{q}\boldsymbol{\nabla}, \\
\Sigma_{JM}(q) \equiv& \mathbf{M}_{JJ}^{M}(q) \cdot \boldsymbol{\sigma}, \\
\Sigma'_{JM}(q) \equiv& -\mathrm{i}\left[\frac{1}{q}\boldsymbol{\nabla} \times \mathbf{M}_{JJ}^{M}(q)\right] \cdot \boldsymbol{\sigma} = \\
& \widehat{J}^{-1}\left[-\sqrt{J}\mathbf{M}_{JJ+1}^{M}(q) + \sqrt{J+1}\mathbf{M}_{JJ-1}^{M}(q)\right] \cdot \boldsymbol{\sigma}, \\
\Sigma''_{JM}(q) \equiv& \left[\frac{1}{q}\boldsymbol{\nabla} M_{JM}(q)\right] \cdot \boldsymbol{\sigma} = \\
& \widehat{J}^{-1}\left[\sqrt{J+1}\mathbf{M}_{JJ+1}^{M}(q) + \sqrt{J}\mathbf{M}_{JJ-1}^{M}(q)\right] \cdot \boldsymbol{\sigma}, \\
\Omega_{JM}(q) \equiv& M_{JM}(q)\boldsymbol{\sigma} \cdot \frac{1}{q}\boldsymbol{\nabla}.
\end{aligned} \tag{53}$$

It is then seen from (32), (52) and (53) that the multipole operators (20) can be constructed from the reduced single-particle matrix elements of the operators $M_{JM}(q)$, $\mathbf{M}_{JL}^{M}(q) \cdot (1/q)\boldsymbol{\nabla}$, $\mathbf{M}_{JL}^{M}(q) \cdot (1/q)\boldsymbol{\sigma}$ and $M_{JM}(q)\boldsymbol{\sigma} \cdot (1/q)\boldsymbol{\nabla}$ where L can take the values $L = J-1, J, J+1$. The reduced single-particle matrix elements of these operators are given by [11]

$$\begin{aligned}
(a\|M_J(q)\|b) =& \frac{(-1)^{J+j_b+1/2}}{\sqrt{4\pi}}\widehat{l_a}\widehat{l_b}\widehat{j_a}\widehat{j_b}\widehat{J}\begin{Bmatrix} l_a & j_a & \frac{1}{2} \\ j_b & l_b & J \end{Bmatrix}\begin{pmatrix} l_a & J & l_b \\ 0 & 0 & 0 \end{pmatrix}\mathcal{R}_{abJ}^{(0)}, \\
(a\|\mathbf{M}_{JL}(q)\cdot\boldsymbol{\sigma}\|b) =& \frac{(-1)^{l_a}}{\sqrt{4\pi}}\sqrt{6}\widehat{l_a}\widehat{l_b}\widehat{j_a}\widehat{j_b}\widehat{J}\widehat{L}\begin{Bmatrix} l_a & l_b & L \\ \frac{1}{2} & \frac{1}{2} & 1 \\ j_a & j_b & J \end{Bmatrix}\begin{pmatrix} l_a & L & l_b \\ 0 & 0 & 0 \end{pmatrix}\mathcal{R}_{abL}^{(0)}, \\
(a\|\mathbf{M}_{JL}(q)\cdot\frac{1}{q}\boldsymbol{\nabla}\|b) =& \frac{(-1)^{L+j_b+1/2}}{\sqrt{4\pi}}\widehat{l_a}\widehat{j_a}\widehat{j_b}\widehat{J}\widehat{L}\begin{Bmatrix} l_a & j_a & \frac{1}{2} \\ j_b & l_b & J \end{Bmatrix}\Big[-\sqrt{(l_b+1)(2l_b+3)} \\
& \times\begin{Bmatrix} L & 1 & J \\ l_b & l_a & l_b+1 \end{Bmatrix}\begin{pmatrix} l_a & L & l_b+1 \\ 0 & 0 & 0 \end{pmatrix}\mathcal{R}_{abL}^{(-)} \\
& + \sqrt{l_b(2l_b-1)}\begin{Bmatrix} L & 1 & J \\ l_b & l_a & l_b-1 \end{Bmatrix}\begin{pmatrix} l_a & L & l_b-1 \\ 0 & 0 & 0 \end{pmatrix}\mathcal{R}_{abL}^{(+)}\Big], \\
(a\|M_J(q)\boldsymbol{\sigma}\cdot\frac{1}{q}\boldsymbol{\nabla}\|b) =& \frac{(-1)^{l_a}}{\sqrt{4\pi}}\widehat{l_a}\widehat{j_a}\widehat{j_b}\widehat{J}\sqrt{2(2j_b-l_b)+1}\begin{Bmatrix} l_a & j_a & \frac{1}{2} \\ j_b & 2j_b-l_b & J \end{Bmatrix}\times \\
& \begin{pmatrix} l_a & J & 2j_b-l_b \\ 0 & 0 & 0 \end{pmatrix}\left[-\delta_{j_b,l_b+1/2}\mathcal{R}_{abJ}^{(-)} + \delta_{j_b,l_b-1/2}\mathcal{R}_{abJ}^{(+)}\right],
\end{aligned} \tag{54}$$

where we have defined the radial matrix elements ($\rho = q|\mathbf{x}|$)

$$\begin{aligned}\mathcal{R}^{(0)}_{abL} \equiv& \langle n_a l_a | j_L(\rho) | n_b l_b \rangle, \\ \mathcal{R}^{(-)}_{abL} \equiv& \langle n_a l_a | j_L(\rho) (\frac{\mathrm{d}}{\mathrm{d}\rho} - \frac{l_b}{\rho}) | n_b l_b \rangle, \\ \mathcal{R}^{(+)}_{abL} \equiv& \langle n_a l_a | j_L(\rho) (\frac{\mathrm{d}}{\mathrm{d}\rho} + \frac{l_b + 1}{\rho}) | n_b l_b \rangle.\end{aligned} \tag{55}$$

2.1.3. Analytical Expressions for the Radial Integrals in a Harmonic Oscillator Basis

In the present work the single-particle energies are calculated from the Coulomb-corrected Woods-Saxon potential. The radial wave functions can then be expanded in terms of the harmonic oscillator wave functions $g_{nl}(r)$ as

$$f_{nlj}(r) = \sum_k A_k^{(nlj} g_{kl}(r). \tag{56}$$

The sum in (56) is usually dominated by one large harmonic oscillator component [20] and we can therefore to a fair approximation use

$$f_{nlj}(r) \approx g_{nl}(r). \tag{57}$$

The radial integrals defined in (55) can then be performed analytically and explicit expressions have been given in e.g. [10]. We discuss next a method to construct analytical expressions for the radial matrix elements (55) in the case of harmonic oscillator wave functions.

The radial matrix element $\mathcal{R}^{(0)}_{abL}$ now takes the form

$$\mathcal{R}^{(0)}_{abL} \equiv \langle n_a l_a | j_L(\rho) | n_b l_b \rangle = \int_0^\infty g_{n_a l_a}(r) j_L(\rho) g_{n_b l_b}(r) r^2 \mathrm{d}r, \tag{58}$$

where $r = |\mathbf{r}|$ and

$$g_{nl}(r) = \sqrt{\frac{2n!}{b'^3 \Gamma(n + l + \frac{3}{2})}} (\frac{r}{b'})^l \mathrm{e}^{-\frac{1}{2}(r/b')^2} L_n^{(l+\frac{1}{2})}((r/b')^2). \tag{59}$$

The associated Laguerre polynomials are here defined as

$$L_n^{(k)}(x) \equiv \sum_{i=0}^{n} \frac{(-1)^i}{i!} \binom{k+n}{n-i} x^i, \tag{60}$$

and they obey the relations (see e.g. [1])

$$\begin{aligned}L_n^{(k)}(x) =& L_n^{(k+1)}(x) - L_{n-1}^{k+1}(x), \\ L_n^{(k+1)}(x) =& \frac{1}{x}\big[(n + k + 1) L_n^{(k)}(x) - (n + 1) L_{n+1}^{(k)}(x)\big].\end{aligned} \tag{61}$$

In (59) the oscillator strength b' is defined by [20]

$$b' \equiv \sqrt{\frac{1}{m_{\mathrm{N}}\omega}}, \tag{62}$$

where the harmonic oscillator frequency ω for a nucleus with A nucleons is approximately given by

$$\omega = \frac{1}{\hbar c}\left(45A^{-1/3} - 25A^{-2/3}\right)\,(\mathrm{fm})^{-1}. \tag{63}$$

With the substitution $x = r/b'$ we can write (58) as

$$\mathcal{R}^{(0)}_{abL} = \int_0^\infty \tilde{g}_{n_a l_a}(x) j_L(qb'x)\tilde{g}_{n_b l_b}(x)x^2\mathrm{d}x, \tag{64}$$

where $\tilde{g}_{nl}(x) \equiv b^{3/2}g_{nl}(x)$. The spherical Bessel functions $j_L(r)$ are given in terms of Bessel functions of the first kind as

$$j_L(r) = \sqrt{\frac{\pi}{2r}}J_{L+1/2}(r) \tag{65}$$

and by exploiting the integral representation of the associated Laguerre polynomials [22]

$$\mathrm{e}^{-x}x^{k/2}L_n^{(k)}(x) = \frac{1}{n!}\int_0^\infty \mathrm{e}^{-t}t^{n+k/2}J_k(2\sqrt{tx})dt \quad \text{for } n = 0, 1, \ldots \text{ and } k > -1, \tag{66}$$

one can derive the result

$$\mathcal{R}^{(0)}_{abL} = \sqrt{\frac{n_a!n_b!\pi}{4\Gamma(n_1 + l_1 + \frac{3}{2})\Gamma(n_b + l_b + \frac{3}{2})}}\mathrm{e}^{-y}y^{L/2} \times \sum_{\kappa_a=0}^{n_a}\sum_{\kappa_b=0}^{n_b} \Lambda_{\kappa_a}(n_a l_a)\Lambda_{\kappa_b}(n_b l_b)L_n^{(L+1/2)}(y)n!, \tag{67}$$

where we have introduced $y \equiv (qb'/2)^2$, $n \equiv (l_a + l_b - L)/2 + \kappa_a + \kappa_b$ and

$$\Lambda_\kappa(nl) \equiv \frac{(-1)^\kappa}{\kappa!}\binom{n + l + \frac{1}{2}}{n - \kappa}. \tag{68}$$

The sums in (67) can subsequently be inverted by using the relation

$$\sum_{\alpha=0}^{\gamma+m}\sum_{\beta=0}^{\alpha+m} C^\alpha_\beta x^\beta = \sum_{\beta=0}^{\gamma+m}\sum_{\alpha=\lambda}^{\gamma} C^\alpha_\beta x^\beta, \tag{69}$$

where $\lambda = \max(0, \beta - m)$ and one then obtains that the matrix elements $\mathcal{R}^{(0)}_{abL}$ can be written on the form

$$\mathcal{R}^{(0)}_{abL} = \langle n_a l_a | j_L(\rho) | n_b l_b\rangle = \mathrm{e}^{-y}y^{L/2}\sum_{\kappa=0}^{n_a+n_b+m} \varepsilon^L_\kappa(n_a l_a n_b l_b)y^\kappa, \tag{70}$$

with the coefficients

$$\varepsilon_{\kappa}^{L}(n_a l_a n_b l_b) \equiv \sqrt{\frac{n_a! n_b! \pi}{4\Gamma(n_a + l_a + \frac{3}{2})\Gamma(n_b + l_b + \frac{3}{2})}} \times \sum_{\kappa_a=\alpha}^{n_a} \sum_{\kappa_b=\beta}^{n_b} \Lambda_{\kappa}(nL)\Lambda_{\kappa_a}(n_a l_a)\Lambda_{\kappa_b}(n_b l_b) n!, \tag{71}$$

and

$$\begin{aligned} m =& (l_a + l_b - L)/2, \\ \alpha =& \max(0, \kappa - m - n_b), \\ \beta =& \max(0, \kappa - m - \kappa_a). \end{aligned} \tag{72}$$

The remaining matrix elements of (55) are given in the harmonic-oscillator basis as

$$\mathcal{R}_{abL}^{(-)} = \langle n_a l_a | j_L(\rho) \big(\frac{\mathrm{d}}{\mathrm{d}\rho} - \frac{l_b}{\rho}\big) | n_b l_b \rangle = \int_0^{\infty} \tilde{g}_{n_a, l_a}(x) j_L(qb'x) \big(\frac{\mathrm{d}}{\mathrm{d}(qb'x)} - \frac{l_b}{qb'x}\big) \tilde{g}_{n_b, l_b}(x) dx, \tag{73}$$

and

$$\mathcal{R}_{abL}^{(+)} = \langle n_a l_a | j_L(\rho) \big(\frac{\mathrm{d}}{\mathrm{d}\rho} + \frac{l_b + 1}{\rho}\big) | n_b l_b \rangle = \int_0^{\infty} \tilde{g}_{n_a, l_a}(x) j_L(qb'x) \big(\frac{\mathrm{d}}{\mathrm{d}(qb'x)} + \frac{l_b + 1}{qb'x}\big) \tilde{g}_{n_b, l_b}(x) dx. \tag{74}$$

With the relations (61) for the associated Laguerre polynomials one can show that

$$\big(\frac{\mathrm{d}}{\mathrm{d}(qb'x)} - \frac{l_b}{qb'x}\big) \tilde{g}_{n_b, l_b}(x) = -\frac{1}{2\sqrt{y}} \big(\sqrt{n_b} \tilde{g}_{n_b - 1, l_b + 1}(x) + \sqrt{n_b + l_b + \tfrac{3}{2}} \tilde{g}_{n_b, l_b + 1}(x)\big), \tag{75}$$

and

$$\big(\frac{\mathrm{d}}{\mathrm{d}(qb'x)} + \frac{l_b + 1}{qb'x}\big) \tilde{g}_{n_b, l_b}(x) = \frac{1}{2\sqrt{y}} \big(\sqrt{n_b + l_b + \tfrac{1}{2}} \tilde{g}_{n_b, l_b - 1}(x) + \sqrt{n_b + 1} \tilde{g}_{n_b + 1, l_b - 1}(x)\big). \tag{76}$$

The radial matrix elements $\mathcal{R}_{abL}^{(\pm)}$ are therefore given by

$$\mathcal{R}_{abL}^{(\pm)} = y^{(L-1)/2} \mathrm{e}^{-y} \sum_{\kappa=0}^{\kappa_{\max}} \zeta_{\kappa, L}^{(\pm)}(n_a l_a n_b l_b) y^{\kappa}, \tag{77}$$

where

$$\begin{aligned} \zeta_{\kappa, L}^{(-)}(n_a l_a n_b l_b) =& -\frac{1}{2} \big[\sqrt{n_b} \varepsilon_{\kappa}^{L}(n_a l_a n_b - 1 l_b + 1) + \\ & \sqrt{n_b + l_b + \tfrac{3}{2}} \varepsilon_{\kappa}^{L}(n_a l_a n_b l_b + 1)\big]; \qquad \text{for } \kappa < \kappa_{\max}, \\ \zeta_{\kappa_{\max}, L}^{(-)}(n_a l_a n_b l_b) =& -\frac{1}{2} \sqrt{n_a + l_b + \tfrac{3}{2}} \varepsilon_{\kappa}^{L}(n_a l_a n_b l_b + 1), \end{aligned} \tag{78}$$

and

$$\begin{aligned}
\zeta^{(+)}_{\kappa,L}(n_a l_a n_b l_b) =& \frac{1}{2}\Big[\sqrt{n_b+1}\varepsilon^L_\kappa(n_a l_a n_b + 1 l_b - 1) + \\
& \sqrt{n_b + l_b + \tfrac{1}{2}}\varepsilon^L_\kappa(n_a l_a n_b l_b - 1)\Big]; \qquad \text{for } \kappa < \kappa_{\max}, \\
\zeta^{(+)}_{\kappa_{\max},L}(n_a l_a n_b l_b) =& \frac{1}{2}\sqrt{n_b+1}\varepsilon^L_\kappa(n_a l_a n_b + 1 l_b - 1).
\end{aligned} \tag{79}$$

Here we have introduced $\kappa_{\max} \equiv n_1 + n_2 + (l_1 + l_2 + 1 - L)/2$ and the coefficients $\varepsilon^L_\kappa(n_a l_a n_b l_b)$ are defined by (71).

2.1.4. Nucleon Form Factors

As discussed in Sec. 2.1.2. the involved single-particle operators given by (52) depend on the neutral-current Dirac ($F_1^{\rm NC:p(n)}(Q^2)$) and Pauli ($F_2^{\rm NC:p(n)}(Q^2)$) nucleon form factors and the axial-vector form factors $F_{\rm A}^{\rm NC:p(n)}(Q^2)$. The form factors for protons (p) and neutrons (n) respectively take the forms (see e.g. [4])

$$\begin{aligned}
F_{1,2}^{\rm NC:p(n)}(Q^2) =& \pm\frac{1}{2}\left[F_{1,2}^{\rm EM:p}(Q^2) - F_{1,2}^{\rm EM:n}(Q^2)\right] - 2\sin^2\theta_{\rm W} F_{1,2}^{\rm EM:p(n)}(Q^2) \\
& - \frac{1}{2}F_{1,2}^{\rm s}(Q^2), \\
F_{\rm A}^{\rm NC:p(n)}(Q^2) =& \pm\frac{1}{2}G_{\rm A}(Q^2) - \frac{1}{2}G_{\rm A}^{\rm s}(Q^2),
\end{aligned} \tag{80}$$

where $F_{1,2}^{\rm EM:p(n)}(Q^2)$ are the corresponding form factors for the electromagnetic current. In (80) the (+) and (−) signs refer to protons and neutrons respectively and θ_W denotes the so-called Weinberg angle. The axial form factor $G_{\rm A}(Q^2)$ in (80) is of the form

$$G_{\rm A}(Q^2) = G_{\rm A}(0)G_{\rm D}(Q^2), \tag{81}$$

with the static value $G_{\rm A}(0) = -1.267$ [8]. In (81) the dipole form factor $G_{\rm D}(Q^2)$ is given by

$$G_{\rm D}(Q^2) = \frac{1}{\left(1+\frac{Q^2}{M^2}\right)^2}, \tag{82}$$

with the axial mass $M = M_{\rm A} = 1.014$ MeV. For the strange-quark form factors $F_1^{\rm s}(Q^2)$, $F_2^{\rm s}(Q^2)$ and $G_{\rm A}^{\rm s}(Q^2)$ in (80) we assume that

$$\begin{aligned}
F_1^{\rm s}(Q^2) =& \frac{F_1^{\rm s}Q^2}{(1+\tau)\left(1+\frac{Q^2}{M_{\rm V}^2}\right)^2}, \\
F_2^{\rm s}(Q^2) =& \frac{\mu_{\rm s}}{(1+\tau)\left(1+\frac{Q^2}{M_{\rm V}^2}\right)^2}, \\
G_{\rm A}^{\rm s} =& \frac{G_{\rm A}^{\rm s}(0)}{\left(1+\frac{Q^2}{M_{\rm A}^{\rm s2}}\right)^2},
\end{aligned} \tag{83}$$

where $\tau = Q^2/4m_{\rm N}^2$ and the fit parameters are given by [4]

$$\begin{aligned} \mu_{\rm s} &= -0.39 \pm 0.70, \\ F_1^{\rm s} &= 0.49 \pm 0.70\,\text{GeV}^{-2}, \\ G_{\rm A}^{\rm s}(0) &= -0.13 \pm 0.09, \\ M_{\rm A}^{\rm s} &= 1.049 \pm 0.023\,\text{GeV}. \end{aligned} \tag{84}$$

The electromagnetic Dirac ($F_1^{\rm EM:p(n)}(Q^2)$) and Pauli ($F_2^{\rm EM:p(n)}(Q^2)$) form factors can be expressed in terms of the commonly used electric ($G_{\rm E}^{\rm p(n)}(Q^2)$) and magnetic ($G_{\rm M}^{\rm p(n)}(Q^2)$) Sachs form factors as

$$\begin{aligned} F_1^{\rm EM:p(n)}(Q^2) &= \frac{G_{\rm E}^{\rm p(n)}(Q^2) + \tau G_{\rm M}^{\rm p(n)}(Q^2)}{1+\tau}, \\ F_2^{\rm EM:p(n)}(Q^2) &= \frac{G_{\rm M}^{\rm p(n)}(Q^2) - G_{\rm E}^{\rm p(n)}(Q^2)}{1+\tau}. \end{aligned} \tag{85}$$

The neutral-current form factors $F_1^{\rm NC:p(n)}(Q^2)$ and $F_2^{\rm NC:p(n)}(Q^2)$ of (80) can consequently be written as

$$\begin{aligned} F_1^{\rm NC:p(n)}(Q^2) &= \pm\frac{1}{2}\frac{\left[G_{\rm E}^{\rm p}(Q^2) - G_{\rm E}^{\rm n}(Q^2)\right] + \tau\left[G_{\rm M}^{\rm p}(Q^2) - G_{\rm M}^{\rm n}(Q^2)\right]}{1+\tau} \\ &\quad - 2\sin^2\theta_{\rm W}\frac{G_{\rm E}^{\rm p(n)}(Q^2) + \tau G_{\rm M}^{\rm p(n)}(Q^2)}{1+\tau} - \frac{1}{2}F_1^{\rm s}(Q^2), \\ F_2^{\rm NC:p(n)}(Q^2) &= \pm\frac{1}{2}\frac{\left[G_{\rm M}^{\rm p}(Q^2) - G_{\rm M}^{\rm n}(Q^2)\right] - \left[G_{\rm E}^{\rm p}(Q^2) - G_{\rm E}^{\rm n}(Q^2)\right]}{1+\tau} \\ &\quad - 2\sin^2\theta_{\rm W}\frac{G_{\rm M}^{\rm p(n)}(Q^2) - G_{\rm E}^{\rm p(n)}(Q^2)}{1+\tau} - \frac{1}{2}F_2^{\rm s}(Q^2). \end{aligned} \tag{86}$$

The proton form factors are usually extracted from elastic electron-proton scattering experiments, while the neutron form factors are fitted to reproduce electron-nucleus scattering data (see e.g. [3]). In this chapter we adopt the parametrizations of [3] and the magnetic form factors therefore take the forms

$$\frac{G_{\rm M}^{\rm p(n)}(Q^2)}{\mu_{\rm p(n)}} = \frac{1 + a_{\rm p(n),1}^{\rm M}\tau}{1 + b_{\rm p(n),1}^{\rm M}\tau + b_{\rm p(n),2}^{\rm M}\tau^2 + b_{\rm p(n),3}^{\rm M}\tau^3}, \tag{87}$$

where $\mu_{\rm p(n)}$ is the magnetic moment of the proton (neutron), and the electric form factor of the proton is given by

$$G_{\rm E}^{\rm p}(Q^2) = \frac{1 + a_{\rm p,1}^{\rm E}\tau}{1 + b_{\rm p,1}^{\rm E}\tau + b_{\rm p,2}^{\rm E}\tau^2 + b_{\rm p,3}^{\rm E}\tau^3}. \tag{88}$$

In the present calculations the fit parameters are those of [3] and they are presented in Table 1. For the electric neutron form factor $G_{\rm E}^{\rm n}(Q^2)$ of (86) we use the so-called Galster-like parametrization

$$G_{\rm E}^{\rm n}(Q^2) = \frac{A\tau}{1+B\tau}G_{\rm D}(Q^2), \tag{89}$$

with $A = 1.68$ and $B = 3.63$ [3]. In (89) $G_{\rm D}(Q^2)$ is of the form (82) with the cut-off mass $M^2 = M_{\rm V}^2 = 0.71\,\text{GeV}^2$.

Table 1. Adopted values [3] of the fit parameters for the proton electric and proton and neutron magnetic form factors.

k	$a^{\mathrm{M}}_{\mathrm{p},k}$	$b^{\mathrm{M}}_{\mathrm{p},k}$	$a^{\mathrm{M}}_{\mathrm{n},k}$	$b^{\mathrm{M}}_{\mathrm{n},k}$	$a^{\mathrm{E}}_{\mathrm{p},k}$	$b^{\mathrm{E}}_{\mathrm{p},k}$
1	1.09	12.31	8.28	21.30	-0.19	11.12
2	-	25.57	-	77	-	15.16
3	-	30.61	-	238	-	21.25

2.2. QRPA and MQPM

In the present work the QRPA (quasiparticle random-phase approximation) is used to construct the final and initial nuclear states of the even-even isotopes and the MQPM (microscopic quasiparticle-phonon model) is adopted for ^{95}Mo and ^{97}Mo. In this section the formalisms of these two models are summarized. For more detailed treatments we refer to [20] (QRPA) and [23] (MQPM). In the MQPM the states of an odd nucleus are formed by coupling together one-quasiparticle states and QRPA phonons. The first step is to generate the quasiparticles in a BCS calculation. These quasiparticles are defined via the Bogolyubov-Valatin transformation as

$$\begin{cases} a^\dagger_\alpha = u_a c^\dagger_\alpha + v_a \tilde{c}_\alpha, \\ \tilde{a}_\alpha = u_a \tilde{c}_\alpha - v_a c^\dagger_a, \end{cases} \tag{90}$$

where the conventions introduced in Sec. 2.1.2. are used and we thus have e.g. $\alpha = (a, m_\alpha)$ with $a = (n_a, l_a, j_a)$

In the calculations the parameters of the BCS calculation are fitted so that the experimental pairing gaps for protons and neutrons respectively are reproduced. The experimental pairing gaps are determined from the three-point formulae [20]

$$\begin{cases} \Delta_{\mathrm{p}} = \frac{1}{4}(-1)^{Z+1}(S_{\mathrm{p}}(A+1, Z+1) - 2S_{\mathrm{p}}(A, Z) + S_{\mathrm{p}}(A-1, Z-1)), \\ \Delta_{\mathrm{n}} = \frac{1}{4}(-1)^{N+1}(S_{\mathrm{n}}(A+1, Z) - 2S_{\mathrm{n}}(A, Z) + S_{\mathrm{n}}(A-1, Z)), \end{cases} \tag{91}$$

where A is the mass number and Z and N are the proton and neutron numbers of the nucleus. In (91) S_{p} and S_{n} denote the experimental separation energies for protons and neutrons respectively.

In the next step the QRPA phonons for the even-even nuclei are constructed by using two-quasiparticle operators coupled to good angular momentum. The QRPA creation operator for an excited state $\omega = \{J_\omega, \pi_\omega, k_\omega\}$ of the even-even nucleus is then given by

$$Q^\dagger_\omega = \sum_{a \leqslant a'} (X^\omega_{aa'} \sigma^{-1}_{aa'} [a^\dagger_a a^\dagger_{a'}]_{J_\omega M_\omega} + Y^\omega_{aa'} \sigma^{-1}_{aa'} [\tilde{a}_a \tilde{a}_{a'}]_{J_\omega M_\omega}), \tag{92}$$

where J_ω and π_ω are the angular momentum and parity respectively and k_ω enumerates the phonons. In (92) we define $\sigma_{aa'} = \sqrt{1 + \delta_{aa'}}$ and the sum runs over all two-proton and two-neutron configurations in such a way that none of them is counted twice. The amplitudes $X^\omega_{aa'}$ and $Y^\omega_{aa'}$ are then solved from the QRPA equations which can be written on the form [20]

$$\begin{pmatrix} \mathbf{A} & \mathbf{B} \\ -\mathbf{B}^* & -\mathbf{A} \end{pmatrix} \begin{pmatrix} \mathbf{X}^\omega \\ \mathbf{Y}^\omega \end{pmatrix} = E_\omega \begin{pmatrix} \mathbf{X}^\omega \\ \mathbf{Y}^\omega \end{pmatrix}, \tag{93}$$

where $\mathbf{A}$ is the QTDA (Quasiparticle Tamm-Dancoff approximation) matrix and the ground-state correlations are taken into account by the matrix $\mathbf{B}$. In (93) the excitation energy of the state ω is given by E_ω.

The states of the odd nuclei are produced by application of the MQPM formalism. In the MQPM the quasiparticles and QRPA phonons of the reference even-even nucleus are coupled together to form three-quasiparticle configurations which together with the one-quasiparticle components represent the states of the odd-A nucleus next to the reference nucleus. The creation operator for the kth MQPM state of angular momentum j has then the form

$$\Gamma_k^\dagger(jm) = \sum_n X_n^k a_{njm}^\dagger + \sum_{a\omega} X_{a\omega}^k [a_a^\dagger Q_\omega^\dagger]_{jm}, \tag{94}$$

where the amplitudes X_n^k and $X_{a\omega}^k$ are obtained by solving the corresponding equations of motion [23]. While solving these equations special care is to be taken to account for the overcompleteness and non-orthogonality of the quasiparticle-phonon basis.

3. Results

In this chapter we calculate cross sections for neutral-current neutrino-scattering off the stable ($A = 92, 94, 95, 96, 97, 98, 100$) molybdenum isotopes. The QRPA (quasiparticle random-phase approximation [20]) is used to calculate the nuclear wave functions of the initial and final nuclear states for the even-even isotopes. The MQPM (microscopic quasiparticle-phonon model [23]) was used in [25] to construct the nuclear states of ^{95}Mo and ^{95}Mo. These wave functions are adopted for the odd nuclei.

In Fig. 3 we show the computed energy spectra for ^{94}Mo and ^{98}Mo together with experimental data [2, 19]. It is seen in the figure that the calculated results are in fair agreement with the experimental data. The calculated results are also in good agreement with the higher-QRPA calculations which were performed in [17]. The same conclusions are also true for the other even-even Mo isotopes. The MQPM calculated energy spectra for ^{95}Mo and ^{97}Mo are compared with experimental data [9, 6] in Fig. 4. It is seen in the figure that except for the $3/2_1^+$ state in ^{95}Mo the agreement between theory and experiment is almost perfect.

In the next step we apply the formalism outlined in Sec. 2.1. to calculate the double-differential cross section (28). The total cross section is then obtained by integrating over the scattering angle θ numerically and summing over all the discrete final states. The interesting quantity from the experimental point-of-view is the averaged cross section $\langle\sigma\rangle$, which is obtained by folding the cross section $\sigma(E_k)$ with a given spectrum of neutrino energies. In this chapter we focus on neutrinos emitted from a core-collapse supernova. The distance from the center at which the neutrinos are detached from the supernova depends on the neutrino flavor. For each neutrino flavor a neutrino sphere is therefore defined characterized by a neutrino temperature at the surface T_ν. The neutrinos are then to a crude approximation radiated away from the supernova as black-body radiation. The energy spectrum for

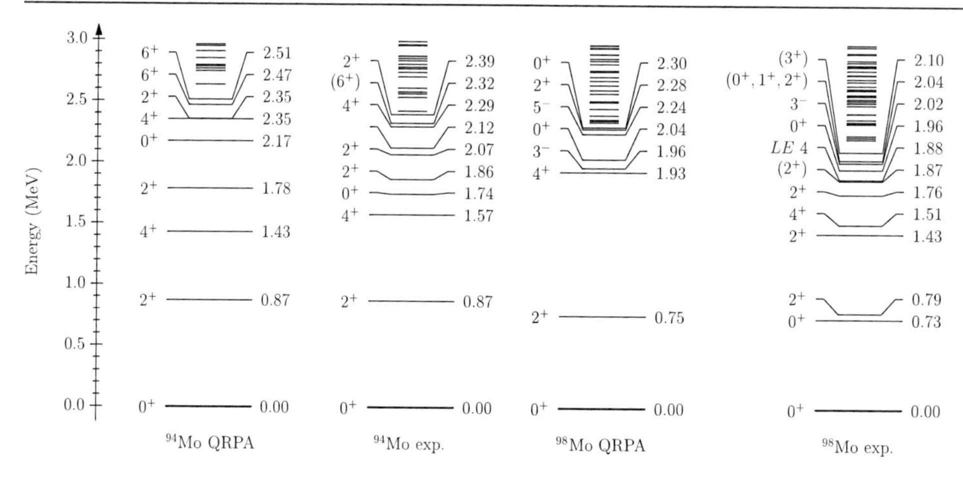

Figure 3. Comparison between the QRPA calculated energy spectra and experimental energies [2, 19] for ^{94}Mo and ^{98}Mo.

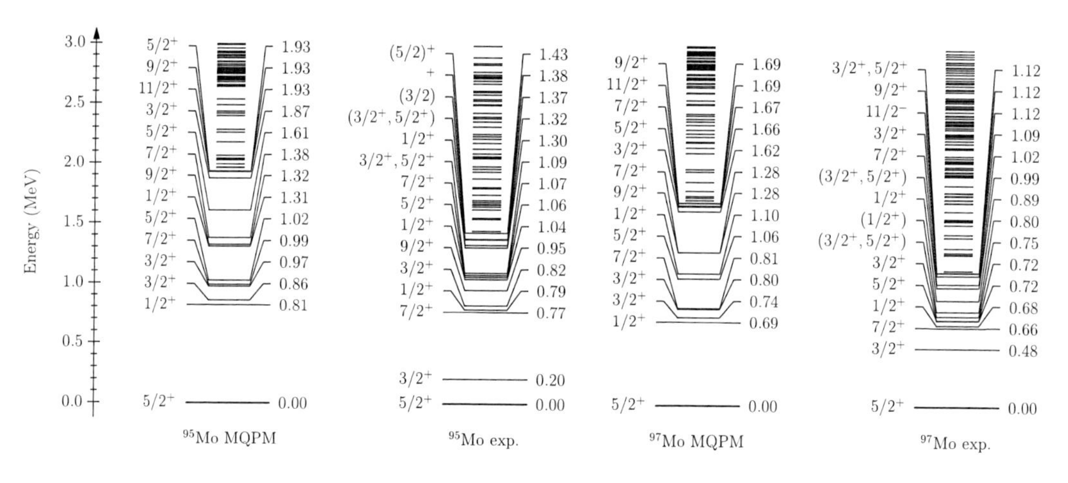

Figure 4. Comparison between the MQPM calculated energy spectra [25] and experimental energies [9, 6] for ^{95}Mo and ^{97}Mo.

neutrinos of a given flavor is in this naive picture described by a thermal spectrum, i.e. a Fermi-Dirac energy spectrum. It has however been recognized [15, 14] that the high-energy tail of the spectra are damped compared to a thermal energy spectrum. In this chapter we consequently adopt a pinched Fermi-Dirac spectrum to describe the distribution of neutrino species. We therefore have that

$$\langle \sigma \rangle = \frac{1}{F_2(\alpha) T^3} \int_0^{\infty} \frac{\sigma(E_k) E_k^2 \mathrm{d}E_k}{1 + \exp(E_k/T - \alpha)}, \tag{95}$$

where T is the neutrino temperature and α is the so-called degeneracy parameter. In (95) the neutrino spectrum is normalized to unit flux by the constant $F_2(\alpha)$.

The outer core of the supernova has a lower proton density than neutron density leading to the situation that reactions of the type $\nu_\mathrm{e}(\mathrm{n},\mathrm{p})\mathrm{e}^-$ dominate over the $\bar{\nu}_\mathrm{e}(\mathrm{p},\mathrm{n})\mathrm{e}^+$ ones. The electron-neutrinos are consequently detached from the supernova at a larger distance from the center compared to electron anti-neutrinos. Neutrinos of the other flavors interact with matter only via neutral-current reactions and are therefore released from the supernova at the smallest radius. The supernova is hotter closer to the center and this leads to the usual hierarchy $T_{\nu_e} < T_{\bar{\nu}_e} < T_{\nu_x} \approx T_{\bar{\nu}_x}$ where ν_x is either a tau- or muon-neutrino. Typical values for the averaged neutrino energies are $\langle E_{\nu_e} \rangle \approx 11$ MeV, $\langle E_{\bar{\nu}_e} \rangle \approx 16$ MeV and $\langle E_{\nu_x} \rangle \approx \langle E_{\bar{\nu}_x} \rangle \approx 25$ MeV [18].

In Table 2 we present the calculated averaged cross sections for incoherent scattering off the ^{92,94,95,96,97,98,100}Mo isotopes for $\alpha = 0.0$ and $\alpha = 3.0$ respectively and neutrino temperatures corresponding to the aforementioned neutrino energies. The results for coherent scattering are similarly shown in Table 3. Only the incoherent cross section can be measured but coherent neutrino-nucleus scattering reactions play an important role in supernova explosions and the coherent cross sections are therefore important inputs for supernova simulations. It is seen in the tables that the coherent cross sections are much larger than the incoherent ones for all nuclei under consideration. We can also conclude that both the incoherent and coherent cross sections increase considerably with increasing neutrino temperature. In Table 2 and Table 3 it is also visible that the cross sections for both incoherent and coherent scattering are notably larger for the even-even isotopes than for the odd ones. By (28) this is due to the non-zero angular momenta of the ground states of the odd nuclei ($J_i = 5/2$) that lead to a statistical factor that suppresses the cross sections of the odd nuclei by a factor of six relative to the cross sections of the even-even nuclei.

Table 2. Averaged incoherent cross sections in units of 10^{-42} cm^2 for different values of the neutrino temperature T and the parameter α for the nuclei under consideration. The same energy spectra are assumed for tau- and muon neutrinos and thus $x = \mu, \tau$.

flavor	α	T (MeV)	$\langle\sigma\rangle^{A=92}_{\text{incoh}}$	$\langle\sigma\rangle^{A=94}_{\text{incoh}}$	$\langle\sigma\rangle^{A=95}_{\text{incoh}}$	$\langle\sigma\rangle^{A=96}_{\text{incoh}}$	$\langle\sigma\rangle^{A=97}_{\text{incoh}}$	$\langle\sigma\rangle^{A=98}_{\text{incoh}}$	$\langle\sigma\rangle^{A=100}_{\text{incoh}}$
ν_e	0.0	3.49	8.66	9.33	3.39	9.58	3.19	8.11	12.3
ν_e	3.0	2.76	6.28	6.93	2.94	7.15	2.77	5.99	9.94
$\bar{\nu}_e$	0.0	5.08	28.4	29.7	7.69	30.1	7.25	26.6	35.1
$\bar{\nu}_e$	3.0	4.01	22.7	24.0	6.67	24.4	6.27	21.2	29.4
ν_x	0.0	7.94	130	130	31.8	130	30.3	117	135
ν_x	3.0	6.27	105	106	26.1	107	24.8	95.6	113
$\bar{\nu}_x$	0.0	7.94	106	107	24.2	108	22.9	98.4	116
$\bar{\nu}_x$	3.0	6.27	89.3	91.3	20.6	91.9	19.4	83.1	101

For both ^{95}Mo and ^{97}Mo the incoherent cross section is dominated by a one-quasiparticle $1d_{5/2} \longrightarrow 1d_{3/2}$ transition which is visualised for ^{95}Mo in Fig. 5. The angular momentum of the odd particle is then changed from $j = 5/2$ to $j = 3/2$ by the action of the Pauli spin operator $\boldsymbol{\sigma}$. There are additionally important transitions to excited states of multipolarity $5/2^+$. We have found that these excitations consist of transitions of either the form $1d_{5/2} \longrightarrow 1d_{5/2} \otimes \omega$ or $1d_{5/2} \longrightarrow 1d_{3/2} \otimes \omega$ where ω is a QRPA phonon. The QRPA phonons can be interpreted as collective vibrations of the core even-even nucleus (^{94}Mo and ^{96}Mo respectively). We can consequently simply interpret these final states as states where an odd 1d-particle is coupled to a vibrating core. This physical interpretation is shown for

Table 3. Averaged coherent cross sections in units of 10^{-40} cm^2 for different values of the neutrino temperature T and the parameter α for the nuclei under consideration. The same energy spectra are assumed for tau- and muon neutrinos and thus $x = \mu, \tau$.

flavor	α	T (MeV)	$\langle\sigma\rangle_{\rm coh}^{A=92}$	$\langle\sigma\rangle_{\rm coh}^{A=94}$	$\langle\sigma\rangle_{\rm coh}^{A=95}$	$\langle\sigma\rangle_{\rm coh}^{A=96}$	$\langle\sigma\rangle_{\rm coh}^{A=97}$	$\langle\sigma\rangle_{\rm coh}^{A=98}$	$\langle\sigma\rangle_{\rm coh}^{A=100}$
ν_e	0.0	3.49	4.64	5.31	0.993	6.03	1.02	6.79	7.60
ν_e	3.0	2.76	4.35	4.98	0.931	5.65	0.959	6.37	7.13
$\bar{\nu}_e$	0.0	5.08	9.38	10.7	2.00	12.2	2.06	13.7	15.3
$\bar{\nu}_e$	3.0	4.01	8.86	10.1	1.89	11.5	1.95	13.0	14.5
ν_x	0.0	7.94	20.5	23.4	4.37	26.5	4.50	29.8	33.3
ν_x	3.0	6.27	19.8	22.7	4.23	25.7	4.36	28.9	32.3
$\bar{\nu}_x$	0.0	7.94	20.5	23.4	4.35	26.5	4.48	29.8	33.3
$\bar{\nu}_x$	3.0	6.27	19.8	22.7	4.21	25.7	4.34	28.9	32.3

^{95}Mo in Fig. 6.

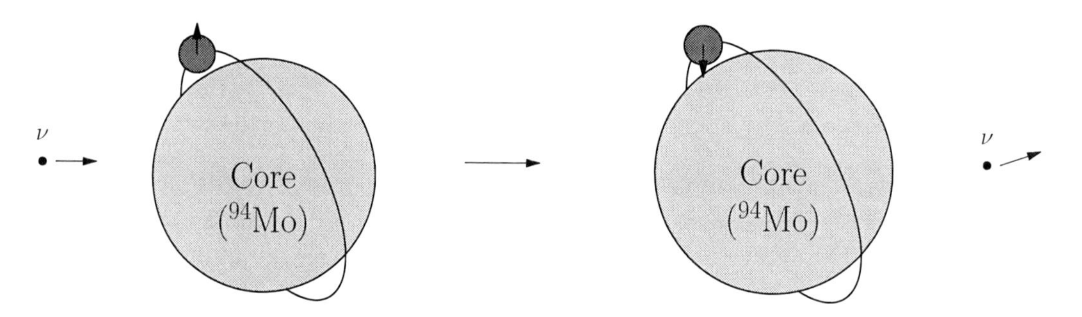

Figure 5. Neutral-current neutrino-nucleus scattering to the $3/2_1^+$ state in ^{95}Mo. The vertical arrows indicate the initial and final spins of the odd particle.

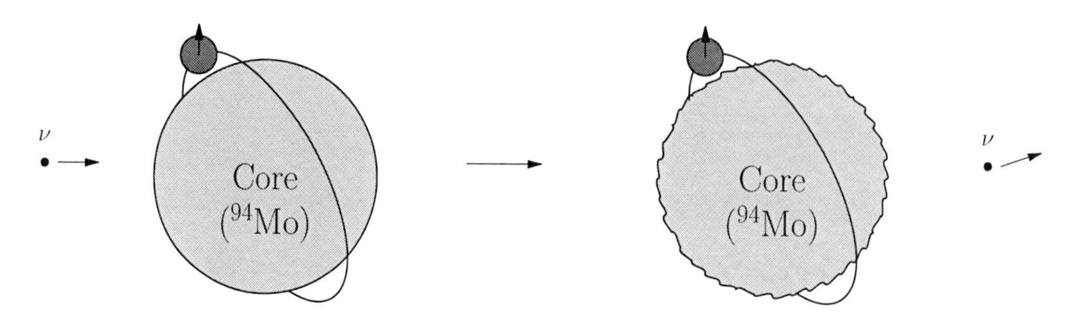

Figure 6. Physical interpretation of neutral-current neutrino-nucleus scattering to excited states of multipolarity $5/2^+$ in ^{95}Mo. The vertical arrows indicate the initial and final spins of the odd particle.

4. Conclusion

We have in this chapter reviewed the theoretical framework for calculations of neutral-current neutrino scattering off nuclei. We have then used the formalism to calculate cross sections for the stable ($A = 92, 94, 95, 96, 97, 98, 100$) molybdenum isotopes. Realistic estimates of the nuclear responses to supernova neutrinos have been computed by folding the obtained cross sections with a two-parameter Fermi-Dirac energy distribution. For the even-even nuclei the QRPA (quasiparticle random-phase approximation) has been employed to compute the wave functions of the initial and final nuclear states. The nuclear wave functions of ^{95}Mo and ^{97}Mo have been constructed with the MQPM (microscopic quasiparticle-phonon model).

For all the aforementioned nuclear targets the coherent cross sections are dominant. The calculated nuclear responses for the even-even nuclei are considerably larger than the ones for the odd nuclei. This result is a consequence of the non-zero angular momenta of the ground states of the odd nuclei. We have additionally found that for the neutral-current neutrino scattering off the odd isotopes (^{95}Mo and ^{97}Mo) the prominent final states can simply be interpreted as either a spin flip of a nucleon in the 1d-shell or as a 1d-nucleon coupled to a vibrational excitation of the core.

Acknowledgements

This work was supported by the Academy of Finland under the Finnish Center Of Excellence Program 2006-2011 (Nuclear and Accelerator Based Program at JYFL). E. Ydrefors has been partly supported by a grant from the Ellen and Artturi Nyyssönen Foundation.

References

[1] M. Abramowitz and I. A. Stegun. *Handbook of Mathematical Functions with Formulas, Graphs and Mathematical Tables*. U.S. Department of Commerce, New York, 2002.

[2] D. Abriola and A. A. Sonzogni. Nuclear Data Sheets for $A = 94$. *Nucl. Data Sheets*, **107**:2423–2578, 2006.

[3] W. M. Alberico, S. M. Bilenky, C. Giunti, and K. M. Graczyk. *Phys. Rev. C*, **79**:065204, 2009.

[4] W. M. Alberico, S. M. Bilenky, and C. Maieron. *Phys. Rep.*, **358**:227–308, 2002.

[5] R. Arnold, C. Augier, J. Baker, A. Barabash, G. Broudin, V.Brudanin, A. J. Caffrey, E. Caurier, V. Egorov, K. Errahmane, A. I. Etienvre, J. L. Guyonnet, F. Hubert, Ph. Hubert, C. Jollet, S. Jullian, O. Kochetov, V. Kovalenko, S. Konovalov, D. Lalanne, F. Leccia, C. Longuemare, G. Lutter, Ch. Marquet, F. Mauger, F. Nowacki, H. Ohsumi, F. Piquemal, J. L. Reyss, R. Saakyan, X. Sarazin, L. Simard, F. Şimkovic, Yu. Shitov, A. Smolnikov, L. Ştekl, J. Suhonen, C. S. Sutton, G. Szklarz, J. Thomas, V. Timkin, V. Tretyak, V. Umatov, L. Vàla, I. Vanushin, V. Vasilyev, V. Vorobel, and Ts. Vylov. *Phys Rev. Lett.*, **95**:182302, 2005.

[6] A. Artna-Cohen. Nuclear Data Sheets Update for $A = 97$. *Nucl. Data Sheets*, **70**:85–172, 1993.

[7] R. M. Bionta, G. Blewitt, C. B. Bratton, D. Casper, A. Ciocio, R. Claus, B. Cortez, M. Crouch, S. T. Dye, S. Errede, G. W. Foster, W. Gajewski, K. S. Ganezer, M. Goldhaber, T. J. Haines, T. W. Jones, D. Kielczewska, W. R. Kropp, J. G. Learned, J. M. LoSecco, J. Matthews, R. Miller M. S. Mudan, H. S. Park, L. R. Price, F. Reines, J. Schultz, S. Seidel, E. Shumard ad D. Sinclair, H. W. Sobel, J. L. Stone, L. R. Sulak, R. Svoboda, G. Thornton, J. C. van der Velde, and C. Wuest. Observation of a neutrino burst in coincidence with supernova 1987 a in the large magellanic cloud. *Phys. Rev. Lett.*, **58**:1494–1496, 1987.

[8] H. Budd, A. Bodek, and J. Arrington. Modeling quasi-elastic form factors for electron and neutrino scattering. *arXiv:hep-ex/0308005v2*, 2010.

[9] T. W. Burrows. Nuclear Data Sheets Update for $A = 95$. *Nucl. Data Sheets*, **68**:635–746, 1993.

[10] V. C. Chasioti and T. S. Kosmas. A unified formalism for the basic nuclear matrix elements in semi-leptonic processes. *Nucl. Phys A*, **829**:234–252, 2009.

[11] T. W. Donnelly and R. D. Peccei. Neutral current effects in nuclei. *Phys. Rep.*, **50**:1–85, 1979.

[12] H. Ejiri. Nuclear isospin responses for low-energy neutrinos. *Phys. Rep.*, **338**:265–351, 2000.

[13] H. Ejiri, P. Doe, S. R. Elliot, M. Finger Jr., M. Finger, K. Fushimi, V. Gehman, M. Greenfield, R. Hazama, P. Kavitov, V. Kekelidze, H. Nakamura, M. Nomachi, R. G. H. Robertson, T. Shima, M. Slunecka, G. Shirkov, A. Sissakian, A. Titov, S. Umehara, V. Vaturin, V. Voronov, J.F. Wilkerson, D. I. Will, S. Yoshida, and V. Vrba. MOON for neutrino-less double beta decays. *Eur. Phys. J. Special Topics*, **162**:239–250, 2008.

[14] Hans-Thomas Janka and W. Hillebrandt. Monte carlo simulations of neutrino transport in type ii supernovae. *Astron. Astrophys. Suppl. Ser.*, **78**:375–397, 1989.

[15] Mathias Th. Keil and Georg G. Raffelt. Monte Carlo study of supernova neutrino spectra formation. *Astrophys. J.*, **590**:971–991, 2003.

[16] E. Kolbe, K. Langanke, G. Martínez-Pinedo, and P. Vogel. Neutrino-nucleus reactions and nuclear structure. *J. Phys. G*, **29**:2569–2596, 2003.

[17] J. Kotila, J. Suhonen, and D. S. Delion. *Nucl. Phys. A*, **765**:354, 2006.

[18] Y.-Z. Qian, W. C. Haxton, K. Langanke, and P. Vogel. Neutrino-induced neutron spallation and supernova r-process nucleosynthesis. *Phys. Rev. C*, **55**:1532–1544, 1997.

[19] B. Singh and Z. Hu. Nuclear Data Sheets for $A = 98$. *Nucl. Data Sheets*, **98**:335–513, 2003.

[20] J. Suhonen. *From Nucleons to Nucleus: Concepts of Microscopic Nuclear Theory*. Springer, Berlin, 2007.

[21] Jouni Suhonen and Osvaldo Civitarese. Weak-interaction and nuclear-structure aspects of nuclear double beta decay. *Phys. Rep.*, **300**:123–214, 1998.

[22] Gabor Szegö. *Orthogonal Polynomials*. American Mathematical Society Colloquium Publications, Providence, 1975.

[23] Jussi Toivanen and Jouni Suhonen. Microscopic quasiparticle-phonon description of odd-mass $^{127-133}$Xe isotopes and their β decay. *Phys. Rev. C*, **57**:1237–1246, 1998.

[24] John Dirk Walecka. *Theoretical Nuclear and Subnuclear Physics*. Imperial College Press, London, 2004.

[25] E. Ydrefors, M. T. Mustonen, and J. Suhonen. MQPM description of the structure and beta decays of the odd $A = 95, 97$ Mo and Tc isotopes. *Nucl. Phys A*, **842**:33–47, 2010.

In: Neutrinos: Properties, Sources and Detection
Editor: Joshua P. Greene, pp. 177-184
ISBN: 978-1-61209-650-6

Chapter 8

NEUTRINO MASS IN A SIX DIMENSIONAL E_6 MODEL

Snigdha Mishra and Sarita Mohanty
Department of Physics, Berhampur University,
Berhampur-760 007, Orissa, India

Abstract

We consider a supersymmetric E_6 GUT in a six dimesional $M^4 \otimes T^2$ space. E_6 breaking is achieved by orbifolding the extra two dimensional torus T^2 via a $T^2/(Z_2 \times Z_2' \times Z_2'')$. In effective four dimension we obtain an extended Pati-Salam group with N = 1 SUSY, as a result of orbifold compactification. We then discuss the problem of neutrino mass with imposed parity assignment. A light Dirac neutrino is predicted with mass of the order of 10^{-2} eV.

1. Introduction

There are several evidences in favour of supersymmetric grand unified theories (SUSYGUT) [1] to describe physics beyond the standard model. However it suffers serious problem like the problem of doublet-triplet mass splitting and the closely related problem of dimensional-five proton decay operators. This suggests that SUSY GUTs may require some drastic modification to reconcile symmetry breaking with the doublet-triplet mass splitting. Recently it has been shown [2] that in models with one or two extra dimensions, orbifold compactification can break grand unified symmetries in such a manner that it resolves the splitting problem as well as proton decay problem automatically without any fine-tunning. These ideas have been used to build models based on SU(5) and SO(10) [3] in five and six dimensions. The exceptional groups like E_6, E_7 and E_8 have been examined by Haba et al [4] in five dimension with N = 1 SUSY. In the present paper we wish to examine the idea in case of E_6 with two extra dimensions. The effective four dimensional theory is obtained by orbifolding the extra dimensional space (which is a torus T^2) with Z_2' s. The motivation to study E_6 is that it may be viewed as remnants of superstring theory with E_8 symmetry in ten dimensions.

The paper is organized as follows; In the next section we shall discuss the technical details of the model showing the gauge symmetry breaking explicitly. Section (3) is devoted to the calculation of the four dimensional Yukawa Lagrangian in the neutral sector predicting a light Dirac neutrino. The result and concluding remarks are discussed in the last section.

2. The Model

In the present model, we take an E_6 gauge theory coupled with N=1 SUSY in a six dimensional space as $M^4 \otimes T^2$ where the extra dimensional space is taken as a Torus with radii R_1 and R_2. The breaking of E_6 symmetry with N=1 SUSY is then realised by orbifolding the two dimensional torus T^2 by two discrete groups Z_2' and Z_2'' leading to two different symmetric subgroups $SU(2)\otimes SU(6)$ (G_{26}) and $SO(10)\otimes U(1)_\chi$ ($G_{10,\ 1}$) in the two orthogonal compact dimensions. The combination $Z_2' \otimes Z_2''$ breaks E_6 to the maximal common subgroups of G_{26} and $G_{10,1}$ i.e. the extended Pati-Salam group ($SU(2)_L\otimes SU(2)_R\otimes SU(4) \otimes U(1)_\chi$). Another Z_2 is needed to break the extended supersymmetry (as N=1 SUSY in d=6 is equivalent to N=2 SUSY in four dimensions). The GUT symmetry is broken on the branes which are fixed points of transformations. Here we may note that, we have four fixed points at $Z = (0, 0), \left(\frac{\pi R_1}{2}, 0\right), \left(0, \frac{\pi R_2}{2}\right)$ and $\left(\frac{\pi R_1}{2}, \frac{\pi R_2}{2}\right)$. The corresponding Gauge symmetries are E_6 at (0, 0), $SU(2)\otimes SU(6)$ at $\left(\frac{\pi R_1}{2}, 0\right)$, $SO(10)\otimes U(1)$ at $\left(0, \frac{\pi R_2}{2}\right)$, and the intersection group $SU(2)\otimes SU(2)\otimes SU(4) \otimes U(1)_\chi$ at the $\left(\frac{\pi R_1}{2}, \frac{\pi R_2}{2}\right)$ brane respectively.

Now to visualise the symmetry breaking explicitly we consider N=1 SUSY Yang-Mills theory in six dimensions.

$$\mathcal{L}_{6d}^{YM} = Tr\left[-\frac{1}{2}V_{MN}V^{MN} + i\bar{\Lambda}\Gamma^M D_M \Lambda\right] \tag{1}$$

with $V_M = T^A V_M^A$, T^A are the E_6 generators. M = (μ, 5, 6) for μ = 1, …4 are the four dimensional indices with coordinates x^μ, and 5, 6 being the two extra dimensional indices with coordinates y and z. Λ is the gaugino consisting of (Λ_1,-iΛ_2), the chiral fermions with opposite 4d chirality. $D_M\Lambda = \partial_M\Lambda - ig[V_M, \Lambda]$ and $V_{MN} = \frac{1}{ig}[D_M, D_N]$. Γ's are the six dimensional Γ-matrices. The operation Z_2 on T^2 reduces the unwanted N = 2 SUSY to N=1, by suitably grouping the vector and chiral multiplets, i.e. V = (V_μ, Λ_1) and Σ=($V_{5,6}$, Λ_2) where V and Σ are the matrices in the adjoint representation of E_6. This is realized by the operation of Z_2 on V and Λ such that,

$$P V_\mu(\text{and } \lambda_1)(x,-y,-z)P^{-1} = +V_\mu(\lambda_1)(x,y,z) \quad (2)$$

$$P V_{5,6}(\text{and } \lambda_2)(x,-y,-z)P^{-1} = -V_{5,6}(\lambda_2)(x,y,z) \quad (3)$$

Here P = ± 1 being the eigen value of Z_2. Further the gauge symmetry breaking of E_6 is done by Z_2' and Z_2'' by operation on the two compact dimensions given as,

$$P' V_\mu(\text{and } \lambda_1)\left(x,-y+\frac{\pi R_1}{2},-z\right)P'^{-1} = +V_\mu(\lambda_1)\left(x,y+\frac{\pi R_1}{2},z\right) \quad (4)$$

$$P'' V_\mu(\text{and } \lambda_1)\left(x,-y,-z+\frac{\pi R_2}{2}.\right)P''^{-1} = V_\mu(\lambda_1)\left(x,y,z+\frac{\pi R_2}{2}\right) \quad (5)$$

Here P′ and P″ are the eigen values of Z_2' and Z_2'' respectively. Here we may note that Z'_2 and Z''_2 operate oppositely for $V_{5,6}$ and λ_2 as shown in equation (3). The component fields belonging to E_6 are again divided according to different parity assignments. Thus the orbifolding to $T^2/(Z_2 \otimes Z'_2 \otimes Z_2'')$ can be viewed through the following parity assignments in the subspaces $E_6 \to SU(2) \otimes SU(6)$ (Z'_2) and $E_6 \to SO(10) \otimes U(1)_\chi$ (Z_2'') as noted in Table 1.

Table 1. Parity assignments for the vector multiplets $V_M^A \in$ 78 of E_6

G_{2241}	G_{26}	$G_{10,1}$	$V = (V_\mu, \lambda_1)$			$\Sigma = (V_{5,6}, \lambda_2)$		
			Z_2	Z_2'	Z_2''	Z_2	Z_2'	Z_2''
$(3, 1, 1)_0$	(3, 1)	45_0	+	+	+	-	-	-
$(1, 3, 1)_0$	(1, 35)	45_0	+	+	+	-	-	-
$(1, 1, 15)_0$	(1, 35)	45_0	+	+	+	-	-	-
$(2, 2, 6)_0$	(2, 20)	45_0	+	-	+	-	+	-
$(1, 1, 1)_0$	(1, 35)	1_0	+	+	+	-	-	-
$(2, 1, 4)_{-3}$	(2, 20)	16_{-3}	+	-	-	-	+	+
$(1, 2, \bar{4})_{-3}$	(1, 35)	16_{-3}	+	+	-	-	-	+
$(2, 1, \bar{4})_{-3}$	(2, 20)	$\overline{16}_3$	+	-	-	-	+	+
$(1, 2, 4)_3$	(1, 35)	$\overline{16}_3$	+	+	-	-	-	+

We observe that the vector bosons belonging to $(3, 1, 1)_0 + (1, 3, 1_0) + (1, 1, 15)_0 + (1,1,1)_0$ in the subspace G_{2241} ($SU(2) \otimes SU(2) \otimes SU(4) \otimes U(1)_\chi$) have all parities positive. These vector bosons remain massless at the compactification scale, which will be obvious from the mode expansion of the fields [5] given by

$$\phi_{+++}(x,y,z)=\frac{1}{\pi\sqrt{R_1 R_2}}\Sigma_{m,n}\ \frac{1}{2^{\delta m_0 \delta n_0}}\phi_{+++}^{2m,2n}(x)\cos\left(\frac{2m}{R_1}y+\frac{2n}{R_2}z\right) \tag{6}$$

Here R_1, R_2 are the radii of the Torus T^2. Thus in effective four-dimension, the residual group will be (G_{2241}) ($SU(2) \otimes SU(2) \otimes SU(4) \otimes U(1)_\chi$ i.e. Pati-Salam group extended by $U(1)_\chi$, coupled with N = 1 SUSY.

We shall next consider the matter and the Higgs sector to break G_{2241} model in order to have a viable model consistent with phenomenology. We take two sets of brane localised matter as well as Higgs field in $27_{1,2}$ and $\overline{27}_{1,2}$ representation. The basic idea is to put all matter and Higgs field on a brane such that the structure of original four dimensional E_6 group is left essentially in tact. In other words, wherever matter resides, the Higgs should live on fixed points i.e. in contact with matter fields to give Yukawa couplings.

We shall now choose the parities for the Higgs 27 to project out the zero modes. In the subspaces G_{26}, $G_{10,1}$ the parities Z_2 Z'_2, Z''_2 are assigned corresponding to the following symmetry breaking pattern.

$$E_6 \rightarrow^{M_c} SU(2)\otimes SU(2)\otimes SU(4)\otimes U(1)_\chi \rightarrow^{M_U}$$

$$SU(2)\otimes SU(2)\otimes SU(4) \rightarrow^{M_R} SU(3)\otimes SU(2)\otimes U(1) \rightarrow^{M_W} SU(3)\otimes U(1) \tag{7}$$

Here the 1st stage of breaking is done by orbifolding at the compactification scale $M_c \sim 10^{19}$ GeV and subsequent stages of breaking are to be achieved at $M_U(\sim 10^{15}$ GeV), M_R ($\sim 10^{12}$ GeV) and M_W ($\sim 10^2$ GeV) scales respectively. The parities of the multiplet, $27 = L(2, 1, 4)_1 \oplus R(1, 2, \bar{4})_1 \oplus D(1, 1, 6)_{-2} \oplus H(2, 2, 1)_{-2} \oplus S(1, 1, 1)_4$ are assigned in Table 2. We shall now take the cubic superpotential on (2, 2, 4, 1) brane, given by:

$$W = 27.27.27 = \lambda_1 \, HHS + \lambda_2 \, LRH + \lambda_3 LLD + \lambda_4 RRD + \lambda_5 DDS \tag{8}$$

Following Table 2, we thus have the zero mode sectors as $D(1, 1, 6)_{-2}$ and $S(1, 1, 1)_4$ which have all parities positive. We may have alternate possibilities by choosing Z''_2 of $L(2, 1, 4)_1$ and $R(1, 2, \bar{4})_1$ belonging to $(16)_1$ as +ve instead of $(1, 1, 6)_{-2}$ and $H(2, 2, 1)_{-2}$ belonging to 10_{-2} as + ve. We then have $S(1, 1, 1)_4$ and $L(2, 1, 4)_1$ to be the zero mode sector. In that case, one has to seek for other zero modes for breaking of SU(4) and other SU(2). This will necessiate the enlargement of Higgs sector. Instead we may use the higher dimensional terms more precisely a four fermi interaction term in the Lagrangian. These terms may come as a result of integrating out the heavy modes which attain mass due to orbifold compactification. We then assume the vevs of Higgs scalar (necessary for symmetry breaking at intermediate as well as weak scales) are formed out of heavy fermion (Higgsinos) bilinear condensates, suppressed by the square of compactification scale [6]. We shall show that this proposition will be advantageous to predict light Dirac neutrino [6,7]. We therefore take the parity assignments of Table (2) to communicate with low energy phenomenology.

In order to break the $U(1)_x$ symmetry we assign vev to $S(1,1,1)_4 \sim M_U$ (as shown in equation (7)). But it will automatically give super heavy mass to the exotic colour triplets D, via the term DDS, so that rapid proton decay is nicely avoided. We next consider the intermediate as well as weak scale breaking through the formation of condensates [6] $\langle\overline{\chi}\chi\rangle$ where $\chi\langle\overline{\chi}\rangle$ are the fermion (anti fermion) partner of scalars arising out of higher dimensional terms. The condensates automatically imply the existence of corresponding scalar modes yielding an effective potential of Ginzberg Landau type [8]. We may here note that the presence of condensates could be through some hidden dynamics at the Planck scale as has been considered by Pati [9]. This alternate mechanism is necessary as the doublets H is heavy at the compactification scale. It may be noted that, presence of condensate mechanism, decouples the generation of mass from the generation of vevs [6] allowing a see-saw mechanism to operate, with the condensate scale ~ $\mu \sim \left(M_C^2 M_W\right)^{1/3} \sim 10^{12}$ Gev. We take the following condensates:

$$< \phi^a >_{(2,2,1)} \sim \frac{1}{M_C^2} < \overline{\chi}_{\overline{H}_a} X_{\overline{S}_a} > \qquad (9)$$

$$< \Delta_R >_{(1,3,\overline{10})} \sim \frac{1}{M_C^2} < \overline{\chi}_{\overline{R}_1} X_{\overline{R}_2} > \qquad (10)$$

$$< \Delta_L >_{(3,1,10)} \sim \frac{1}{M_C^2} < \overline{\chi}_{\overline{L}_1} X_{\overline{L}_2} > \qquad (11)$$

Here the superscript a = 1, 2 stand for the two sets of condensates corresponding to the two sets 27_1 and 27_2. $\overline{R}_{1,2}, \overline{L}_{1,2}$ corresponds to the $\overline{27}_{1,2}$ respectively. Here we have taken $< \phi > \sim k \sim (0(M_W))$, $< \Delta_R > \sim \vartheta_R \sim O(M_R)$ and $< \Delta_L > \sim \vartheta_L$, such that $\vartheta_R >> k >> \vartheta_L$ [7].

Table 2. Parity assignments for brane-localised Higgs 27

G'_{ps}	G_{26}	$G_{10,1}$	Z_2	Z'_2	Z''_2
$L(2,1,4)_1$	(1, 15)	16_1	+	+	-
$R(1,2,\overline{4})_1$	$(2,\overline{6})$	16_1	+	-	-
$D(1,1,6)_{-2}$	(1, 15)	10_{-2}	+	+	+
$H(2,2,1)_{-2}$	$(2,\overline{6})$	10_{-2}	+	-	+
$S(1,1,1)_4$	(1, 15)	1_4	+	+	+

Here we may note that the condensates are formed out of heavy Higgsino fields, similar to the attempt for super symmetry breaking through gaugino condensates.

3. Neutrino Mass Matrix

We can now associate the condensate mechanism to generate a light Dirac neutrino by writing down the Yukawa Lagrangian.

$$L = h_1\left[\bar{f}_L^1 f_R^1 \phi_1 + \bar{f}_L^2 f_R^2 \phi_2\right] + h.c. + h_2\left[\bar{f}_L^{c1} f_L^2 \Delta_L + \bar{f}_R^{c1} f_R^2 \Delta_R\right] + h.c. \qquad (12)$$

Here the brane localized matter sector $F^1\{f_L^1, f_L^{c1}\}$ and $F^2\{f_L^2, f_L^{c2}\}$ correspond to qarks and leptons belonging to 27^1, 27^2 respectively. Obviously in the subspace G_{2241}, $f_L^1 = (2,1,4)_1$ and $f_L^{c1} = (1,2,\bar{4})_1$. Similarly the right handed sectors $\{f_R, f_R^c\}$ belong to the conjugate representation $\overline{27}$ such that $f_R^1 = (2,1,\bar{4})_{-1}$ and $f_R^c = (1,2,4)_{-1}$. We have further assumed that one generation of fermion is heaver than the other set. $\phi_{1,2}$, Δ_L, Δ_R are the condensates as taken in equations (9 – 10). Here ϕ_1 (ϕ_2) is formed out of Higgsinos $\bar{\chi}_1 \chi_1 (\bar{\chi}_2 \chi_2)$ belonging to $F^1(F^2)$. However Δ_L, (Δ_R) is formed by $\bar{\chi}_1 \chi_2$ with $\chi_1 \in F^1$ and $\chi_2 \in F^2$. Here we may note that L_y permits Dirac mass term (1st term) as well as Majorana mass term (2nd term) as a result of an imposed parity .The Lagrangian is invariant under the transformation, $F_1 \rightarrow F^1$, $F^2 \rightarrow -F^2$, $\Delta_{L,R} \rightarrow -\Delta_{L,R}$. The interesting thing to observe that such a transformation can be obtained if one takes parity of F^1 (27^1) as +ve with that of F^2 (27^2) as –ve. As has been mentioned before, the condensates $\Delta_{L,R}$ are formed out of the Higgsinos belonging to (27^1) and (27^2), so that $\Delta_{L,R} \rightarrow -\Delta_{L,R}$ can be achieved. In fact this discrete symmetry changes the scenario of conventional GUTs with Majorana mass. On the other hand it introduces a mass mixing between the basic and heavy set of fermions which will yield Dirac neutrino.

We shall now write the mass matrix in the neutral sector (ν_a, ν_a^c) with a = 1, 2 for two sets of neutrinos, taking into account the Yukawa part [10]. Here ν_a and ν_a^c are the left handed neutrino and its conjugate respectively.

	ν_1	ν_2^c	$\nu_1^c\ V_1^c$	ν_2
ν_1		0		A
ν_2^c				
$\nu_1^c\ V_1^c$		A^T		
ν_2				0

where $A = \begin{pmatrix} h_1 k & h_2 \vartheta_L \\ h_2 \vartheta_R & h_1 k' \end{pmatrix}$. Here k and k′ are the vevs corresponding to ϕ_1 and ϕ_2 respectively, with the hierarchy of scales $\vartheta_R >> k >> k' >> \vartheta_L$. On diagonalising the mass matrix we get four eigen values each of which consists of degenerate members (mass with opposite sign) yielding two Dirac particles [7]. The masses in the leading order occur as $m_1 \sim h_2 \vartheta_R$ and $m_2 \sim \frac{h_2^2 \vartheta_L \vartheta_R - h_1^2 kk'}{h_2 \vartheta_R}$. Phenomenologically we choose $\vartheta_R \sim 10^{12}$ Gev [11], $h_1 k \sim 10^2$ Gev which gives mass to heavier fermion and $h_1 k' \sim 10^{-3}$ Gev which gives masses to ordinary fermions. Further $\vartheta_L \sim \frac{h_1^2 k^2}{\vartheta_R} \sim 10^{-8}$ Gev [7] which is obtained by potential minimization. h_2 is taken to be 10^{-3}, so that we have a super heavy neutrino $m_1 \sim 10^9$ Gev and an ultra light Dirac neutrino as $m_2 \sim h_2 \vartheta_L \sim 10^{-2}$ ev.

4. Conclusion

We shall now briefly emphasize the salient features and its relevance to phenomenology. We have studied an E_6 gauge theory on $M^4 \otimes T^2 / (Z_2 \otimes Z'_2 \otimes Z''_2)$ coupled with N = 1 SUSY to obtain an extended Pati-Salam group (G_{2241}) as a result of orbifold compactification. We have taken the E_6 group in six dimension with the point in mind that it may bear the memory of E_8 group in d = 10. Due to this reason, we are also confined to the brane-localised Higgs sector belonging to $27(\overline{27})$ only. The question of doublet-triplet splitting is absent here as the doublet (H) is super heavy at the compactification scale and the triplet (D) attains mass of the order of M_U from Vev of "S". The suppression of rapid proton decay is automatically achieved. We have made use of non-zero of bilinear fermion condensates, which are responsible for symmetry breaking at two different levels. The condensate mechanism is conjectured in such a way that the condensates may be there in an intermediate scale and the corresponding Higgs particle may have a mass in GUT scale and still it has a vev in weak scale. The possibility of fermion condensates may be relevant in case of non observation of Higgs particles. A light Dirac neutrino is predicted in neutral sector with imposed parity assignments. One open question is still to be investigated about the super symmetry breaking which may occur via gaugino condensates.

Acknowledgments

One of the authors (S. Mishra) is thankful to Prof. Asim Ku. Ray for useful discussions on this work.

References

[1] H.P. Nilles, *Phys. Rep.* 110, 2(1984).
[2] H.E. Haber and G.L. Kane, *Nucl. Phys.* B232, 333, (1984).
[3] Y. Kawamura, *Prog. Thor. Phys.* 105, 999 (2001).
[4] L.J. Hall and Y. Nomura, *Phys. Rev.* D64, 055003 (2001).
[5] Hebecker and J. March-Russel, *Nucl. Phys.* B613, 3 (2001).

[6] T. Asaka et. al., *Phys. Lett.* B523, 199 (2001).
[7] G. Altarelli and F. Feruglio, *Phys. Lett.* B511, 257 (2001).
[8] C.H. Albright and S.M. Barr, *Phys. Rev.* D67, 013002 (2003).
[9] L. Hall et al., *Phys. Rev.* D65, 035008 (2002)
[10] Y. Nomura, et. al., *Nucl. Phys.* B613, 147 (2001).
[11] R. Dermisek and A. Mati, *Phys. Rev.* D65, 0552002 (2002).
[12] N. Haba and Y. Shimizu, *Phys. Rev.* D67, 095001 (2003).
[13] T. Asaka et. Al., *Phys, Lett.* B523, 199 (2002).
[14] S. Mahapatra and S.P. Misra, *Phys. Rev.* D33, 464 (1986).
[15] S. Mishra and S.P. Misra, *Phys. Lett.* B186, 99 (1987).
[16] S. Misra and S.P. Misra, *Phys. Lett.* B217, 66 (1989).
[17] S. Mishra et. al., *Phys. Rev.* D35, 975 (1986).
[18] R. N. Mohapatra and G.N. Senjanovic, *Phys. Rev.* D23, 165 (1981).
[19] J.C. Pati, *Phys. Lett.* B144, 375 (1984).
[20] J.C. Pati, *Phys. Rev.* D30, 1144 (1984).
[21] S. Mishra and S.P. Misra, *Phys. Lett.* B217, 66 (1989).
[22] K. Bhattacharya et. Al., *Phys. Rev.* D74, 015003(2006).

In: Neutrinos: Properties, Sources and Detection
Editor: Joshua P. Greene, pp. 185-198
ISBN 978-1-61209-650-6

Chapter 7

ELECTROMAGNETIC PROPERTIES OF THE TAU AND THE TAU-NEUTRINO

***A. Gutiérrez-Rodríguez*[1]*, M.A. Hernández-Ruíz*[2] *and J.M. Rivera-Juárez*[1]**
[1] Facultad de Física, Universidad Autónoma de Zacatecas
Apartado Postal C-580, 98060 Zacatecas, Zacatecas México
[2] Facultad de Ciencias Químicas, Universidad Autónoma de Zacatecas
Apartado Postal 585, 98060 Zacatecas, Zacatecas México

Abstract

Measuremets on the anomalous magnetic moment and the electric dipole moment of the tau and its neutrino are calculated through the reactions $e^+e^- \to \tau^+\tau^-\gamma$ and $e^+e^- \to \nu_\tau\bar{\nu}_\tau\gamma$ at the Z_1-pole and in the framework of a left-right symmetric model. The results are based on the recent data reported by the L3 and OPAL Collaborations at CERN LEP. Due to the stringent limit of the model mixing angle ϕ, the effect of this angle on the dipole moments is quite small.

1. Introduction

In the Standard Model (SM) [1], the electromagnetic interactions of each of the three charged leptons are identical. However, there is no experimentally verified explanation for the existence of three generations of leptons nor for why they have such different masses. New insight might be forthcoming if the leptons were observed to have a substructure which could manifest itself in deviations from the SM values for the anomalous magnetic or electric dipole moments. The anomalous moments for the electron and muon have been measured with very high precision [2] compared to those of tau for which there are only upper limits [3, 4, 5, 6, 7].

In general, an on-shell photon may couple to a tau through its electric charge, magnetic dipole moment or electric dipole moment. This coupling may be parametrised using a matrix element in which the usual γ^α is replaced by a more general Lorentz-invariant form

$$\Gamma^\alpha = eF_1(q^2)\gamma^\alpha + \frac{ie}{2m_\tau}F_2(q^2)\sigma^{\alpha\mu}q_\mu + eF_3(q^2)\gamma_5\sigma^{\alpha\mu}q_\mu, \quad (1)$$

where m_τ is the mass of the τ lepton and $q = p' - p$ is the momentum transfer. The q^2-dependent form-factors, $F_i(q^2)$, have familiar interpretations for $q^2 = 0$: $F_1(0) \equiv Q_\tau$ is the electric charge; $F_2(0) \equiv a_\tau = (g-2)/2$ is the anomalous magnetic moment; and $F_3 \equiv d_\tau/Q_\tau$, where d_τ is the electric dipole moment.

The analysis of radiative τ pair production provides a means to determine the anomalous magnetic and electric dipole moments of the τ lepton at $q^2 = 0$. An anomalous magnetic dipole moment at $q^2 = 0$ ($F_2(0)$) or an electric dipole moment ($F_3(0)$) affects the total cross section for the process $e^+e^- \rightarrow \tau^+\tau^-\gamma$ as well as the shape of energy and angular distributions of the three final state particles [5, 8, 9]. Previous experimental limits [5, 6, 7, 10, 11] on $F_2(0)$ and $F_3(0)$ have been based on approximate calculations of the $e^+e^- \rightarrow \tau^+\tau^-\gamma$ cross section and photon energy distribution.

The first analysis of the anomalous magnetic moment of the τ lepton, *i.e.* of $F_2(0)$, is due to Grifols and Méndez using L3 data [5]. They derived a limit for $F_2(q^2 = 0) \leq 0.11$ and $F_3(q^2 = 0) \leq 6 \times 10^{-16}$ecm at $q^2 = 0$. More recently, Escribano and Massó [4] have used electroweak data to find $d_\tau \leq 1.1 \times 10^{-17}$ecm and $-0.004 \leq a_\tau \leq 0.006$ at the 2σ confidence level.

On the Z_1 peak, where a large number of Z_1 events are collected at e^+e^- colliders, one may hope to constrain or eventually measure the anomalous magnetic moment and electric dipole moment of the τ by selecting $\tau^+\tau^-$ events accompanied by a hard photon. The Feynman diagrams which give the most important contribution to the cross section are shown in Fig. 1.

Our aim is analyze the reaction $e^+e^- \rightarrow \tau^+\tau^-\gamma$. We use recent data collected with the L3 and OPAL detector at CERN LEP [6, 7] in the Z_1 boson resonance. The analysis is carried out in the context of a left-right symmetric model [12, 13] and we attribute a magnetic moment and an electric dipole moment to the tau lepton. Processes measured in the resonance serve to set limits on the tau magnetic moment and electric dipole moment. In this work, we take advantage of this fact to set limits for a_τ and d_τ for different values of the mixing angle ϕ [14, 15, 16], which is consistent with other constraints previously reported [4, 5, 6, 7, 17].

We do our analysis on the Z_1 peak ($s = M^2_{Z_1}$). Our results are therefore independent of the mass of the additional heavy Z_2 gauge boson which appears in these kind of models. Thus, we have the mixing angle ϕ between the left and the right bosons as the only additional parameter apart from the SM parameters.

In the case of the tau neutrino, in many extensions of the Standard Model the neutrino acquires a nonzero mass, a magnetic moment and an electric dipole moment [18]. In this manner the neutrinos seem to be likely candidates for carrying features of physics beyond the Standard Model [1]. Apart from masses and mixings, magnetic moments and electric dipole moments are also signs of new physics and are of relevance in terrestrial experiments, in the solar neutrino problem, in astrophysics and in cosmology [19].

At the present time, all of the available experimental data for electroweak processes can be understood in the context of the Standard Model, with the exception of the results of the SUPER-KAMIOKANDE experiment on the neutrino-oscillations [20], as well as the GALLEX, SAGE, GNO, HOMESTAKE and LSND [21] experiments. Nonetheless, the SM is still the starting point for all the extended gauge models. In other words, any gauge group with physical characteristics must have as a subgroup the $SU(2)_L \times U(1)$

group of the standard model in such a way that their predictions agree with those of the SM at low energies. The purpose of the extended theories is to explain some fundamental aspects which are not clarified in the frame of the SM. One of these aspects is the origin of the parity violation at current energies. The Left-Right Symmetric Models (LRSM), based on the $SU(2)_R \times SU(2)_L \times U(1)$ gauge group [22], give an answer to this problem by restoring the parity symmetry at high energies and giving their violations at low energies as a result of the breaking of gauge symmetry. Detailed discussions on LRSM can be found in the literature [12, 13, 23].

In 1994, T. M. Gould and I. Z. Rothstein [24] reported a bound on the tau neutrino magnetic moment which they obtained through the analysis of the process $e^+e^- \to \nu\bar{\nu}\gamma$, near the Z_1-resonance, by considering a massive tau neutrino and using Standard Model $Z_1e^+e^-$ and $Z_1\nu\bar{\nu}$ couplings.

At low center of mass energy $s \ll M_{Z_1}^2$, the dominant contribution to the process $e^+e^- \to \nu\bar{\nu}\gamma$ involves the exchange of a virtual photon[25]. The dependence on the magnetic moment comes from a direct coupling to the virtual photon, and the observed photon is a result of initial state Bremsstrahlung.

At higher s, near the Z_1 pole $s \approx M_{Z_1}^2$, the dominant contribution for $E_\gamma > 10\ GeV$ [26] involves the exchange of a Z_1 boson. The dependence on the magnetic moment and the electric dipole moment now comes from the radiation of the photon observed by the neutrino or antineutrino in the final state. The Feynman diagrams which give the most important contribution to the cross section are shown in Fig. 2. We emphasize here the importance of the final state radiation near the Z_1 pole, which occurs preferentially at high E_γ compared to conventional Bremsstrahlung.

As in the case of the tau lepton, our aim is to analyze the reaction $e^+e^- \to \nu\bar{\nu}\gamma$. We use recent data collected with the L3 detector at CERN LEP [6, 26] near the Z_1 boson resonance in the framework of a left-right symmetric model and we attribute an anomalous magnetic moment and an electric dipole moment to a massive tau neutrino. Processes measured near the resonance serve to set limits on the tau neutrino magnetic moment and electric dipole moment. In this work, we take advantage of this fact to set bounds for μ_{ν_τ} and d_{ν_τ} for different values of the mixing angle ϕ [14, 15, 16], which is consistent with other constraints previously reported [6, 24, 25, 27, 28] and [4, 29].

We do our analysis near the resonance of the Z_1 ($s \approx M_{Z_1}^2$). Thus, our results are independent of the mass of the additional heavy Z_2 gauge boson which appears in these kinds of models. Therefore, we have the mixing angle ϕ between the left and the right bosons as the only additional parameter besides the SM parameters.

The L3 Collaboration evaluated the selection efficiency using detector-simulated $e^+e^- \to \nu\bar{\nu}\gamma(\gamma)$ events, random trigger events, and large-angle $e^+e^- \to e^+e^-$ events. A total of 14 events were found by the selection. The distributions of the photon energy and the cosine of its polar angle are consistent with SM predictions. The total number of events expected from the SM is 14.1. If the photon energy is greater than half the beam energy, 2 events are selected from the data and 2.4 events are expected from the SM in the $\nu\bar{\nu}\gamma$ channel.

2. The Left-Right Symmetric Model (LRSM)

We consider a Left-Right Symmetric Model (LRSM) consisting of one bidoublet Φ and two doublets χ_L, χ_R. The vacuum expectation values of χ_L, χ_R break the gauge symmetry to give mass to the left and right heavy gauge bosons. This is the origin of the parity violation at low energies [13] *i.e.*, at energies produced in actual accelerators. The Lagrangian for the Higgs sector of the LRSM is [12]

$$\mathcal{L}_{LRSM} = (D_\mu \chi_L)^\dagger (D^\mu \chi_L) + (D_\mu \chi_R)^\dagger (D^\mu \chi_R) + Tr(D_\mu \Phi)^\dagger (D^\mu \Phi). \tag{2}$$

The covariant derivatives are written as

$$\begin{aligned} D_\mu \chi_L &= \partial_\mu \chi_L - \frac{1}{2} i g \tau \cdot \mathbf{W}_L \chi_L - \frac{1}{2} i g' B \chi_L, \\ D_\mu \chi_R &= \partial_\mu \chi_R - \frac{1}{2} i g \tau \cdot \mathbf{W}_R \chi_R - \frac{1}{2} i g' B \chi_R, \\ D_\mu \Phi &= \partial_\mu \Phi - \frac{1}{2} i g (\tau \cdot \mathbf{W}_L \Phi - \Phi \tau \cdot \mathbf{W}_R). \end{aligned} \tag{3}$$

There are seven gauge bosons in this model: the charged $W^1_{L,R}$, $W^2_{L,R}$ and the neutral $W^3_{L,R}$, B. The gauge couplings constants g_L and g_R of the $SU(2)_L$ and $SU(2)_R$ subgroups, respectively, are equal: $g_L = g_R = g$, since manifest left-right symmetry is assumed [30]. g' is the gauge coupling for the $U(1)$ group.

The transformation properties of the Higgs bosons under the group $SU(2)_L \times SU(2)_R \times U(1)$ are $\chi_L \sim (1/2, 0, 1)$, $\chi_R \sim (0, 1/2, 1)$ and $\Phi \sim (1/2, 1/2, 0)$. After spontaneous symmetry breaking, the ground states are of the form

$$\langle \chi_L \rangle = \frac{1}{\sqrt{2}} \begin{pmatrix} 0 \\ v_L \end{pmatrix}, \quad \langle \chi_R \rangle = \frac{1}{\sqrt{2}} \begin{pmatrix} 0 \\ v_R \end{pmatrix}, \quad \langle \Phi \rangle = \frac{1}{\sqrt{2}} \begin{pmatrix} k & 0 \\ 0 & k' \end{pmatrix}, \tag{4}$$

which break the symmetry group to form the $U(1)_{em}$, giving mass to the gauge bosons and fermions with the photon remaining massless. In Eq. (4), v_L, v_R, k and k' are the vacuum expectation values. The part of the Lagrangian that contains the mass terms for the charged boson is

$$\mathcal{L}^C_{mass} = (W^+_L \; W^+_R) M^C \begin{pmatrix} W^-_L \\ W^-_R \end{pmatrix}, \tag{5}$$

where $W^\pm = \frac{1}{\sqrt{2}}(W^1 \mp W^2)$.

The mass matrix M^C is

$$M^C = \frac{g^2}{4} \begin{pmatrix} v_L^2 + k^2 + k'^2 & -2kk' \\ -2kk' & v_R^2 + k^2 + k'^2 \end{pmatrix}. \tag{6}$$

This matrix is diagonalized by an orthogonal transformation parametrized [30] by the angle ζ. This angle has a very small value because of the hyperon β decay data [31].

Similarly, the part of the Lagrangian that contains the mass terms for the neutral bosons is

$$\mathcal{L}^N_{mass} = \frac{1}{8} (W^3_L \; W^3_R \; B) M^N \begin{pmatrix} W^3_L \\ W^3_R \\ B \end{pmatrix}, \tag{7}$$

where the matrix M^N is given by

$$M^N = \frac{1}{4}\begin{pmatrix} g^2(v_L^2 + k^2 + k'^2) & -g^2(k^2 + k'^2) & -gg'v_L^2 \\ -g^2(k^2 + k'^2) & g^2(v_R^2 + k^2 + k'^2) & -gg'v_R^2 \\ -gg'v_L^2 & -gg'v_R^2 & g'^2(v_L^2 + v_R^2) \end{pmatrix}. \quad (8)$$

Since the process $e^+e^- \to \tau^+\tau^-\gamma$ is neutral, we center our attention on the mass terms of the Lagrangian for the neutral sector as shown in Eq. (1.7).

The matrix M^N for the neutral gauge bosons is diagonalized by an orthogonal transformation which can be written in terms of the angles θ_W and ϕ [32]

$$U^N = \begin{pmatrix} c_W c_\phi & -s_W t_W c_\phi - r_W s_\phi / c_W & t_W(s_\phi - r_W c_\phi) \\ c_W s_\phi & -s_W t_W s_\phi + r_W c_\phi / c_W & -t_W(c_\phi + r_W s_\phi) \\ s_W & s_W & r_W \end{pmatrix}, \quad (9)$$

where $c_W = \cos\theta_W$, $s_W = \sin\theta_W$, $t_W = \tan\theta_W$ and $r_W = \sqrt{\cos 2\theta_W}$, and θ_W is the electroweak mixing angle. Here, $c_\phi = \cos\phi$ and $s_\phi = \sin\phi$. The angle ϕ is considered as the angle that mixes the left and right handed neutral gauge bosons $W^3_{L,R}$. The expression that relates the left and right handed neutral gauge bosons $W^3_{L,R}$ and B with the physical bosons Z_1, Z_2 and the photon is:

$$\begin{pmatrix} Z_1 \\ Z_2 \\ A \end{pmatrix} = U^N \begin{pmatrix} W_L^3 \\ W_R^3 \\ B \end{pmatrix}. \quad (10)$$

The diagonalization of (1.6) and (1.8) gives the mass of the charged $W^\pm_{1,2}$ and neutral $Z_{1,2}$ physical fields:

$$M^2_{W_{1,2}} = \frac{g^2}{8}\left[v_L^2 + v_R^2 + 2(k^2 + k'^2) \mp \sqrt{(v_R^2 - v_L^2)^2 + 16(kk')^2}\right], \quad (11)$$

$$M^2_{Z_1,Z_2} = B \mp \sqrt{B^2 - 4C}, \quad (12)$$

respectively, with

$$B = \frac{1}{8}\Big[(g^2 + g'^2)(v_L^2 + v_R^2) + 2g^2(k^2 + k'^2)\Big],$$

$$C = \frac{1}{64} g^2(g^2 + 2g'^2)\Big[v_L^2 v_R^2 + (k^2 + k'^2)(v_L^2 + v_R^2)\Big].$$

Taking into account that $M^2_{W_2} \gg M^2_{W_1}$, we conclude from the expressions for the masses of M_{Z_1} and M_{Z_2}, that the relation $M^2_{W_1} = M^2_{Z_1}\cos^2\theta_W$ still holds in this model.

From the Lagrangian of the LRSM, we extract the terms for the neutral interaction of a fermion with the gauge bosons $W^3_{L,R}$ and B:

$$\mathcal{L}^N_{int} = g(J_L^3 W_L^3 + J_R^3 W_R^3) + \frac{g'}{2} J_Y B. \quad (13)$$

Specifically, the Lagrangian interaction for $Z_1 \to f\bar{f}$ [33] is

$$\mathcal{L}^N_{int} = \frac{g}{c_W} Z_1 \left[\left(c_\phi - \frac{s_W^2}{r_W} s_\phi\right) J_L - \frac{c_W^2}{r_W} s_\phi J_R\right], \quad (14)$$

where the left (right) current for the fermions are

$$J_{L,R} = J^3_{L,R} - \sin^2\theta_W J_{em}$$

and

$$J_{em} = J^3_L + J^3_R + \frac{1}{2}J_Y$$

is the electromagnetic current. From (1.14), we find that the amplitude $\mathcal{M}$ for the decay of the Z_1 boson with polarization ϵ^λ into a fermion-antifermion pair is:

$$\mathcal{M} = \frac{g}{c_W}\left[\bar{u}\gamma^\mu\frac{1}{2}(ag_V - bg_A\gamma_5)v\right]\epsilon^\lambda_\mu, \tag{15}$$

with

$$a = c_\phi - \frac{s_\phi}{r_W} \quad \text{and} \quad b = c_\phi + r_W s_\phi, \tag{16}$$

where ϕ is the mixing parameter of the LRSM [14, 15].

In the following section, we make the calculations for the reaction $e^+e^- \to \tau^+\tau^-\gamma$, $\nu_\tau\bar{\nu}_\tau\gamma$ by using the expression (1.15) for the transition amplitude.

3. The Total Cross Section of $e^+e^- \to \tau^+\tau^-\gamma, \nu_\tau\bar{\nu}_\tau\gamma$

We calculate the total cross section of the processes $e^+e^- \to \tau^+\tau^-\gamma, \nu_\tau\bar{\nu}_\tau\gamma$, using the Breit-Wigner resonance form [34, 35]:

$$\sigma(e^+e^- \to f^+f^-\gamma) = \frac{4\pi(2J+1)\Gamma_{e^+e^-}\Gamma_{f^+f^-\gamma}}{(s - M^2_{Z_1})^2 + M^2_{Z_1}\Gamma^2_{Z_1}}, \tag{17}$$

where $\Gamma_{e^+e^-}$ is the decay rate of Z_1 to the channel $Z_1 \to e^+e^-$ and $\Gamma_{f^+f^-\gamma}$ is the decay rate of Z_1 to the channel $Z_1 \to f^+f^-\gamma$.

In the next subsections, we calculate the widths of Eq. (1.17) to both processes $e^+e^- \to \tau^+\tau^-\gamma, \nu_\tau\bar{\nu}_\tau\gamma$.

3.1. Width of $Z_1 \to e^+e^-$

In this section we calculate the total width of the reaction

$$Z_1 \to e^+e^-, \tag{18}$$

in the context of the left-right symmetric model which is described in Section 1.2.

The expression for the total width of the process $Z_1 \to e^+e^-$, due only to the Z_1 boson exchange according to the diagrams depicted in Fig. 1 and using the expression for the amplitude given in Eq. (1.15), is

$$\Gamma_{(Z_1\to e^+e^-)} = \frac{G_F M^3_{Z_1}}{6\pi\sqrt{2}}\sqrt{1-4\eta}\big[a^2(g^e_V)^2(1+2\eta) + b^2(g^e_A)^2(1-4\eta)\big], \tag{19}$$

where $\eta = m^2_e/M^2_{Z_1}$.

We take $g_V^e = -\frac{1}{2} + 2\sin^2\theta_W$ and $g_A^e = -\frac{1}{2}$, the couplings of the standard model [34] so that the total width is

$$\Gamma_{(Z_1\to e^+e^-)} = \frac{\alpha M_{Z_1}}{24}\left[\frac{\frac{1}{2}(a^2+b^2) - 4a^2x_W + 8a^2x_W^2 + (a^2 - 2b^2 - 8a^2x_W + 16a^2x_W^2)\eta}{x_W(1-x_W)}\right], \tag{20}$$

where $x_W = \sin^2\theta_W$ and $\alpha = e^2/4\pi$ is the fine structure constant. In the limit when $m_e = 0$, $\eta = 0$ and Eq. (1.20) is reduced to the expression (19) given in the Ref. [27].

3.2. Widths of $Z_1 \to \tau^+\tau^-\gamma, \nu_\tau\bar{\nu}_\tau\gamma$

The expression for the Feynman amplitude $\mathcal{M}$ of the processes $Z_1 \to \tau^+\tau^-\gamma, \nu_\tau\bar{\nu}_\tau\gamma$ is due only to the Z_1 boson exchange, as shown in the diagrams in Figs. 1, 2. We use the expression for the amplitude given in Eq. (1.15) and assume that a tau lepton is characterized by the following phenomenological parameters: a charge radius $\langle r^2\rangle$, a magnetic moment a_τ (μ_{ν_τ}), and an electric dipole moment d_τ (d_{ν_τ}). Therefore, the expression for the Feynman amplitude $\mathcal{M}$ of the processes $Z_1 \to \tau^+\tau^-\gamma, \nu_\tau\bar{\nu}_\tau\gamma$ is given by

$$\mathcal{M}_1 = \begin{cases} \left[\bar{u}(p_{\tau^-})\Gamma^\alpha \frac{i}{(\not{l} - m_\tau)}\left(-\frac{ig}{2\cos\theta_W}\gamma^\beta(ag_V^\tau - bg_A^\tau\gamma_5)\right)v(p_{\tau^+})\right]\epsilon_\alpha^\lambda(\gamma)\epsilon_\beta^\lambda(Z_1), & \tau^+\tau^-\gamma \\ \left[\bar{u}(p_\nu)\Gamma^\alpha \frac{i}{(\not{l} - m_\nu)}\left(-\frac{ig}{4\cos\theta_W}\gamma^\beta(a - b\gamma_5)\right)v(p_{\bar{\nu}})\right]\epsilon_\alpha^\lambda(\gamma)\epsilon_\beta^\lambda(Z_1), & \nu_\tau\bar{\nu}_\tau\gamma \end{cases} \tag{21}$$

and

$$\mathcal{M}_2 = \begin{cases} \left[\bar{u}(p_{\tau^-})\left(-\frac{ig}{2\cos\theta_W}\gamma^\beta(ag_V^\tau - bg_A^\tau\gamma_5)\right)\frac{i}{(\not{k} - m_\tau)}\Gamma^\alpha v(p_{\tau^+})\right]\epsilon_\alpha^\lambda(\gamma)\epsilon_\beta^\lambda(Z_1), & \tau^+\tau^-\gamma \\ \left[\bar{u}(p_\nu)\left(-\frac{ig}{4\cos\theta_W}\gamma^\beta(a - b\gamma_5)\right)\frac{i}{(\not{k} - m_\nu)}\Gamma^\alpha v(p_{\bar{\nu}})\right]\epsilon_\alpha^\lambda(\gamma)\epsilon_\beta^\lambda(Z_1), & \nu_\tau\bar{\nu}_\tau\gamma \end{cases} \tag{22}$$

where Γ^α is given by the Eq. (1.1). ϵ_α^λ and ϵ_β^λ are the polarization vectors of photon and of the boson Z_1, respectively. l (k) stands by the momentum of the virtual tau (antitau), neutrino (antineutrino) and the coupling constants a and b are given in Eq. (1.16).

After applying some of the trace theorems of the Dirac matrices and of sum and average over the initial and final spin, the square of the matrix elements becomes

$$\sum_s |\mathcal{M}_T|^2 = \begin{cases} \frac{g^2}{\cos^2\theta_W}\left[\frac{e^2a_\tau^2}{4m_\tau^2} + d_\tau^2\right]\Big[(a^2(g_V^\tau)^2 + b^2(g_A^\tau)^2)(s - 2\sqrt{s}E_\gamma) \\ \qquad + a^2(g_A^\tau)^2E_\gamma^2\sin^2\theta_\gamma\Big], & \tau^+\tau^-\gamma \\ \frac{g^2}{4\cos^2\theta_W}(\mu_{\nu_\tau}^2 + d_{\nu_\tau}^2)\Big[(a^2+b^2)(s - 2\sqrt{s}E_\gamma) + a^2E_\gamma^2\sin^2\theta_\gamma\Big], & \nu_\tau\bar{\nu}_\tau\gamma \end{cases} \tag{23}$$

Now that we know the square of Eq. (1.23) transition amplitude, our final step is to calculate the total widths of $Z_1 \to \tau^+\tau^-\gamma, \nu_\tau\bar{\nu}_\tau\gamma$:

$$\Gamma = \begin{cases} \int \frac{\alpha}{12\pi^2 M_{Z_1} x_W(1-x_W)} \left[\frac{e^2 a_\tau^2}{4m_\tau^2} + d_\tau^2\right] \Big[(a^2(g_V^\tau)^2 + b^2(g_A^\tau)^2)(s - 2\sqrt{s}E_\gamma) \\ \qquad + a^2(g_A^\tau)^2 E_\gamma^2 \sin^2\theta_\gamma\Big] E_\gamma dE_\gamma d\cos\theta_\gamma, \quad Z_1 \to \tau^+\tau^-\gamma \\ \int \frac{\alpha(\mu_{\nu_\tau}^2 + d_{\nu_\tau}^2)}{96\pi^2 M_{Z_1} x_W(1-x_W)} \Big[(a^2+b^2)(s-2\sqrt{s}E_\gamma) + a^2 E_\gamma^2 \sin^2\theta_\gamma\Big] E_\gamma dE_\gamma d\cos\theta_\gamma, \\ \qquad Z_1 \to \nu_\tau\bar{\nu}_\tau\gamma \end{cases} \tag{24}$$

where E_γ and $\cos\theta_\gamma$ are the energy and scattering angle of the photon.

The substitution of (1.20) and (1.24) in (1.17) gives

$$\sigma = \begin{cases} \int \frac{\alpha^2}{48\pi}\left[\frac{e^2 a_\tau^2}{4m_\tau^2} + d_\tau^2\right] \\ \left[\frac{\frac{1}{2}(a^2+b^2) - 4a^2 x_W + 8a^2 x_W^2 + (a^2 - 2b^2 - 8a^2 x_W + 16a^2 x_W^2)\eta}{x_W^2(1-x_W)^2}\right] \\ \left[\frac{\left[\frac{1}{2}(a^2+b^2) - 4a^2 x_W + 8a^2 x_W^2\right](s - 2\sqrt{s}E_\gamma) + \frac{1}{2}a^2 E_\gamma^2 \sin^2\theta_\gamma}{(s-M_{Z_1}^2)^2 + M_{Z_1}^2\Gamma_{Z_1}^2}\right] E_\gamma dE_\gamma d\cos\theta_\gamma, \\ \qquad \text{for} \quad e^+e^- \to \tau^+\tau^-\gamma \\ \int \frac{\alpha^2(\mu_{\nu_\tau}^2 + d_{\nu_\tau}^2)}{192\pi}\left[\frac{\frac{1}{2}(a^2+b^2) - 4a^2 x_W + 8a^2 x_W^2}{x_W^2(1-x_W)^2}\right] \\ \left[\frac{(a^2+b^2)(s - 2\sqrt{s}E_\gamma) + a^2 E_\gamma^2 \sin^2\theta_\gamma}{(s-M_{Z_1}^2)^2 + M_{Z_1}^2\Gamma_{Z_1}^2}\right] E_\gamma dE_\gamma d\cos\theta_\gamma, \\ \qquad \text{for} \quad e^+e^- \to \nu_\tau\bar{\nu}_\tau\gamma \end{cases} \tag{25}$$

It is useful to consider the smallness of the mixing angle ϕ, as indicated in the Eq. (1.31), to approximate the cross section in Eq. (1.25) by its expansion in ϕ up to the linear term: $\sigma = [\frac{e^2 a_\tau^2}{4m_\tau^2}(\mu_{\nu_\tau}^2) + d_\tau^2(d_{\nu_\tau}^2)][A(A_1) + B(B_1)\phi + O(\phi^2)]$, where $A(A_1)$ and $B(B_1)$ are constants which can be evaluated. Such an approximation for deriving the bounds of $a_\tau(\mu_{\nu_\tau})$ and $d_\tau(d_{\nu_\tau})$ is more illustrative and easier to manipulate.

For $\phi < 1$, the total cross section for the processes $e^+e^- \to \tau^+\tau^-\gamma, \nu_\tau\bar{\nu}_\tau\gamma$ is given by

$$\sigma = \begin{cases} \left(\frac{e^2 a_\tau^2}{4m_\tau^2} + d_\tau^2\right)\left[A + B\phi + O(\phi^2)\right], & \text{for} \quad e^+e^- \to \tau^+\tau^-\gamma \\ (\mu_{\nu_\tau}^2 + d_{\nu_\tau}^2)\left[A_1 + B_1\phi + O(\phi^2)\right], & \text{for} \quad e^+e^- \to \nu_\tau\bar{\nu}_\tau\gamma \end{cases} \tag{26}$$

where explicitly A and B are:

$$A = \int \frac{\alpha^2}{48\pi}\left[\frac{1 - 4x_W + 8x_W^2 + (-1 - 4x_W + 8x_W^2)\eta}{x_W^2(1-x_W)^2}\right] \left[\frac{(1 - 4x_W + 8x_W^2)(s - 2\sqrt{s}E_\gamma) + \frac{1}{2}E_\gamma^2\sin^2\theta_\gamma}{(s-M_{Z_1}^2)^2 + M_{Z_1}^2\Gamma_{Z_1}^2}\right] E_\gamma dE_\gamma d\cos\theta_\gamma, \tag{27}$$

$$B = \int \frac{\alpha^2}{48\pi}\Bigg\{\left[\frac{1 - 4x_W + 8x_W^2 - (1 + 4x_W - 8x_W^2)\eta}{r_W x_W^2(1-x_W)^2}\right]$$

$$\left[\frac{(6x_W - 16x_W^2)(s - 2\sqrt{s}E_\gamma) - E_\gamma^2 \sin^2\theta_\gamma}{(s - M_{Z_1}^2)^2 + M_{Z_1}^2\Gamma_{Z_1}^2}\right]$$
$$+ \left[\frac{6x_W - 16x_W^2 + (-6 + 16x_W - 16x_W^2)\eta}{r_W x_W^2 (1 - x_W)^2}\right]$$
$$\left[\frac{(1 - 4x_W + 8x_W^2)(s - 2\sqrt{s}E_\gamma) + \frac{1}{2}E_\gamma^2 \sin^2\theta_\gamma}{(s - M_{Z_1}^2)^2 + M_{Z_1}^2\Gamma_{Z_1}^2}\right]\Bigg\} E_\gamma dE_\gamma d\cos\theta_\gamma, \quad (28)$$

while A_1 and B_1 are given by

$$A_1 = \int \frac{\alpha^2}{96\pi}\left[\frac{1 - 4x_W + 8x_W^2 + (-1 - 4x_W + 8x_W^2)\eta}{x_W^2(1 - x_W)^2}\right]$$
$$\left[\frac{s - 2\sqrt{s}E_\gamma + \frac{1}{2}E_\gamma^2 \sin^2\theta_\gamma}{(s - M_{Z_1}^2)^2 + M_{Z_1}^2\Gamma_{Z_1}^2}\right] E_\gamma dE_\gamma d\cos\theta_\gamma, \quad (29)$$

$$B_1 = \int \frac{\alpha^2}{48\pi}\Bigg\{\left[\frac{3x_W - 8x_W^2 + (-3 + 8x_W - 8x_W^2)\eta}{r_W x_W^2(1 - x_W)^2}\right]\left[\frac{s - 2\sqrt{s}E_\gamma + \frac{1}{2}E_\gamma^2 \sin^2\theta_\gamma}{(s - M_{Z_1}^2)^2 + M_{Z_1}^2\Gamma_{Z_1}^2}\right]$$
$$- \left[\frac{1 - 4x_W + 8x_W^2 + (-1 - 4x_W + 8x_W^2)\eta}{r_W x_W^2(1 - x_W)^2}\right]$$
$$\left[\frac{x_W(s - 2\sqrt{s}E_\gamma) + \frac{1}{2}E_\gamma^2 \sin^2\theta_\gamma}{(s - M_{Z_1}^2)^2 + M_{Z_1}^2\Gamma_{Z_1}^2}\right]\Bigg\} E_\gamma dE_\gamma d\cos\theta_\gamma. \quad (30)$$

In the case of the reaction $e^+e^- \to \tau^+\tau^-\gamma$, the expression given for A corresponds to the cross section previously reported by A. Grifols and A. Méndez [5] with $\eta = 0$, while B comes from the contribution of the LRSM.

For the case of $e^+e^- \to \nu_\tau\bar{\nu}_\tau\gamma$, the expression given for A_1 corresponds to the cross section previously reported by T.M. Gould and I.Z. Rothstein [24] with $\eta = 0$, while B_1 comes from the contribution of the LRSM.

Evaluating the limit when the mixing angle is $\phi = 0$, the second term with $B(B_1)$ in (1.26) is zero and Eq. (1.26) is reduced to the expression (4 (3)) given in Refs. [5], [24].

4. Limits on the Dipole Moments of the τ and the ν_τ

In practice, detector geometry imposes a cut on photon polar angle with respect to the electron direction, and further cuts must be applied on the photon energy and minimum opening angle between the photon and tau in order to suppress background from tau decay products. Therefore, to evaluate the integral of the total cross section as a function of mixing angle ϕ, we require cuts on the photon angle and energy to avoid divergences when the integral is evaluated at the important intervals of each experiment.

4.1. Magnetic Moment and Electric Dipole Moment of the Tau

We integrate over $\cos\theta_\gamma$ from -0.74 to 0.74 and E_γ from 5 GeV to 45.5 GeV for various fixed values of the mixing angle $\phi = -0.009, -0.005, 0, 0.004$. Using the numerical values: $\sin^2\theta_W = 0.2314$, $m_\tau = 1.776\ GeV$, $M_{Z_1} = 91.187\ GeV$, and $\Gamma_{Z_1} = 2.49\ GeV$, we obtain the cross section $\sigma = \sigma(\phi, a_\tau, d_\tau)$.

For the mixing angle ϕ between Z_1 and Z_2, we use the reported data of M. Maya *et al.* [14]:

$$-9 \times 10^{-3} \leq \phi \leq 4 \times 10^{-3}, \tag{31}$$

with a 90 % C. L. Other limits on the mixing angle ϕ reported in the literature are given in the Refs. [15, 16].

As was discussed in Refs. [6, 7], $N \approx \sigma(\phi, a_\tau, d_\tau)\mathcal{L}$, where N and $\mathcal{L}$ are the number of events and the luminosity respectively. Using the Poisson statistic [6, 36], we require that $N \approx \sigma(\phi, a_\tau, d_\tau)\mathcal{L}$ be less than 1559 (1429), with $\mathcal{L} = 100\ pb^{-1}$ ($180\ pb^{-1}$), according to the data reported by the L3 (OPAL) Collaboration Ref. [6, 7]. Taking this into consideration, we put a bound for the tau lepton magnetic moment as a function of the ϕ mixing parameter with $d_\tau = 0$. We show the value of this bound for values of the ϕ parameter in Tables 1 and 2.

Table 1. Limits on the a_τ magnetic moment and d_τ electric dipole moment of the τ-lepton for different values of the mixing angle ϕ before the Z_1 resonance, *i.e.*, $s \approx M^2_{Z_1}$.

ϕ	L3 [6]		OPAL [7]	
	a_τ	$d_\tau(10^{-16}ecm)$	a_τ	$d_\tau(10^{-16}ecm)$
-0.009	0.084	4.61	0.094	5.15
-0.005	0.083	4.59	0.093	5.12
0	0.082	4.55	0.092	5.09
0.004	0.081	4.53	0.091	5.06

These results differ from the limits obtained in the references [6, 7]. However, the derived limits in Table 1 could be improved by including data from the entire Z_1 resonance as shown in Table 2.

Table 2. Limits on the a_τ magnetic moment and d_τ electric dipole moment of the τ-lepton for different values of the mixing angle ϕ in the Z_1 resonance, *i.e.*, $s = M^2_{Z_1}$.

ϕ	L3 [6]		OPAL [7]	
	a_τ	$d_\tau(10^{-16}ecm)$	a_τ	$d_\tau(10^{-16}ecm)$
-0.009	0.06	3.27	0.067	3.66
-0.005	0.059	3.25	0.066	3.64
0	0.058	3.22	0.065	3.61
0.004	0.057	3.21	0.064	3.59

The above analysis and comments can readily be translated to the electric dipole moment of the τ-lepton. The resulting limit for the electric dipole moment as a function of the

ϕ mixing parameter is shown in Table 1.

The results in Table 2 for the electric dipole moment are in agreement with those found by the L3 (OPAL) Collaboration [6, 7].

Fig. 3 shows the total cross section as a function of the mixing angle ϕ for the limits of the magnetic moment given in Tables 1, 2. We observe in Fig. 3 that for $\phi = 0$, we reproduce the data previously reported in the literature. Also, we observe that the total cross section increases constantly and reaches its maximum value for $\phi = 0.004$.

Our results for the dependence of the differential cross section on the photon energy versus the cosine of the opening angle between the photon and the beam direction (θ_γ) are presented in Fig. 4, for $\phi = 0.004$ and $a_\tau = 0.057$. We observe in this figure that the energy and angular distributions are consistent with those reported in the literature. In addition, the form of the distributions does not change significanty for the values ϕ and a_τ because ϕ and a_τ are very small in value, as shown in Table 2.

Finally in this subsection we plotting the differential cross section in Fig. 5 as a function of the photon energy for the limits of the magnetic moment given in Tables 1, 2. We observed in this figure that the energy distributions are consistent with those reported in the literature.

4.2. Magnetic Moment and Electric Dipole Moment of the Tau Neutrino

In order to evaluate the integral of the total cross section as a function of mixing angle ϕ, we require cuts on the photon angle and energy to avoid divergences when the integral is evaluated at the important intervals of each experiment. We integrate over θ_γ from 44.5^o to 135.5^o and E_γ from 15 GeV to 100 GeV for various fixed values of the mixing angle $\phi = -0.009, -0.005, 0, 0.004$. Using the following numerical values: $\sin^2\theta_W = 0.2314$, $M_{Z_1} = 91.187\ GeV, \Gamma_{Z_1} = 2.49\ GeV$, we obtain the cross section $\sigma = \sigma(\phi, \mu_{\nu_\tau}, d_{\nu_\tau})$.

As was discussed in Ref. [24], $N \approx \sigma(\phi, \mu_{\nu_\tau}, d_{\nu_\tau})\mathcal{L}$. Using the Poisson statistic [26, 36], we require that $N \approx \sigma(\phi, \mu_{\nu_\tau}, d_{\nu_\tau})\mathcal{L}$ be less than 14, with $\mathcal{L} = 137\ pb^{-1}$, according to the data reported by the L3 Collaboration Ref. [26] and references therein. Taking this into consideration, we put a bound for the tau neutrino magnetic moment as a function of the ϕ mixing parameter with $d_{\nu_\tau} = 0$. We show the value of this bound for values of the ϕ parameter in Tables 3 and 4.

Table 3. Bounds on the μ_{ν_τ} magnetic moment and d_{ν_τ} electric dipole moment for different values of the mixing angle ϕ before the Z_1 resonance, *i.e.*, $s \approx M^2_{Z_1}$.

	L3 [26]	
ϕ	$\mu_{\nu_\tau}(10^{-6}\mu_B)$	$d_{\nu_\tau}(10^{-17}e\text{cm})$
-0.009	4.48	8.64
-0.005	4.44	8.56
0	4.40	8.49
0.004	4.37	8.43

These results are comparable with the bounds obtained in the references [24, 25].

However, the derived bounds in Table 3 could be improved by including data from the entire Z_1 resonance as is shown in Table 4.

Table 4. Bounds on the μ_{ν_τ} magnetic moment and d_{ν_τ} electric dipole moment for different values of the mixing angle ϕ in the Z_1 resonance, *i.e.*, $s = M^2_{Z_1}$.

	L3 [26]	
ϕ	$\mu_{\nu_\tau}(10^{-6}\mu_B)$	$d_{\nu_\tau}(10^{-17}e\mathrm{cm})$
-0.009	3.37	6.50
-0.005	3.34	6.44
0	3.31	6.38
0.004	3.28	6.32

The above analysis and comments can readily be translated to the electric dipole moment of the τ-neutrino with $\mu_{\nu_\tau} = 0$. The resulting bound for the electric dipole moment as a function of the ϕ mixing parameter is shown in Table 3.

The results in Table 4 for the electric dipole moment are in agreement with those found by the L3 Collaboration [26].

We end this subsection by plotting the total cross section in Fig. 6 as a function of the mixing angle ϕ for the bounds of the magnetic moment given in Tables 3, 4. We observe in Fig. 6 that for $\phi = 0$, we reproduce the data previously reported in the literature. Also, we observe that the total cross section increases constantly and reaches its maximum value for $\phi = 0.004$.

5. Conclusions

We have determined a limit on the magnetic moment and the electric dipole moment of the tau lepton and the tau-neutrino in the framework of a left-right symmetric model as a function of the mixing angle ϕ, as shown in Tables 1-4.

In summary, we conclude that the estimated limit for the tau lepton magnetic moment and the electric dipole moment as well as for the tau-neutrino, are almost independent of the experimental allowed values of the ϕ parameter of the model. In the limit $\phi = 0$, our bound takes the value previously reported in the literature [6, 7, 26].

Acknowledgments

This work was supported in part by SEP-CONACYT (México) (**Proyects: 2003-01-32-001-057** and **40729-F**), *Sistema Nacional de Investigadores* (SNI) (México) and Programa de Mejoramiento al Profesorado (PROMEP). The authors would also like to thank Nova-Science for publishing this manuscript.

References

[1] S.L. Glashow, *Nucl. Phys.* **22**, 579 (1961); S. Weinberg, *Phys. Rev. Lett.* **19**, 1264 (1967); A. Salam, in *Elementary Particle Theory*, Ed. N. Svartholm (Almquist and Wiskell, Stockholm, 1968), p. 367.

[2] E.R. Cohen and B.N. Taylor, *Rev. Mod. Phys.* **59** (1987), 1121.

[3] D.J. Silverman and G.L. Shaw, *Phys. Rev.* **D27** (1983), 1196.

[4] R. Escribano and E. Massó, *Phys. Lett.* **B395**(1997), 369.

[5] J.A. Grifols and A. Méndez, *Phys. Lett.* **B255** (1991), 611; Erratum ibid B259 (1991), 512.

[6] The L3 Collaboration, *Phys. Lett.* **B434** (1998), 169, and references therein.

[7] The OPAL Collaboration, *Phys. Lett.* **B431** (1998), 188, and references therein.

[8] S.S. Gau, T. Paul, J. Swain, L. Taylor, *Nucl. Plys.* **B523** (1998), 439.

[9] J. Biebel and T Riemann, *Z. Phys.* **C76** (1997), 53.

[10] J. Swain, *Nucl. Phys. B (Proc. Suppl.)* **98** (2001), 351.

[11] L. Taylor, *Nucl. Phys. B (Proc. Suppl.)* **76** (1999), 237.

[12] G. Senjanovic, *Nucl. Phys.* **B 153** (1979), 334.

[13] G. Senjanovic and R. N. Mohapatra, *Phys. Rev.* **D12** (1975), 1502.

[14] M. Maya and O. G. Miranda, *Z. Phys.* **C68** (1995), 481.

[15] J. Polak, M. Zralek, *Phys. Rev.* **D46** (1992), 3871.

[16] L3 Collab., O. Adriani et al., *Phys. Lett.* **B306** (1993), 187.

[17] A.Gutiérrez-Rodríguez, M. Hernández-Ruíz and L. N. Luis-Noriega, *Mod. Phys. Lett.* **A19** (2004), 2227.

[18] R.N.Mohapatra and P.B.Pal, in "*Massive Neutrinos in Physics and Astrophysics*", World Scientific, Singapore, 1991.

[19] R. Cisneros, *Astrophys. Space Sci.* **10** (1971), 87; M. B. Voloshin and M. I. Vysotskii, *Sov. J. Nucl. Phys.* **44** (1986). 544; L. B. Okun, *Sov. J. Nucl. Phys.* **44** (1986), 546; M. B. Voloshin, M. I. Vysotskii and L. B. Okun, *Sov. J. Nucl. Phys.* **44** (1986), 440; *Sov. Phys. JETP* **64** (1986), 446.

[20] Y. Fukuda *et al.*, *Measurements of the Solar Neutrino Flux from Super-Kamiokande's First 300 Days*, Preprint hep-ex/9805021, 1998.

[21] SAGE Collaboration, J.N. Abdurashitov et al., *Phys. Rev. Lett.* **83** (1999), 4686; *Phys. Rev.* **C60**, (1999) 055801; GALLEX Collaboration, W. Hampel et al., *Phys. Lett.* **B447** (1999), 127; GNO Collaboration, M. Altmann et al., *Phys. Lett.* **B490**, (2000) 16; B.T. Cleveland et al., *Astrophys. J.* **496** (1998), 505; LSND Collaboration, *Phys. Rev. Lett.* **81** (1998), 1774.

[22] J. C. Pati and A. Salam, *Phys. Rev.* **D10** (1974), 275; R. N. Mohapatra and J.C. Pati, ibid. 11 (1975), 566; **11** (1975), 2558; R. N. Mohapatra, *in Quarks, Leptons and Beyond* edited by H. Fritzsch et al., Nato Advanced Studies Institute Series B: Series B Vol. 122 (Plenum, New York, 1985), p. 219.

[23] R. N. Mohapatra, *Prog. Part. Nucl. Phys.* **26** (1992), 1.

[24] T. M. Gould and I. Z. Rothstein, *Phys. Lett.* **B333** (1994), 545.

[25] H. Grotch and R. Robinet, *Z. Phys.* **C39** (1988), 553.

[26] L3 Collab. M. Acciarri et al., *Phys. Lett.* **B412** (1997), 201, and references therein.

[27] A. Gutiérrez-Rodríguez, M. Hernández-Ruíz and A. Del Rio-De Santiago, *Phys. Rev.* **D69** (2004), 073008; M. A. Pérez, G. Tabares-Velasco, J.J. Toscano, *Int. J. Mod. Phys.* **A19** (2004), 159; F. Larios, M. A. Pérez, G. Tabares-Velasco, *Phys. Lett.* **B531** (2002), 231; A. Gutiérrez, M. A. Hernández, M. Maya and A. Rosado, *Phys. Rev.* **D58** (1998), 117302; A. Gutiérrez, M. A. Hernández, M. Maya and A. Rosado, *Rev. Mex. Fis.* **45** (1998), 249; M. Maya, M.A.Pérez, G. Tabares-Velasco, B. Vega, *Phys. Lett.* **B434** (1998), 354;

[28] Aytekin Aydemir and Ramazan Sever, *Modern Phys. Lett.* **A16** (2001), 7, 457. and references therein.

[29] Keiichi Akama, Takashi Hattori and Kazuo Katsuura, *Phys. Rev. Lett.* **88** (2002), 201601.

[30] M. A. B. Beg, R. V. Budny, R. Mohapatra, A. Sirlin, *Phys. Rev. Lett.* **22** (1977), 1252.

[31] M. Aquino, A. Fernández, A. García, *Phys. Lett.* **B261** (1991), 280.

[32] J. Polak, M. Zralek, *Nucl. Phys.* **B363** (1991), 385.

[33] P. Langacker, M. Luo, A. K. Mann, *Rev. Mod. Phys.* **64** (1992), 87.

[34] Particle Data Group, *Phys. Rev.* **D66** (2002), 1.

[35] Peter Renton, *An Introduction to the Physics of Quarks and Leptons*, Cambridge University Press, Cambridge, England, 1990.

[36] R. M. Barnett et al. *Phys. Rev.* **D54** (1996)+, 166.

INDEX

A

B

C

D

E

F

G

H

T

U

V

W

X

Y